Multicomponent Ultrafine Microstructures

MATERIALS RESEARCH SOCIETY SYMPOSIUM PROCEEDINGS VOLUME 132

Multicomponent Ultrafine Microstructures

Symposium held November 30-Decmber 1, 1988, Boston, Massachusetts, U.S.A.

EDITORS:

L.E. McCandlish
Rutgers University, Piscataway, New Jersey, U.S.A.

D.E. Polk
Office of Naval Research, Arlington, Virginia, U.S.A.

R.W. Siegel
Argonne National Laboratories, Argonne, Illinois, U.S.A.

B.H. Kear
Rutgers University, Piscataway, New Jersey, U.S.A.

MRS MATERIALS RESEARCH SOCIETY
Pittsburgh, Pennsylvania

This work was supported in part by the U.S. Army Research Office under Grant Number DAAL03-89-G-0001. The views, opinions, and/or findings contained in this report are those of the authors and should not be construed as an official Department of the Army position, policy, or decision unless so designated by other documentation.

CODEN: MRSPDH

Copyright 1989 by Materials Research Society.
All rights reserved.

This book has been registered with Copyright Clearance Center, Inc. For further information, please contact the Copyright Clearance Center, Salem, Massachusetts.

Published by:

Materials Research Society
9800 McKnight Road, Suite 327
Pittsburgh, Pennsylvania 15237
Telephone (412) 367-3003

Library of Congress Cataloging in Publication Data

Printed in the United States of America

Contents

PREFACE ... ix

ACKNOWLEDGMENTS ... xi

MATERIALS RESEARCH SOCIETY SYMPOSIUM PROCEEDINGS ... xii

PART I: CERAMICS

*SYNTHESIS, CHARACTERIZATION, AND PROPERTIES OF NANOPHASE CERAMICS ... 3
    R.W. Siegel and J.A. Eastman

SINTERING OF NANOPHASE TiO_2 AT 550°C ... 15
    J.E. Epperson, R.W. Siegel, J.W. White, T.E. Klippert, A. Narayanasamy, J.A. Eastman, and F. Trouw

SMALL ANGLE NEUTRON SCATTERING FROM NANOPHASE TITANIUM AS A FUNCTION OF OXIDATION ... 21
    J.A. Eastman, J.E. Epperson, H. Hahn, T.E. Klippert, A. Narayanasamy, S. Ramasamy, R.W. Siegel, J.W. White, and F. Trouw

PHASE CONTROL IN NANOPHASE MATERIALS FORMED FROM ULTRA-FINE Ti OR Pd POWDERS ... 27
    J.A. Eastman

MICROSTRUCTURE OF NANOCRYSTALLINE CERAMICS ... 35
    H. Hahn, J. Logas, H.J. Höfler, Th. Bier, and R.S. Averback

PHYSICAL-CHEMICAL PROPERTIES OF TiO_2 MEMBRANES CONTROLLED BY SOL-GEL PROCESSING ... 41
    Qunyin Xu and Marc A. Anderson

PART II: COMPOSITES

*CHEMICAL VAPOR DEPOSITION OF ULTRAFINE CERAMIC STRUCTURES ... 49
    B.M. Gallois, R. Mathur, S.R. Lee, and J.Y. Yoo

ULTRAFINE COMPOSITE SYNTHESIS BY LASER-REACTIONS AND RAPID CONDENSATION ... 61
    Gan-Moog Chow and Peter R. Strutt

METASTABLE NANOCRYSTALLINE CARBIDES IN CHEMICALLY SYNTHESIZED W-Co-C TERNARY ALLOYS ... 67
    L.E. McCandlish, B.H. Kear, B.K. Kim, and L.W. Wu

*Invited Paper

FABRICATION OF FERROMAGNETIC Fe/MULLITE COMPOSITE
MATERIALS VIA SOL-GEL PROCESSING 73
    Deanne P. Yamato, Abraham L. Landis, and
    Teh S. Kuan

CRYOMILLING OF NANO-PHASE DISPERSION STRENGTHENED
ALUMINUM 79
    M.J. Luton, C.S. Jayanth, M.M. Disko, S. Matras,
    and J. Vallone

MATERIAL WITH NOVEL COMPOSITIONS AND FINE MICRO-
STRUCTURES PRODUCED VIA THE MIXALLOY PROCESS 87
    Arthur K. Lee, Luis E. Sanchez-Caldera,
    Jung-Hoon Chun, and Nam P. Suh

MICROSTRUCTURE OF AN Al_2O_3/METAL COMPOSITE CONTAINING
AN Al_2O_3 FILLER MATERIAL 93
    E. Breval and A.S. Nagelberg

INFRARED PROPERTIES OF THIN Pt/Al_2O_3 GRANULAR METAL-
INSULATOR COMPOSITE FILMS 99
    M.F. MacMillan, R.P. Devaty, and J.V. Mantese

PROPERTIES OF TRANSPARENT SILICA GEL - PMMA COMPOSITES 105
    Edward J.A. Pope, Minuo Asami, and John D.
    Mackenzie

PREPARATION OF ULTRAFINE COPPER PARTICLES IN POLY
(2-VINYLPYRIDINE) 111
    Alan M. Lyons, S. Nakahara, and E.M. Pearce

OPTICAL PROPERTIES OF FINELY-STRUCTURED PARTICULATES 119
    Ping Sheng, Min-Yao Zhou, Zhe Chen, and S.T. Chui

PART III: ALLOYS, METALS, AND MAGNETIC MATERIALS

*A NEW LOOK AT AMORPHOUS VS MICROCRYSTALLINE STRUCTURE 127
    Frans Spaepen

MECHANISM OF ACHIEVING NANOCRYSTALLINE AlRu BY BALL
MILLING 137
    E. Hellstern, H.J. Fecht, C. Garland, and
    W.L. Johnson

FABRICATION AND PROPERTIES OF GRANULAR Fe-Ni ALLOYS 143
    A. Gavrin and C.L. Chien

SMALL ANGLE SCATTERING FROM NANOCRYSTALLINE Pd 149
    G. Wallner, E. Jorra, H. Franz, J. Peisl,
    R. Birringer, H. Gleiter, T. Haubold, and
    W. Petry

CHARACTERIZATION OF ULTRAFINE MICROSTRUCTURES USING A
POSITION-SENSITIVE ATOM PROBE (POSAP) 155
    Alfred Cerezo, Chris R.M. Grovenor, Mark G.
    Hetherington, Barbara A. Shollock, and
    George D.W. Smith

*Invited Paper

STUDY OF THE LATTICE DYNAMICS OF IRON NANOCRYSTALS BY
MÖSSBAUER SPECTROSCOPY   161
    J. Childress, A. Levy, and C.L. Chien

MICROSTRUCTURE AND MAGNETO-OPTICAL PROPERTIES OF
TbFeCo FILMS PREPARED BY FACING TARGETS SPUTTERING   167
    H. Ito, T. Hirata, and M. Naoe

STUDY OF MULTICOMPONENT MAGNETIC NANOSTRUCTURES WITH
DIGITAL IMAGE PROCESSING TECHNIQUE   173
    A.P. Valanju, I.S. Jeong, D.Y. Kim, and
    R.M. Walser

MAGNETIC PROPERTIES OF IRON/SILICA GEL NANOCOMPOSITES   179
    Robert D. Shull, Joseph J. Ritter, Alexander J.
    Shapiro, Lydon J. Swartzendruber, and Lawrence H.
    Bennett

COERCIVITY IN GRANULAR Fe-Al$_2$O$_3$ 185
    F.H. Streitz and C.L. Chien

ON THE COERCIVITY OF GRANULAR Fe-SiO$_2$ FILMS   191
    S.H. Liou, C.H. Chen, H.S. Chen, A.R. Kortan,
    and C.L. Chien

          PART IV: MULTILAYERS AND SUPERLATTICES

*CRITICAL PHENOMENA IN NANOSCALE MULTILAYER MATERIALS   199
    T. Tsakalakos and A. Jankowski

MICROSTRUCTURE AND MAGNETIC BEHAVIOR OF Fe/Ti MULTI-
LAYERED FILMS   213
    Satoshi Ono, Michio Nitta, and Masahiko Naoe

ELECTRODEPOSITED METALLIC SUPERLATTICES   219
    D.S. Lashmore, Robert Oberle, Moshe P. Dariel,
    L.H. Bennett, and Lydon Swartzendruber

PREPARATION AND STRUCTURE OF Cu-W MULTILAYERS   225
    K.M. Unruh, B.M. Patterson, S.I. Shah,
    G.A. Jones, Y.-W. Kim, and J.E. Greene

X-RAY DIFFRACTION ANALYSIS OF Au/Ni MULTILAYERS   231
    J. Chaudhuri, S. Shah, and A.F. Jankowski

AUTHOR INDEX   239

SUBJECT INDEX   241

*Invited Paper

Preface

In 1986, a small group of researchers, recognizing the growing importance of ultrafine microstructures as a scientific discipline, decided that it was time to provide a forum for those interested in this interdisciplinary field. This led to the first symposium on the subject, Multicomponent Ultrafine Microstructures, sponsored by the Office of Naval Research at the Fall Meeting of the Materials Research Society. The attendance at this one and one-half day meeting was small but sufficient to justify a continuation of the dialogue at future symposia. The 1988 meeting represents the second in this new series. As before, the scope of the meeting was limited to metals, ceramics, polymers and their composites. Semiconductor, superconductor, and opto-electronic materials were excluded because they are being adequately covered by other symposia.

The 1988 meeting was a two-day symposium addressing the preparation, characterization, and properties of materials with ultrafine or "nanophase" structures. These materials are characterized by microstructures in the 1 to 100 nanometer range. Indicative of the rapid growth and intellectual excitement in the field, the attendance at the symposium averaged 150 and at one stage overflowed the conference room.

In these proceedings, the 33 articles are organized under four main headings: ceramics; composites; alloys, metals, and magnetic materials; multilayers and superlattices. At this stage in the development of the field, research emphasis is on the synthesis and characterization of novel nanophase materials. Chemical and physical vapor deposition techniques remain one of the most active research areas in the synthesis of nanophase materials. A growing activity in the chemical synthesis of nanophases is apparent. Such methods depend on controlled thermochemical conversion of precursor molecules to achieve the desired ultrafine microstructures. A major effort is being made to characterize the intergranular and interphase interfaces using x-ray and neutron scattering, Mössbauer spectroscopy and coercivity measurements, atom probe, and high resolution electron microscopy. Controversy still exists regarding the nature of grain boundaries in nanophase materials produced by compaction of fine powders. Some authors claim that these boundaries have a random gas-like structure, whereas others contend that the boundaries are of conventional type, although perhaps containing a distribution of small pores. Major contributions have been made in the determination of magnetic and mechanical properties. Mechanical property data is extensive in a few areas, e.g., materials produced by mechanical reduction of metal based systems or by cold compaction and sintering of ultrafine ceramics. Confirmation of the supermodulus effect has been obtained in many different multilayered materials. Understanding of critical phenomena in nanoscale multilayer structures has also advanced.

February 1989

L.E. McCandlish
D.E. Polk
R.W. Siegel
B.H. Kear

Acknowledgments

The symposium organizers would like to acknowledge the participants and speakers, who contributed greatly to the overall success of the meeting. A special thanks goes to Donald Polk who is responsible for recognizing the importance of ultrafine microstructures and who initiated the 1986 and 1988 symposia. The MRS staff was especially helpful at the meeting and in the preparation of this book. Their dedication to their work is deeply appreciated. Finally we would like to thank the Office of Naval Research and the Army Research Office for their generous support.

MATERIALS RESEARCH SOCIETY SYMPOSIUM PROCEEDINGS

ISSN 0272 - 9172

Volume 1—Laser and Electron-Beam Solid Interactions and Materials Processing, J. F. Gibbons, L. D. Hess, T. W. Sigmon, 1981, ISBN 0-444-00595-1

Volume 2—Defects in Semiconductors, J. Narayan, T. Y. Tan, 1981, ISBN 0-444-00596-X

Volume 3—Nuclear and Electron Resonance Spectroscopies Applied to Materials Science, E. N. Kaufmann, G. K. Shenoy, 1981, ISBN 0-444-00597-8

Volume 4—Laser and Electron-Beam Interactions with Solids, B. R. Appleton, G. K. Celler, 1982, ISBN 0-444-00693-1

Volume 5—Grain Boundaries in Semiconductors, H. J. Leamy, G. E. Pike, C. H. Seager, 1982, ISBN 0-444-00697-4

Volume 6—Scientific Basis for Nuclear Waste Management IV, S. V. Topp, 1982, ISBN 0-444-00699-0

Volume 7—Metastable Materials Formation by Ion Implantation, S. T. Picraux, W. J. Choyke, 1982, ISBN 0-444-00692-3

Volume 8—Rapidly Solidified Amorphous and Crystalline Alloys, B. H. Kear, B. C. Giessen, M. Cohen, 1982, ISBN 0-444-00698-2

Volume 9—Materials Processing in the Reduced Gravity Environment of Space, G. E. Rindone, 1982, ISBN 0-444-00691-5

Volume 10—Thin Films and Interfaces, P. S. Ho, K.-N. Tu, 1982, ISBN 0-444-00774-1

Volume 11—Scientific Basis for Nuclear Waste Management V, W. Lutze, 1982, ISBN 0-444-00725-3

Volume 12—In Situ Composites IV, F. D. Lemkey, H. E. Cline, M. McLean, 1982, ISBN 0-444-00726-1

Volume 13—Laser-Solid Interactions and Transient Thermal Processing of Materials, J. Narayan, W. L. Brown, R. A. Lemons, 1983, ISBN 0-444-00788-1

Volume 14—Defects in Semiconductors II, S. Mahajan, J. W. Corbett, 1983, ISBN 0-444-00812-8

Volume 15—Scientific Basis for Nuclear Waste Management VI, D. G. Brookins, 1983, ISBN 0-444-00780-6

Volume 16—Nuclear Radiation Detector Materials, E. E. Haller, H. W. Kraner, W. A. Higinbotham, 1983, ISBN 0-444-00787-3

Volume 17—Laser Diagnostics and Photochemical Processing for Semiconductor Devices, R. M. Osgood, S. R. J. Brueck, H. R. Schlossberg, 1983, ISBN 0-444-00782-2

Volume 18—Interfaces and Contacts, R. Ludeke, K. Rose, 1983, ISBN 0-444-00820-9

Volume 19—Alloy Phase Diagrams, L. H. Bennett, T. B. Massalski, B. C. Giessen, 1983, ISBN 0-444-00809-8

Volume 20—Intercalated Graphite, M. S. Dresselhaus, G. Dresselhaus, J. E. Fischer, M. J. Moran, 1983, ISBN 0-444-00781-4

Volume 21—Phase Transformations in Solids, T. Tsakalakos, 1984, ISBN 0-444-00901-9

Volume 22—High Pressure in Science and Technology, C. Homan, R. K. MacCrone, E. Whalley, 1984, ISBN 0-444-00932-9 (3 part set)

Volume 23—Energy Beam-Solid Interactions and Transient Thermal Processing, J. C. C. Fan, N. M. Johnson, 1984, ISBN 0-444-00903-5

Volume 24—Defect Properties and Processing of High-Technology Nonmetallic Materials, J. H. Crawford, Jr., Y. Chen, W. A. Sibley, 1984, ISBN 0-444-00904-3

Volume 25—Thin Films and Interfaces II, J. E. E. Baglin, D. R. Campbell, W. K. Chu, 1984, ISBN 0-444-00905-1

MATERIALS RESEARCH SOCIETY SYMPOSIUM PROCEEDINGS

Volume 26—Scientific Basis for Nuclear Waste Management VII, G. L. McVay, 1984, ISBN 0-444-00906-X

Volume 27—Ion Implantation and Ion Beam Processing of Materials, G. K. Hubler, O. W. Holland, C. R. Clayton, C. W. White, 1984, ISBN 0-444-00869-1

Volume 28—Rapidly Solidified Metastable Materials, B. H. Kear, B. C. Giessen, 1984, ISBN 0-444-00935-3

Volume 29—Laser-Controlled Chemical Processing of Surfaces, A. W. Johnson, D. J. Ehrlich, H. R. Schlossberg, 1984, ISBN 0-444-00894-2

Volume 30—Plasma Processing and Synthesis of Materials, J. Szekely, D. Apelian, 1984, ISBN 0-444-00895-0

Volume 31—Electron Microscopy of Materials, W. Krakow, D. A. Smith, L. W. Hobbs, 1984, ISBN 0-444-00898-7

Volume 32—Better Ceramics Through Chemistry, C. J. Brinker, D. E. Clark, D. R. Ulrich, 1984, ISBN 0-444-00898-5

Volume 33—Comparison of Thin Film Transistor and SOI Technologies, H. W. Lam, M. J. Thompson, 1984, ISBN 0-444-00899-3

Volume 34—Physical Metallurgy of Cast Iron, H. Fredriksson, M. Hillerts, 1985, ISBN 0-444-00938-8

Volume 35—Energy Beam-Solid Interactions and Transient Thermal Processing/1984, D. K. Biegelsen, G. A. Rozgonyi, C. V. Shank, 1985, ISBN 0-931837-00-6

Volume 36—Impurity Diffusion and Gettering in Silicon, R. B. Fair, C. W. Pearce, J. Washburn, 1985, ISBN 0-931837-01-4

Volume 37—Layered Structures, Epitaxy, and Interfaces, J. M. Gibson, L. R. Dawson, 1985, ISBN 0-931837-02-2

Volume 38—Plasma Synthesis and Etching of Electronic Materials, R. P. H. Chang, B. Abeles, 1985, ISBN 0-931837-03-0

Volume 39—High-Temperature Ordered Intermetallic Alloys, C. C. Koch, C. T. Liu, N. S. Stoloff, 1985, ISBN 0-931837-04-9

Volume 40—Electronic Packaging Materials Science, E. A. Giess, K.-N. Tu, D. R. Uhlmann, 1985, ISBN 0-931837-05-7

Volume 41—Advanced Photon and Particle Techniques for the Characterization of Defects in Solids, J. B. Roberto, R. W. Carpenter, M. C. Wittels, 1985, ISBN 0-931837-06-5

Volume 42—Very High Strength Cement-Based Materials, J. F. Young, 1985, ISBN 0-931837-07-3

Volume 43—Fly Ash and Coal Conversion By-Products: Characterization, Utilization, and Disposal I, G. J. McCarthy, R. J. Lauf, 1985, ISBN 0-931837-08-1

Volume 44—Scientific Basis for Nuclear Waste Management VIII, C. M. Jantzen, J. A. Stone, R. C. Ewing, 1985, ISBN 0-931837-09-X

Volume 45—Ion Beam Processes in Advanced Electronic Materials and Device Technology, B. R. Appleton, F. H. Eisen, T. W. Sigmon, 1985, ISBN 0-931837-10-3

Volume 46—Microscopic Identification of Electronic Defects in Semiconductors, N. M. Johnson, S. G. Bishop, G. D. Watkins, 1985, ISBN 0-931837-11-1

Volume 47—Thin Films: The Relationship of Structure to Properties, C. R. Aita, K. S. SreeHarsha, 1985, ISBN 0-931837-12-X

Volume 48—Applied Materials Characterization, W. Katz, P. Williams, 1985, ISBN 0-931837-13-8

Volume 49—Materials Issues in Applications of Amorphous Silicon Technology, D. Adler, A. Madan, M. J. Thompson, 1985, ISBN 0-931837-14-6

MATERIALS RESEARCH SOCIETY SYMPOSIUM PROCEEDINGS

Volume 50—Scientific Basis for Nuclear Waste Management IX, L. O. Werme, 1986, ISBN 0-931837-15-4

Volume 51—Beam-Solid Interactions and Phase Transformations, H. Kurz, G. L. Olson, J. M. Poate, 1986, ISBN 0-931837-16-2

Volume 52—Rapid Thermal Processing, T. O. Sedgwick, T. E. Seidel, B.-Y. Tsaur, 1986, ISBN 0-931837-17-0

Volume 53—Semiconductor-on-Insulator and Thin Film Transistor Technology, A. Chiang. M. W. Geis, L. Pfeiffer, 1986, ISBN 0-931837-18-9

Volume 54—Thin Films—Interfaces and Phenomena, R. J. Nemanich, P. S. Ho, S. S. Lau, 1986, ISBN 0-931837-19-7

Volume 55—Biomedical Materials, J. M. Williams, M. F. Nichols, W. Zingg, 1986, ISBN 0-931837-20-0

Volume 56—Layered Structures and Epitaxy, J. M. Gibson, G. C. Osbourn, R. M. Tromp, 1986, ISBN 0-931837-21-9

Volume 57—Phase Transitions in Condensed Systems—Experiments and Theory, G. S. Cargill III, F. Spaepen, K.-N. Tu, 1987, ISBN 0-931837-22-7

Volume 58—Rapidly Solidified Alloys and Their Mechanical and Magnetic Properties, B. C. Giessen, D. E. Polk, A. I. Taub, 1986, ISBN 0-931837-23-5

Volume 59—Oxygen, Carbon, Hydrogen, and Nitrogen in Crystalline Silicon, J. C. Mikkelsen, Jr., S. J. Pearton, J. W. Corbett, S. J. Pennycook, 1986, ISBN 0-931837-24-3

Volume 60—Defect Properties and Processing of High-Technology Nonmetallic Materials, Y. Chen, W. D. Kingery, R. J. Stokes, 1986, ISBN 0-931837-25-1

Volume 61—Defects in Glasses, F. L. Galeener, D. L. Griscom, M. J. Weber, 1986, ISBN 0-931837-26-X

Volume 62—Materials Problem Solving with the Transmission Electron Microscope, L. W. Hobbs, K. H. Westmacott, D. B. Williams, 1986, ISBN 0-931837-27-8

Volume 63—Computer-Based Microscopic Description of the Structure and Properties of Materials, J. Broughton, W. Krakow, S. T. Pantelides, 1986, ISBN 0-931837-28-6

Volume 64—Cement-Based Composites: Strain Rate Effects on Fracture, S. Mindess, S. P. Shah, 1986, ISBN 0-931837-29-4

Volume 65—Fly Ash and Coal Conversion By-Products: Characterization, Utilization and Disposal II, G. J. McCarthy, F. P. Glasser, D. M. Roy, 1986, ISBN 0-931837-30-8

Volume 66—Frontiers in Materials Education, L. W. Hobbs, G. L. Liedl, 1986, ISBN 0-931837-31-6

Volume 67—Heteroepitaxy on Silicon, J. C. C. Fan, J. M. Poate, 1986, ISBN 0-931837-33-2

Volume 68—Plasma Processing, J. W. Coburn, R. A. Gottscho, D. W. Hess, 1986, ISBN 0-931837-34-0

Volume 69—Materials Characterization, N. W. Cheung, M.-A. Nicolet, 1986, ISBN 0-931837-35-9

Volume 70—Materials Issues in Amorphous-Semiconductor Technology, D. Adler, Y. Hamakawa, A. Madan, 1986, ISBN 0-931837-36-7

Volume 71—Materials Issues in Silicon Integrated Circuit Processing, M. Wittmer, J. Stimmell, M. Strathman, 1986, ISBN 0-931837-37-5

Volume 72—Electronic Packaging Materials Science II, K. A. Jackson, R. C. Pohanka, D. R. Uhlmann, D. R. Ulrich, 1986, ISBN 0-931837-38-3

Volume 73—Better Ceramics Through Chemistry II, C. J. Brinker, D. E. Clark, D. R. Ulrich, 1986, ISBN 0-931837-39-1

Volume 74—Beam-Solid Interactions and Transient Processes, M. O. Thompson, S. T. Picraux, J. S. Williams, 1987, ISBN 0-931837-40-5

MATERIALS RESEARCH SOCIETY SYMPOSIUM PROCEEDINGS

Volume 75—Photon, Beam and Plasma Stimulated Chemical Processes at Surfaces, V. M. Donnelly, I. P. Herman, M. Hirose, 1987, ISBN 0-931837-41-3

Volume 76—Science and Technology of Microfabrication, R. E. Howard, E. L. Hu, S. Namba, S. Pang, 1987, ISBN 0-931837-42-1

Volume 77—Interfaces, Superlattices, and Thin Films, J. D. Dow, I. K. Schuller, 1987, ISBN 0-931837-56-1

Volume 78—Advances in Structural Ceramics, P. F. Becher, M. V. Swain, S. Sōmiya, 1987, ISBN 0-931837-43-X

Volume 79—Scattering, Deformation and Fracture in Polymers, G. D. Wignall, B. Crist, T. P. Russell, E. L. Thomas, 1987, ISBN 0-931837-44-8

Volume 80—Science and Technology of Rapidly Quenched Alloys, M. Tenhover, W. L. Johnson, L. E. Tanner, 1987, ISBN 0-931837-45-6

Volume 81—High-Temperature Ordered Intermetallic Alloys, II, N. S. Stoloff, C. C. Koch, C. T. Liu, O. Izumi, 1987, ISBN 0-931837-46-4

Volume 82—Characterization of Defects in Materials, R. W. Siegel, J. R. Weertman, R. Sinclair, 1987, ISBN 0-931837-47-2

Volume 83—Physical and Chemical Properties of Thin Metal Overlayers and Alloy Surfaces, D. M. Zehner, D. W. Goodman, 1987, ISBN 0-931837-48-0

Volume 84—Scientific Basis for Nuclear Waste Management X, J. K. Bates, W. B. Seefeldt, 1987, ISBN 0-931837-49-9

Volume 85—Microstructural Development During the Hydration of Cement, L. Struble, P. Brown, 1987, ISBN 0-931837-50-2

Volume 86—Fly Ash and Coal Conversion By-Products Characterization, Utilization and Disposal III, G. J. McCarthy, F. P. Glasser, D. M. Roy, S. Diamond, 1987, ISBN 0-931837-51-0

Volume 87—Materials Processing in the Reduced Gravity Environment of Space, R. H. Doremus, P. C. Nordine, 1987, ISBN 0-931837-52-9

Volume 88—Optical Fiber Materials and Properties, S. R. Nagel, J. W. Fleming, G. Sigel, D. A. Thompson, 1987, ISBN 0-931837-53-7

Volume 89—Diluted Magnetic (Semimagnetic) Semiconductors, R. L. Aggarwal, J. K. Furdyna, S. von Molnar, 1987, ISBN 0-931837-54-5

Volume 90—Materials for Infrared Detectors and Sources, R. F. C. Farrow, J. F. Schetzina, J. T. Cheung, 1987, ISBN 0-931837-55-3

Volume 91—Heteroepitaxy on Silicon II, J. C. C. Fan, J. M. Phillips, B.-Y. Tsaur, 1987, ISBN 0-931837-58-8

Volume 92—Rapid Thermal Processing of Electronic Materials, S. R. Wilson, R. A. Powell, D. E. Davies, 1987, ISBN 0-931837-59-6

Volume 93—Materials Modification and Growth Using Ion Beams, U. Gibson, A. E. White, P. P. Pronko, 1987, ISBN 0-931837-60-X

Volume 94—Initial Stages of Epitaxial Growth, R. Hull, J. M. Gibson, David A. Smith, 1987, ISBN 0-931837-61-8

Volume 95—Amorphous Silicon Semiconductors—Pure and Hydrogenated, A. Madan, M. Thompson, D. Adler, Y. Hamakawa, 1987, ISBN 0-931837-62-6

Volume 96—Permanent Magnet Materials, S. G. Sankar, J. F. Herbst, N. C. Koon, 1987, ISBN 0-931837-63-4

Volume 97—Novel Refractory Semiconductors, D. Emin, T. Aselage, C. Wood, 1987, ISBN 0-931837-64-2

Volume 98—Plasma Processing and Synthesis of Materials, D. Apelian, J. Szekely, 1987, ISBN 0-931837-65-0

MATERIALS RESEARCH SOCIETY SYMPOSIUM PROCEEDINGS

Volume 99—High-Temperature Superconductors, M. B. Brodsky, R. C. Dynes, K. Kitazawa, H. L. Tuller, 1988, ISBN 0-931837-67-7

Volume 100—Fundamentals of Beam-Solid Interactions and Transient Thermal Processing, M. J. Aziz, L. E. Rehn, B. Stritzker, 1988, ISBN 0-931837-68-5

Volume 101—Laser and Particle-Beam Chemical Processing for Microelectronics, D.J. Ehrlich, G.S. Higashi, M.M. Oprysko, 1988, ISBN 0-931837-69-3

Volume 102—Epitaxy of Semiconductor Layered Structures, R. T. Tung, L. R. Dawson, R. L. Gunshor, 1988, ISBN 0-931837-70-7

Volume 103—Multilayers: Synthesis, Properties, and Nonelectronic Applications, T. W. Barbee Jr., F. Spaepen, L. Greer, 1988, ISBN 0-931837-71-5

Volume 104—Defects in Electronic Materials, M. Stavola, S. J. Pearton, G. Davies, 1988, ISBN 0-931837-72-3

Volume 105—SiO_2 and Its Interfaces, G. Lucovsky, S. T. Pantelides, 1988, ISBN 0-931837-73-1

Volume 106—Polysilicon Films and Interfaces, C.Y. Wong, C.V. Thompson, K-N. Tu, 1988, ISBN 0-931837-74-X

Volume 107—Silicon-on-Insulator and Buried Metals in Semiconductors, J. C. Sturm, C. K. Chen, L. Pfeiffer, P. L. F. Hemment, 1988, ISBN 0-931837-75-8

Volume 108—Electronic Packaging Materials Science II, R. C. Sundahl, R. Jaccodine, K. A. Jackson, 1988, ISBN 0-931837-76-6

Volume 109—Nonlinear Optical Properties of Polymers, A. J. Heeger, J. Orenstein, D. R. Ulrich, 1988, ISBN 0-931837-77-4

Volume 110—Biomedical Materials and Devices, J. S. Hanker, B. L. Giammara, 1988, ISBN 0-931837-78-2

Volume 111—Microstructure and Properties of Catalysts, M. M. J. Treacy, J. M. Thomas, J. M. White, 1988, ISBN 0-931837-79-0

Volume 112—Scientific Basis for Nuclear Waste Management XI, M. J. Apted, R. E. Westerman, 1988, ISBN 0-931837-80-4

Volume 113—Fly Ash and Coal Conversion By-Products: Characterization, Utilization, and Disposal IV, G. J. McCarthy, D. M. Roy, F. P. Glasser, R. T. Hemmings, 1988, ISBN 0-931837-81-2

Volume 114—Bonding in Cementitious Composites, S. Mindess, S. P. Shah, 1988, ISBN 0-931837-82-0

Volume 115—Specimen Preparation for Transmission Electron Microscopy of Materials, J. C. Bravman, R. Anderson, M. L. McDonald, 1988, ISBN 0-931837-83-9

Volume 116—Heteroepitaxy on Silicon: Fundamentals, Structures,and Devices, H.K. Choi, H. Ishiwara, R. Hull, R.J. Nemanich, 1988, ISBN: 0-931837-86-3

Volume 117—Process Diagnostics: Materials, Combustion, Fusion, K. Hays, A.C. Eckbreth, G.A. Campbell, 1988, ISBN: 0-931837-87-1

Volume 118—Amorphous Silicon Technology, A. Madan, M.J. Thompson, P.C. Taylor, P.G. LeComber, Y. Hamakawa, 1988, ISBN: 0-931837-88-X

Volume 119—Adhesion in Solids, D.M. Mattox, C. Batich, J.E.E. Baglin, R.J. Gottschall, 1988, ISBN: 0-931837-89-8

Volume 120—High-Temperature/High-Performance Composites, F.D. Lemkey, A.G. Evans, S.G. Fishman, J.R. Strife, 1988, ISBN: 0-931837-90-1

Volume 121—Better Ceramics Through Chemistry III, C.J. Brinker, D.E. Clark, D.R. Ulrich, 1988, ISBN: 0-931837-91-X

Volume 122—Interfacial Structure, Properties, and Design, M.H. Yoo, W.A.T. Clark, C.L. Briant, 1988, ISBN: 0-931837-92-8

MATERIALS RESEARCH SOCIETY SYMPOSIUM PROCEEDINGS

Volume 123—Materials Issues in Art and Archaeology, E.V. Sayre, P. Vandiver, J. Druzik, C. Stevenson, 1988, ISBN: 0-931837-93-6

Volume 124—Microwave-Processing of Materials, M.H. Brooks, I.J. Chabinsky, W.H. Sutton, 1988, ISBN: 0-931837-94-4

Volume 125—Materials Stability and Environmental Degradation, A. Barkatt, L.R. Smith, E. Verink, 1988, ISBN: 0-931837-95-2

Volume 126—Advanced Surface Processes for Optoelectronics, S. Bernasek, T. Venkatesan, H. Temkin, 1988, ISBN: 0-931837-96-0

Volume 127—Scientific Basis for Nuclear Waste Management XII, W. Lutze, R.C. Ewing, 1989, ISBN: 0-931837-97-9

Volume 128—Processing and Characterization of Materials Using Ion Beams, L.E. Rehn, J. Greene, F.A. Smidt, 1989, ISBN: 1-55899-001-1

Volume 129—Laser and Particle-Beam Chemical Processes on Surfaces, G.L. Loper, A.W. Johnson, T.W. Sigmon, 1989, ISBN: 1-55899-002-X

Volume 130—Thin Films: Stresses and Mechanical Properties, J.C. Bravman, W.D. Nix, D.M. Barnett, D.A. Smith, 1989, ISBN: 0-55899-003-8

Volume 131—Chemical Perspectives of Microelectronic Materials, M.E. Gross, J. Jasinski, J.T. Yates, Jr., 1989, ISBN: 0-55899-004-6

Volume 132—Multicomponent Ultrafine Microstructures, L.E. McCandlish, B.H. Kear, D.E. Polk, and R.W. Siegel, 1989, ISBN: 1-55899-005-4

Volume 133—High Temperature Ordered Intermetallic Alloys III, C.T. Liu, A.I. Taub, N.S. Stoloff, C.C. Koch, 1989, ISBN: 1-55899-006-2

Volume 134—The Materials Science and Engineering of Rigid-Rod Polymers, W.W. Adams, R.K. Eby, D.E. McLemore, 1989, ISBN: 1-55899-007-0

Volume 135—Solid State Ionics, G. Nazri, R.A. Huggins, D.F. Shriver, 1989, ISBN: 1-55899-008-9

Volume 136—Fly Ash and Coal Conversion By-Products: Characterization, Utilization, and Disposal V, R.T. Hemmings, E.E. Berry, G.J. McCarthy, F.P. Glasser, 1989, ISBN: 1-55899-009-7

Volume 137—Pore Structure and Permeability of Cementitious Materials, L.R. Roberts, J.P. Skalny, 1989, ISBN: 1-55899-010-0

Volume 138—Characterization of the Structure and Chemistry of Defects in Materials, B.C. Larson, M. Ruhle, D.N. Seidman, 1989, ISBN: 1-55899-011-9

Volume 139—High Resolution Microscopy of Materials, W. Krakow, F.A. Ponce, D.J. Smith, 1989, ISBN: 1-55899-012-7

Volume 140—New Materials Approaches to Tribology: Theory and Applications, L.E. Pope, L. Fehrenbacher, W.O. Winer, 1989, ISBN: 1-55899-013-5

Volume 141—Atomic Scale Calculations in Materials Science, J. Tersoff, D. Vanderbilt, V. Vitek, 1989, ISBN: 1-55899-014-3

Volume 142—Nondestructive Monitoring of Materials Properties, J. Holbrook, J. Bussiere, 1989, ISBN: 1-55899-015-1

Volume 143—Synchrotron Radiation in Materials Research, R. Clarke, J.H. Weaver, J. Gland, 1989, ISBN: 1-55899-016-X

Volume 144—Advances in Materials, Processing and Devices in III-V Compound Semiconductors, D.K. Sadana, L. Eastman, R. Dupuis, 1989, ISBN: 1-55899-017-8

MATERIALS RESEARCH SOCIETY CONFERENCE PROCEEDINGS

Tungsten and Other Refractory Metals for VLSI Applications, R. S. Blewer, 1986; ISSN 0886-7860; ISBN 0-931837-32-4

Tungsten and Other Refractory Metals for VLSI Applications II, E.K. Broadbent, 1987; ISSN 0886-7860; ISBN 0-931837-66-9

Ternary and Multinary Compounds, S. Deb, A. Zunger, 1987; ISBN 0-931837-57-x

Tungsten and Other Refractory Metals for VLSI Applications III, Victor A. Wells, 1988; ISSN 0886-7860; ISBN 0-931837-84-7

Atomic and Molecular Processing of Electronic and Ceramic Materials: Preparation, Characterization and Properties, Ilhan A. Aksay, Gary L. McVay, Thomas G. Stoebe, 1988; ISBN 0-931837-85-5

Materials Futures: Strategies and Opportunities, R. Byron Pipes, U.S. Organizing Committee, Rune Lagneborg, Swedish Organizing Committee, 1988; ISBN 0-55899-000-3

Tungsten and Other Refractory Metals for VLSI Applications IV, Robert S. Blewer, Carol M. McConica, 1989; ISSN: 0886-7860; ISBN: 0-931837-98-7

PART I

Ceramics

SYNTHESIS, CHARACTERIZATION, AND PROPERTIES OF NANOPHASE CERAMICS

R. W. SIEGEL and J. A. EASTMAN
Materials Science Division, Argonne National Laboratory, Argonne, Illinois 60439.

ABSTRACT

Ultrafine-grained ceramics have been synthesized by the production of ultrafine (2-20 nm) particles, using the gas-condensation method, followed by their in-situ, ultra-high vacuum consolidation at room temperature. These new nanophase ceramics have properties that are significantly improved relative to those of their coarser-grained, conventionally-prepared counterparts. For example, nanophase rutile (TiO_2) with an initial mean grain diameter of 12 nm sinters at 400 to 600°C lower temperatures than conventional powders, without the need for compacting or sintering aids. The sintered nanophase rutile exhibits both improved microhardness and fracture characteristics. These property improvements result from the reduced scale of the grains and the increased cleanliness of the particle surfaces and the subsequently-formed grain boundaries. Research completed on the synthesis, characterization, and properties of nanophase ceramics is reviewed and the potential for using the nanophase synthesis method for engineering new and/or improved ceramics and composites is considered.

INTRODUCTION

A means of producing a new class of ultrafine-grained materials having desirable structures and properties has recently been developed. These nanophase materials are composed of clusters of atoms (typically with diameters of 2 to 20 nm) which have been consolidated to form a bulk solid material. They are produced in a two-step process that was first suggested by Gleiter in 1981 [1]. In this process, small particles are formed by the condensation of evaporated atoms in an inert gas and are subsequently consolidated without exposure to air. Because the clusters (or grains) of which these materials are composed are extremely small, depending upon the actual sizes of the grains, on the order of 50% or more of the atoms in the material are found in or very near the interfaces between grains. This ultrafine grain size and large fraction of interfacial volume is responsible for nanophase materials having a variety of novel and potentially advantageous properties and characteristics, which are only now beginning to be discovered. Many of the first results in this new area, particularly with respect to metals, have already been reviewed [2,3].

Several reasons can be cited for interest in the study, development, and use of nanophase materials, in general, and nanophase ceramics in particular. Superior properties are obtained because of the smaller grain sizes and, hence, correspondingly small radii of curvature and larger interface areas. The small size of the particles produced can also be exploited in order to form alloys or compounds by reacting two or more materials. Besides the advantages imparted by being able to produce homogeneous materials by mixing powders on a very fine scale, additional benefits, such as greatly enhanced sinterability, are obtained due to their short diffusion distances and large volume fractions of highly reactive surface material. Likewise, these same conditions make controlled and efficient doping possible. Metastable material production should also be possible, since the relative energies of formation of phases can be altered by the increase in surface area which is obtained with small particles. This direction in nanophase processing, which is only beginning to be explored, holds considerable promise for future materials design and engineering.

One of the primary advantages of ceramic materials processing using the gas-condensation method is that the ultrafine powders produced can be consolidated under pristine conditions. This is a considerable advantage over other processing techniques presently used to form ceramics from small particles, most of which involve forming powders in liquid solutions by precipitation during chemical reactions. Collecting the powders that are produced in the gas-condensation method and transferring them directly under vacuum to a compaction unit, on the other hand, makes it possible in the nanophase processing method to produce ceramics with very clean interfaces. These processes appear to be easily scaled to the production of larger amounts of material as well.

SYNTHESIS METHOD

An apparatus for the synthesis of nanophase ceramics is comprised of an ultrahigh-vacuum (UHV) system fitted with evaporation sources, a cold finger and scraper assembly, and in situ compaction devices for consolidating the powders produced and collected in the chamber. Such an apparatus is depicted schematically in Figure 1. Nanophase ceramics are produced by a processing method that is actually a combination and extension of two techniques which have been known to scientists and engineers for many years; the gas-condensation method for producing ultrafine powders, and powder processing to produce bulk materials. The gas-condensation method [4-6] involves evaporating materials inside a vacuum chamber which is first pumped to a vacuum better than 10^{-5} Pa and then backfilled with a controlled high-purity gas atmosphere. This gas atmosphere is typically a few hundred Pa of an inert gas such as He, if the condensed particles are to be subsequently reacted to form ceramic clusters, or of an appropriate reactive gas or gas mixture. Because the atoms or molecules being evaporated collide with the gas atoms inside the chamber, small discrete gas-borne particles are formed rather than the continuous films which are commonly produced by evaporating materials in a vacuum onto a substrate. The small ceramic clusters thus produced are then collected and consolidated in-situ in order to produce a bulk ceramic.

Much of the nanophase materials research at Argonne has concentrated on producing and

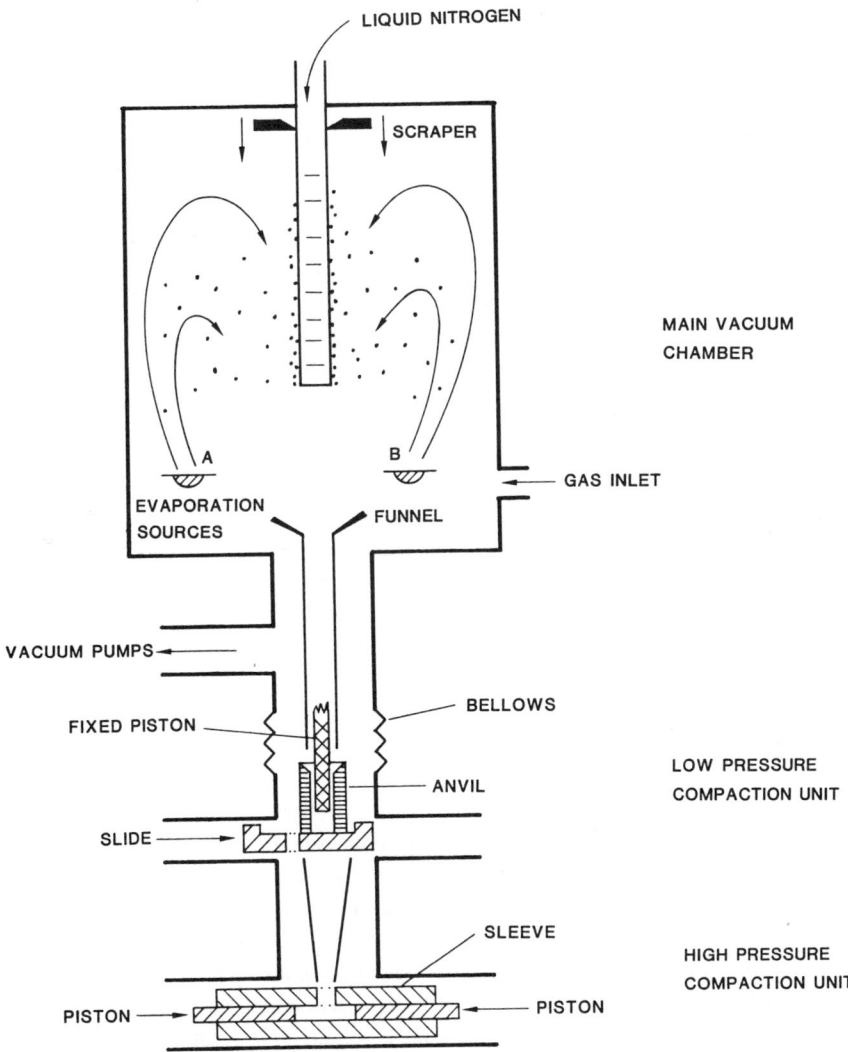

Figure 1. Schematic drawing of a gas-condensation chamber for the synthesis of nanophase materials. The material evaporated from sources A and/or B condenses in the gas and is transported via convection to the liquid-N_2 filled cold finger. The powders are subsequently scraped from the cold finger, collected via the funnel, and consolidated first in the low-pressure compaction device and then in the high-pressure compaction device, all in vacuum.

characterizing ceramic oxides. Three nanophase oxides have been produced to date: TiO_2, Al_2O_3 and MgO. Each requires a different procedure in order to be successfully produced. TiO_2, which has been studied in the most detail, is produced in a two-step process. First, Ti metal is evaporated in about 300 Pa of He to form ultrafine Ti particles, and then these particles are oxidized to the TiO_2 rutile phase by the rapid introduction of pure oxygen into the vacuum chamber at room temperature. The chamber is then evacuated again so that the oxide powders can be collected and consolidated under vacuum conditions, resulting in a nanophase compact with around 12 nm average sized grains. This particular technique is not successful for producing Al_2O_3 or MgO, however. Exposure of Al or Mg particles to oxygen in the chamber at room temperature does not transform the metal clusters to the desired crystalline oxide. Instead, the material remains largely untransformed, with perhaps a very thin amorphous coating on the surfaces (this coating must be extremely thin because it is not visible by x-ray diffraction even though the relative surface area of the particles is very large). Because of this, it has been necessary to search for different techniques to produce these oxides.

In order to produce Al_2O_3, Al metal clusters are produced and then annealed in air at 1000°C, resulting in their transformation to the stable α-phase of Al_2O_3 with very little increase in particle size. Subsequent room temperature consolidation of these powders produces nanophase α-Al_2O_3 with a grain size of approximately 18 nm. The effect of removing the powder from the vacuum chamber, which is a deviation from the standard nanophase processing method, has not been studied. Recently, ultrafine grain size MgO (5 nm diameter) has been produced for the first time by direct heating of MgO in the vacuum chamber in a few hundred Pa of He. Most oxides, including MgO, have very high melting points, and because of this, are not good candidates for evaporation by resistive heating. MgO is special, however, in that it has a very high vapor pressure at temperatures well below its melting point. This makes it possible to sublime sufficient MgO to make nanophase samples by simply heating it in 200 Pa of He to temperatures of around 1600°C (MgO melts at 2852°C). The material which sublimes is oxygen deficient, but is fully converted to stoichiometric MgO by subsequent exposure to oxygen introduced into the vacuum chamber. While most of the nanophase ceramics that have been produced to date have been oxides, the processing technique is by no means limited to such materials. Other compounds have been produced in powder form and it is only a matter of time before researchers begin examining the properties of new nanophase ceramics made from such powders. For example, Iwama and coworkers [7] have succeeded in producing a variety of transition-metal nitrides in ultrafine powder form by evaporating metals in either N_2 or NH_3 gas.

In the gas-condensation process, the main parameters which are controlled in order to produce particles of a desired size are the inert gas pressure, the evaporation rate, and the gas composition. The effects of varying these three processing parameters were first examined in detail by Granqvist and Buhrman [5] in the mid 1970's. They demonstrated that the particle size produced can be decreased by decreasing either the gas pressure in the chamber or the evaporation rate, and by using lighter inert gases such as He rather than heavier ones such as Xe. For example, smaller particles will result by using He than by using the same pressure of Xe. Generally, in producing nanophase materials, these parameters are adjusted to produce clusters having the smallest particle size while still maintaining reasonably fast evaporation rates, so that sufficient quantities of material can be produced in reasonable times. The conditions which are most often used are a few hundred Pa of He and evaporation temperatures which correspond to a vapor pressure of around 10 Pa.

Recently, experiments performed at Argonne have revealed that, at least in one case, control of the inert gas pressure affects not only particle size, but also the phase of the resulting material [8]. Ultrafine powders of Ti are found to react with oxygen to form rutile (TiO_2) particles of average diameter 12 nm, if He pressures of greater than 500 Pa are present during evaporation. However, if the He pressure used is less than 500 Pa and greater than about 10 Pa, small Ti particles are still formed, but these particles do not form rutile when exposed to oxygen. Instead, in this case, an unexpected amorphous phase is formed. This behavior is not observed for all materials, however. Small crystalline particles having the normal bulk material phase are formed, regardless of inert gas pressure, when Pd is evaporated. While the reason for this unusual behavior in Ti (recent experiments indicate that Si also shows the same behavior) is not yet understood, what is clear is that this demonstrates that phase control using nanophase processing is a potential future materials engineering application. Unexpected new phases of erbium oxides have also been produced via the nanophase processing route [9].

Collection of the powders formed in close proximity to the evaporation source is carried out by establishing convective currents inside the vacuum chamber. This is accomplished by designing the chamber so that it contains a collection device (cold finger) that can be cooled with liquid nitrogen. The gas in the chamber is heated by the hot evaporation source and cooled by this collection device, setting up a convective current which carries the condensed ultrafine powders to the cold finger. In the apparatus now in use at Argonne, the collection device is in the form of a hollow tube which can be filled with liquid nitrogen from outside the vacuum chamber. The particles collected on the surface of the cold finger form very open fractal structures as seen by transmission electron microscopy. A typical view of collected TiO_2 particles is shown in Figure 2(a). After completing the evaporation process, these particles are removed by scraping the material from the tube using a Teflon annular ring, which is moved downward along the length of the cold finger. The tube is in a vertical orientation so that gravity can be used in order to transfer the particles which collect on the cold finger to the in situ compaction devices located below the base of the main vacuum chamber.

The consolidation of the powders to form a bulk ceramic is presently accomplished using a two-stage compaction unit, as shown in Figure 1. The upper stage consists of a simple piston and anvil arrangement which operates using very low pressures in order to form a loosely compacted pellet. This pellet is then transferred under vacuum to a second unit, in which the pellet is

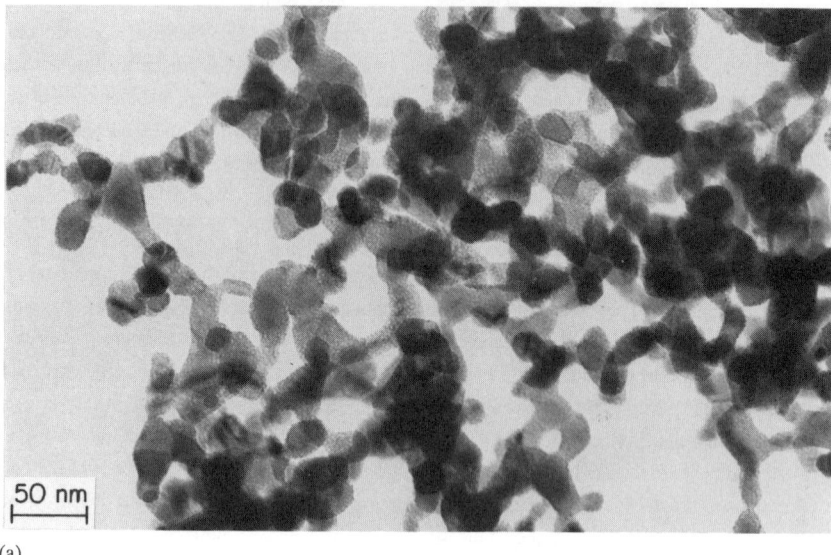

(a)

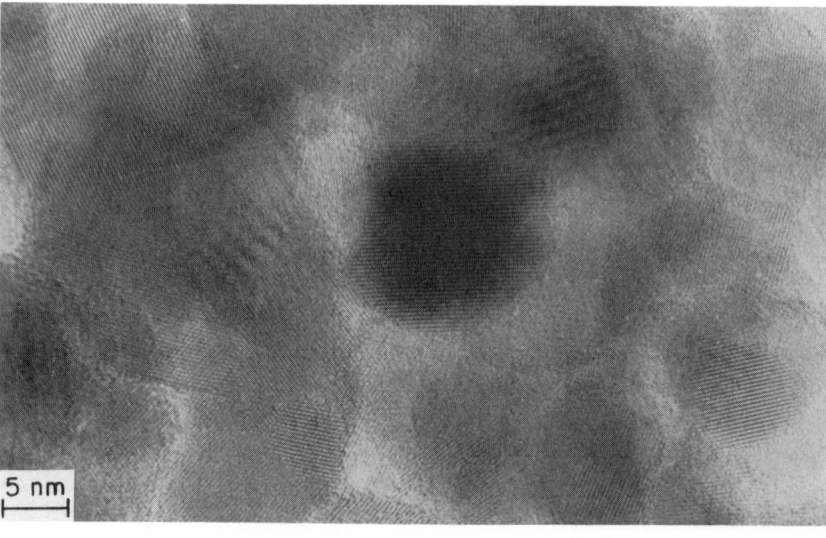

(b)

Figure 2. Transmission electron micrographs of (a) as-collected and oxidized TiO$_2$ particles and (b) nanophase TiO$_2$ (rutile) after in situ consolidation at room temperature and 1.4 GPa pressure and then sintering in air for one-half hour at 500°C. After Reference [10].

consolidated at higher pressures (typically 1.4 GPa) between two tungsten-carbide pistons. The scraping and consolidation is performed under UHV conditions after removal of the inert or reactive gases from the chamber, in order to maximize the cleanliness of the particle surfaces and the interfaces that are subsequently formed in the consolidated material, while minimizing any possibility of trapping remnants of these gases in the nanophase ceramic. Surprisingly high densities of the as-compacted samples have been measured, with values of about 50 to 80% of bulk density for ceramic oxides (TiO_2 and MgO). A typical view of a consolidated nanophase TiO_2 sample is shown in Figure 2(b). The fractal aggregation of the particles on the cold finger is not a hinderance to dense consolidation, in contrast to the usual problems confronted in consolidating conventionally aggregated ceramic powders. The resulting samples are disc shaped and are typically 9 mm in diameter and 0.1 to 0.5 mm in thickness. While these are rather small dimensions by normal industrial standards, the need at present is only to produce samples large enough to facilitate characterization of their structure and properties. If design modifications were to be made, there are no impediments which preclude scaling the process up to produce larger commercial-sized samples.

Despite the encouraging successes in producing nanophase ceramics using the above described techniques, room for future improvements clearly exists. One of the primary limitations at present is due to the standard use of Joule (resistive) heating for the evaporation of materials. This type of heating has several drawbacks, such as: (1) only a limited number of materials may be evaporated due to chemical reactions of many materials with the evaporation sources, (2) the evaporation temperatures that can be obtained are not high enough to evaporate many interesting materials having high melting temperatures or low vapor pressures, and (3) control of the evaporation temperature , which is important for maintaining a narrow size distribution of particles, is not ideal. Alternative methods of evaporating materials are now being examined. For instance, a new evaporation system, which utilizes a specially designed differentially pumped electron gun for evaporation, is currently being built at Argonne. This system will allow a variety of new nanophase materials to be produced. The additional control obtained with this type of system will be especially useful as one moves toward synthesizing more complex multicomponent nanophase ceramics, such as the high-T_c perovskite superconductors. In addition, the transport of the particles via gas convection can be improved upon by using, instead, forced gas motion, which will greatly facilitate the scaling up of the method to the collection of larger amounts of material than are normally produced (less than a few hundred mg) in the type of apparatus now in use.

STRUCTURE AND PROPERTIES

The structures of the nanophase ceramics that have been synthesized to date were investigated by a number of methods. These have included transmission electron microscopy, x-ray and neutron scattering, and positron-annihilation and Raman spectroscopy. The grains in nanophase compacts seen by transmission electron microscopy are typically rather equiaxed and retain the narrow log-normal size distributions representative of the particles formed in the gas-condensation method, as shown in Figure 3. The grain-size distributions are thus relatively unaffected by rection with gases, as in the oxidation process, or the consolidation process itself. Moreover, these distributions remain relatively stable to elevated temperatures. For example, as also shown in Figure 3, the 12 nm initial average grain diameter of an as-compacted TiO_2 sample is unchanged after 1/2-h sintering at 500°C, and only rises to about 20 nm after subsequent sintering to 700°C; grain growth to larger than 100 nm occurs only above 800°C [10]. Similar behavior is found for nanocrystalline metals as well.

A rather unique aspect of nanophase materials is the large fraction of their atoms that reside in grain boundaries, although this feature is shared with ultrafine multilayered materials. Atoms located within a certain distance of a boundary (i.e., within the grain boundatry thickness) are known to be displaced from their normal sites in a perfect crystal, due to the presence of atoms across the boundary. If one assumes a grain-boundary thickness of approximately 1 nm (i.e., about four nearest-neighbor distances), a nanophase material with a mean grain diameter of 5 nm will have nearly 50% of its atoms in its grain boundaries, and an even higher percentage for smaller grain sizes. This falls only to about 30% for a 10 nm grain size, but is as low as 3% at a 100 nm grain

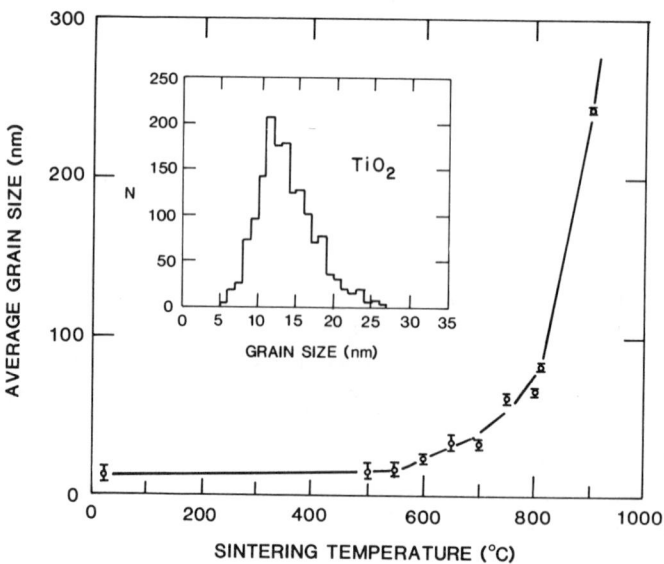

Figure 3. A typical grain-size distribution (inset) for an as-consolidated nanophase TiO_2 (rutile) sample measured from transmission electron micrographs and the variation of the average grain size with sintering temperature. After Reference [10].

size. The properties of these new materials are thus expected to be strongly influenced by their interface volume (which represents such a significant fraction of their total volume) and the changes in their electronic structure that are likely to result from these disturbed regions.

Because of the large fraction of atoms in the grain boundaries of nanophase materials, the interface structures can play a significant role in determining their properties. X-ray scattering studies on nanocrystalline metals have been interpreted by Gleiter and coworkers [11, 12] to indicate that their grain-boundary atomic structures may be random, rather than possessing either the short-range or long-range order found in the grain boundaries of coarser-grained polycrystalline materials [13]. Whether such a conclusion will be sustained in the future in the light of further experiments and calculations of the structure of and the scattering from nanophase boundaries is still an open question. Raman spectroscopy measurements, however, which probe local atomic arrangements by exciting active phonons with incident laser light, recently completed on the nanophase ceramic TiO_2 have shown no evidence for such random structures in the grain boundaries of this material [14]. While it is possible that the nature of the interfaces differs in different classes of materials, small-angle neutron scattering (SANS) measurements reported in these Proceedings indicate similar lower densities in the grain boundaries of both nanophase metals [15] and ceramics [16] than are found for bulk material. These results are not necessarily surprising, however, since grain boundaries are expected to possess more open structures than bulk material owing to their constituent atomic displacements [17]. Another important aspect of the interface microstructure of as-consolidated nanophase ceramics is the ultrafine-scale porosity inherent in their structures. Positron annihilation spectroscopy (PAS), because of its continuous range of sensitivity to open-volume defects ranging in size from single atomic vacancies to macroscopic voids, has been used to study the porosity in nanophase TiO_2 as a function of sintering temperature [10]. The small voids found by PAS to be present in this as-consolidated nanophase ceramic are removed by sintering. The relationships between this porosity and the structures deduced for nanophase interfaces from scattering studies are, however, not well understood at present, but SANS measurements [16] may be beginning to shed light in this area.

The properties of nanophase ceramics that have been measured appear to be rather different and often considerably improved in comparison with those of conventional ceramics. The synthesis of nanophase TiO_2 (rutile) has demonstrated [10] considerable improvements in both the sinterability and resulting mechanical properties of this material relative to conventionally synthesized rutile. Nanophase TiO_2 (rutile), with an initial average grain size of 12 nm, sinters at 400 to 600°C lower temperatures than conventionally prepared, commercial TiO_2 powders with 1.3 μm average diameters, as shown in Figure 4. The nanophase ceramic processing is performed without the need for the compacting and sintering aid polyvinyl alcohol (pva). On the other hand, without pva the commercial TiO_2 does not sinter. Also, the resulting hardness values and fracture characteristics [18] developed for the nanophase TiO_2 are better than those for normal TiO_2. It is expected that nanophase ceramics in general, with their high grain-boundary purity and ultrafine grain sizes, will

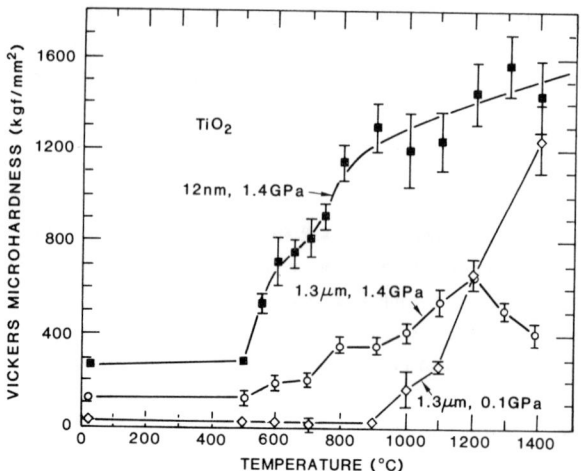

Figure 4. Vickers microhardness of TiO$_2$ (rutile) measured at room temperature as a function of one-half hour sintering at successively higher temperatures in air. Results for a nanophase sample (filled squares) with an initial average grain size of 12 nm consolidated at 1.4 GPa are compared with those for two conventional, coarser-grained samples with 1.3 μm initial average grain sizes consolidated from commercial powder at 0.1 GPa with (diamonds) and at 1.4 GPa without (circles) the aid of polyvinyl alcohol. The enhanced sinterability of the nanophase rutile is clearly demonstrated. After Reference [10].

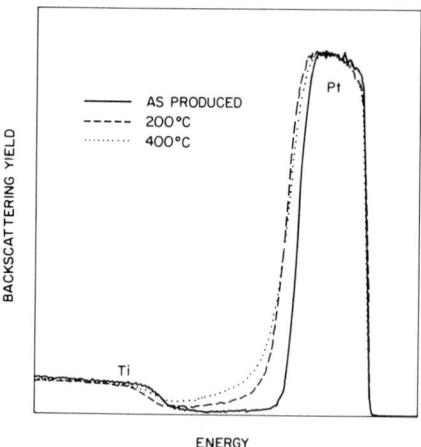

Figure 5. Rutherford (He) backscattering spectra from a nanophase TiO$_2$ sample with an ca. 100-nm thick Pt film deposited onto its surface: before annealing (solid curve), after 2 h at 200°C (dashed curve), and after an additional 2 h at 400°C (dotted curve). The rapid diffusion of Pt into the nanophase TiO$_2$ is demonstrated in the region between the Ti and Pt spectra.

sinter at much lower temperatures than conventional ceramics and will exhibit generally superior properties as well. For example, the lack of impurity-induced brittle phase formation at their interfaces and their small grain sizes may lead to more efficient deformation mechanisms and more effective crack dissipation in nanophase ceramics [19].

Another important aspect of nanophase ceramics (shared with other nanophase materials investigated), and one that should have a major impact on a number of the the eventual applications of these ceramics, is the rapidity of atomic diffusion in them. Experiments carried out on nanophase ceramics, and on metals as well [20], indicate that atomic transport is orders of magnitude faster in these materials than in normal polycrystalline samples and is similar in magnitude and temperature dependence to surface diffusion. This rapid diffusion appears to be intrinsically coupled with the nature of the interfaces (including both atomic structure and porosity) in nanophase materials. Indeed, such measurements could help to elucidate their interface structures along with the results from more direct characterization methods. For example, Rutherford back-scattering (RBS) investigations of Pt diffusion into nanophase TiO_2 have been carried out, and some results are shown in Figure 5. These experiments, which utilize the back-scattering of He ions to locate the relative depths of a sample's atomic constituents below its surface, clearly demonstrate enhanced diffusivities at relatively low temperatures in nanophase ceramics. The ultrafine grain sizes and high grain-boundary diffusivities in nanophase ceramics may even lead to their ductility at ambient temperatures. Indications of such behavior are already evident in the forming of ceramic nanophase pellets [10] and in their deformation behavior at low strain rates [21]. However, more sophisticated mechanical properties measurements are clearly needed before any firm conclusions can be drawn regarding such behavior. The possibilities seem great indeed, nonetheless, for being able to efficiently dope nanophase ceramics via rapid diffusion along their ubiquitous grain-boundary networks or for utilizing solid-state reactions to form stable or metastable phases in order to synthesize new ceramics with tailored optical, electrical, or mechanical properties.

Only limited research has been carried out to date on nanophase ceramics and, clearly, much work remains to be done in this interesting new research area before a full understanding will be developed of the relationships between their synthesis conditions, structures, and properties. In order to develop such an understanding, a number of aspects regarding nanophase ceramics, and other nanophase materials, need to be elucidated. Primary among these aspects is the characterization of the local structure of the grain boundaries and interfaces in these materials, which comprise such a large fraction of their volumes and affect their properties so dramatically. Beyond such structural studies, measurements of the electrical, optical, and mechanical properties of these new materials are certainly needed. Furthermore, the relationships between the nanophase processing parameters and the resulting materials' properties need to be elucidated, and the capabilities of the processing method itself need to be expanded to encompass a broader range of ceramics and ceramic-based composites. A knowledge of the variation of their properties with the structures and synthesis parameters of nanophase ceramics should lead to the ability to design new and improved ceramics with selected properties for a variety of technological applications.

This work was supported by the U.S. Department of Energy, BES-Materials Sciences, under Contract W-31-109-Eng-38.

REFERENCES

1. H. Gleiter, in **Deformation of Polycrystals: Mechanisms and Microstructures**, N. Hansen et al., eds. (Risø National Laboratory, Roskilde, 1981) p. 15.
2. R. Birringer, U. Herr, and H. Gleiter, Suppl. Trans. Jpn. Inst. Met. 27, 43 (1986).
3. R. W. Siegel and H. Hahn, in **Current Trends in the Physics of Materials**, M. Yussouff, ed. (World Scientific Publ. Co., Singapore, 1987) p. 403.
4. K. Kimoto, Y. Kamiya, M. Nonoyama, and R. Uyeda, Jpn. J. Appl. Phys. 2, 702 (1963).
5. C. G. Granqvist and R. A. Buhrman, J. Appl. Phys. 47, 2200 (1976).
6. A. R. Thölén, Acta Metall. 27, 1765 (1979).
7. S. Iwama, K. Hayakawa, and T. Arizumi, J. Cryst. Growth 66, 189 (1984).
8. J. A. Eastman, these Proceedings.
9. Z. Li, H. Hahn, and R. W. Siegel, Mater. Lett. 6, 342 (1988).
10. R. W. Siegel, S. Ramasamy, H. Hahn, Z. Li, T. Lu, and R. Gronsky, J. Mater. Res. 3, 1367 (1988).
11. X. Zhu, R. Birringer, U. Herr, and H. Gleiter, Phys. Rev. B 35, 9085 (1987).
12. T. Haubold, R. Birringer, B. Lengeler, and H. Gleiter, Nature, submitted (1988).
13. S. L. Sass, J. Appl. Cryst. 13, 109 (1980).
14. C. A. Melendres, A. Narayanasamy, V. A. Maroni, and R. W. Siegel, Mater. Res. Soc. Symp. Proc. 153, to be published (1989).
15. G. Wallner, E. Jorra, H. Franz, J. Peisl, R. Birringer, H. Gleiter, T. Haubold, and W. Petry, these Proceedings.
16. J. E. Epperson, R.W. Siegel, J.W. White, T. E. Klippert, A. Narayanasamy, J. A. Eastman, and F. Trouw, these Proceedings.
17. K. L. Merkle, J. F. Reddy, and C. L. Wiley, J. de Physique, Colloque C4, 46, 95 (1985).
18. Z. Li, S. Ramasamy, H. Hahn, and R. W. Siegel, Mater. Lett. 6, 195 (1988).
19. H. Hahn, J. A. Eastman, and R. W. Siegel, in Ceramic Transactions, **Ceramic Powder Science**, Vol. 1, Part B, G. L. Messing et al., eds. (American Ceramic Society, Westerville, 1988) p. 1115.
20. R. Birringer, H. Hahn, H. Höfler, J. Karch, and H. Gleiter, in **Diffusion Processes in High Technology Materials**, D. Gupta et al., eds. (Trans Tech. Publ., Aedermannsdorf, 1988) p. 17.
21. H. Karch, R. Birringer, and H. Gleiter, Nature 330, 556 (1987).

SINTERING OF NANOPHASE TiO$_2$ AT 550°C

J. E. EPPERSON,* R. W. SIEGEL,* J. W. WHITE,[†] T. E. KLIPPERT,*[#]
A. NARAYANASAMY*, J. A. EASTMAN,* and F. TROUW[††]
*Materials Science Division, Argonne National Laboratory, Argonne, Illinois 60439 USA.
[†]Research School of Chemistry, Australian National University, Canberra, Australia and Argonne Fellow, Argonne National Laboratory, Argonne, Illinois 60439 USA.
[††]Intense Pulsed Neutron Source, Argonne National Laboratory, Argonne, Illinois 60439 USA.

ABSTRACT

Samples of nanophase TiO$_2$ were prepared by the condensation of Ti vapors into clusters, their in situ oxidation to TiO$_2$, and their consolidation into thin disks. Small angle neutron scattering was measured in the as-consolidated condition and after selected isothermal sintering anneals of up to 23 h at 550°C. The maximum entropy analysis method was used to obtain the size distributions of the scattering centers from the scattering curves. The results are interpreted in terms of a microstructural model consisting of nanometer sized grains of TiO$_2$ separated by about 0.5 nm wide boundary regions, which contain voids and TiO$_2$ of $\leq$ 60-70% of bulk density.

INTRODUCTION

Nanophase materials have grain sizes in the range of 2 to 20 nm and are assembled from powders. As such, they may be metals, ceramics, or composites. They are usually produced by evaporation of metal vapors in a gas and controlled nucleation of metal clusters under ultraclean conditions, followed by their in situ consolidation in vacuum. Obviously, such materials have a high interface-to-volume ratio, with a significant fraction of their atoms residing in grain boundaries. As a result, their chemical and physical properties can differ markedly from those of respective coarser-grained bulk materials. The work done to date on nanophase materials and their potential for the future has been reviewed [1-3]. Since their ultrafine-scale microstructure is of key importance in determining the properties of nanophase materials, small angle neutron scattering (SANS) has been used in the present investigation to characterize the microstructure of nanophase TiO$_2$, upon which a number of complementary measurements have already been performed [4]. A more complete report of the present work will appear elsewhere.

[#]Now at Advanced Photon Source, Argonne National Laboratory, Argonne, Illinois 60439 USA. This work was supported by the U. S. Department of Energy, BES-DMS, under Contract W-31-109-Eng-38.

SAMPLE PREPARATION AND EXPERIMENTAL PROCEDURE

The nanophase TiO_2 samples used in the present investigation were produced by the gas-condensation method, which has been described elsewhere [4]. Titanium of 99.7% purity was evaporated from a resistance-heated tungsten boat at temperatures between 1550 and 1650°C into a 0.3-0.7 kPa He atmosphere over a period of 15 to 30 minutes. The Ti particles condensed in the He atmosphere were deposited on a cold-finger in the chamber and subsequently oxidized to TiO_2 by the rapid introduction of about 2 kPa of oxygen. After evacuation, these particles were scraped off the cold-finger and consolidated into disks 9 mm in diameter and about 0.5 mm thick.

The SANS measurements were carried out at the Argonne National Laboratory, Intense Pulsed Neutron Source (IPNS), using the SAD (small angle diffractometer) instrument. Neutrons are produced at IPNS when short bursts of energetic protons impact on a heavy metal target (depleted uranium). The resulting neutron spectrum was made more suitable for small angle scattering by use of a solid methane moderator operated at about 22 K. A single-crystal MgO filter maintained at liquid nitrogen temperature removed most of the unwanted fast neutrons. Collimation was achieved by use of a crossed pair of Soller collimators. In the present experiment, neutrons having a wavelength between 2.16 and 14.0 Å were employed. Scattering data were collected in a constant $\Delta\lambda/\lambda = 0.05$ binning scheme, using an area-sensitive, gas-filled proportional counter of the Kopp-Borkowski type [5] and with the sample maintained at room temperature. All scattering spectra were normalized to a common monitor value and corrected for the presence of a high-purity quartz cell, which held the TiO_2 compacts in a fixed position in the neutron beam during the measurements. The net sample scattering was then normalized with the sample thickness, with the distribution of neutron wavelengths incident on the sample, and with the wavelength-dependent transmission coefficients. These time-of-flight scattering data were binned into an S(q) array from $0.009 \leq q \leq 0.248$ Å^{-1}, conversion to absolute cross sections being accomplished by use of a secondary intensity standard obtained from Bates [6]. The present SANS data were analyzed using the maximum entropy method, first applied to small angle scattering by Potton and coworkers [7]. For this investigation, a generic maximum entropy computer program was obtained from Potton [8] and modified by the first author for small angle scattering analysis.

RESULTS AND DISCUSSION

Small angle neutron scattering was measured for three samples in the as-consolidated condition, which indicated excellent reproducibility of the nanophase synthesis method. For one of these samples, SANS was also measured after isothermal sintering in air at 550°C for times ranging from 15 min to 23 h. A survey of the changes produced in the scattering from this sample is shown in Figure 1. The scatter seen in the high-q region is a result of poor counting statistics traceable in part to the short-wavelength cutoff of the MgO filter and in part to our not using any neutrons with wavelengths below 2.16 Å. It is readily apparent from these data that small, but systematic,

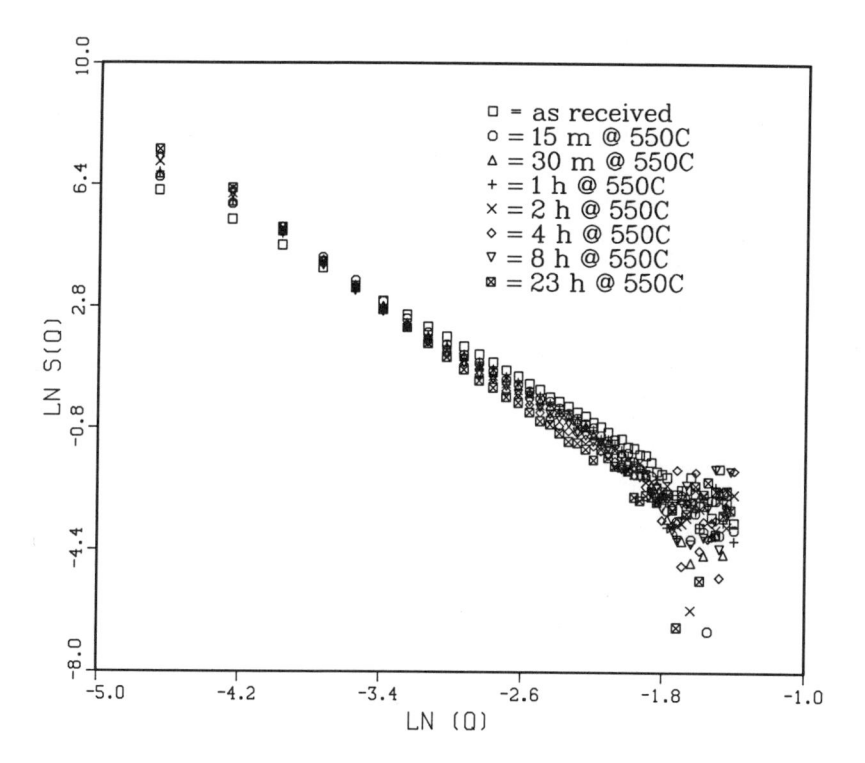

Fig. 1. Survey of changes in the SANS resulting from isothermal sintering of a nanophase TiO$_2$ compact at 550°C for times up to 23 hours. The sintering times are indicated.

changes are occurring as a result of the 550°C sintering. As a first crude means of assessing the magnitude of these changes, the Guinier radius was determined from the low-q limiting slope of a plot of ln S(q) vs. q^2 for each sample condition. The results found were an initial rapid increase in the Guinier radius, which slowed considerably after 2 h sintering. A cursory examination of the Guinier plots from this sintering series reveals marked curvature, however, indicating the presence in the sample of a distribution of sizes of the inhomogeneities responsible for the observed scattering. Since the Guinicr radius is heavily weighted (as r^6) in favor of larger scatterers, a more useful analysis method was sought.

The maximum entropy method (MEM) permits a more sophisticated analysis of these SANS data in terms of the size distribution of scattering centers. By applying this method, one can represent the measured small angle scattering in terms of a size distribution consistent with the intensity observations. Figure 2(a) shows the MEM size distribution determined for the as-consolidated TiO$_2$ sample, and, as an example of the changes produced by sintering at 550°C, that determined

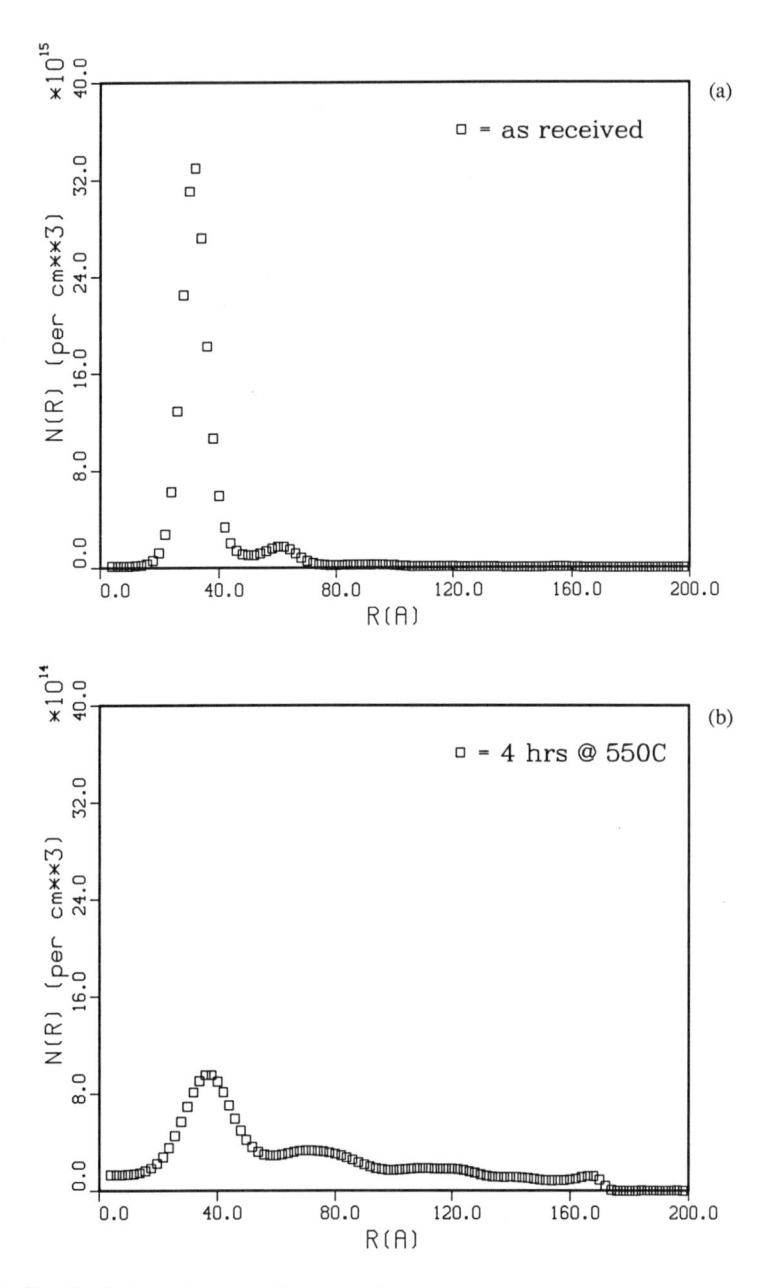

Fig. 2. Size distributions determined by the maximum entropy method for: (a) the as-compacted nanophase TiO$_2$ sample and (b) the same sample after isothermal sintering for 4 h at 550°C.

after 4 h sintering is shown in Fig. 2(b). Clearly, a coarsening process is taking place during 550°C sintering of nanophase TiO_2, and a more detailed report of this behavior will be forthcoming. The presence of well-defined peaks in the size distributions shown here is striking. The nature of the inhomogeneities causing such peaks cannot usually be resolved on the basis of small angle scattering alone. In special cases, however, contrast variation techniques can be utilized to resolve such issues [9].

Extraction of a volume fraction of scatterers from small angle scattering data requires that one know, or assume, the contrast. For the moment, we shall assume that the observed small angle scattering results from the difference in scattering length density between the nanocrystalline grains and homogeneous boundary regions separating such grains. One can then represent the scattering by the expression

$$S(q) = \int_{r=0}^{\infty} [V_p(r) \, \Delta\rho \, \Phi(qr)]^2 \, N(r) \, dr, \tag{1}$$

where $\Delta\rho$ is the difference in scattering length density, $V_p(r)$ is the volume of a particle of radius r, and for homogeneous spherical particles

$$\Phi(qr) = [3/(qr)^3] \, [\sin(qr) - (qr)\cos(qr)]. \tag{2}$$

If one has determined the volume fraction $[f_v(0)]$ of scattering entities, assuming that bulk TiO_2 is contrasted with vacuum (as was done in obtaining the MEM distributions shown in Fig. 2), one can immediately write down the volume fraction (f_v) of scatterers for any other $\Delta\rho$ value

$$f_v = f_v(0) / (1 - \delta_b)^2, \tag{3}$$

where δ_b is the fractional density (of bulk TiO_2) in the boundary region. From this expression, one can readily estimate the maximum fractional density of TiO_2 in the boundary regions, which, at this point, are still assumed to be homogeneous

$$\delta_b(max) = 1 - [f_v(0)]^{1/2}. \tag{4}$$

Using the stated assumptions, the $f_v(0)$ values were obtained from the MEM size distributions for the 550°C sintering series. The $f_v(0)$ value is 0.133 for the as-consolidated sample, goes through a maximum for sintering times up to 2 h (maximum value of 0.147 at 0.5 h) and then decreases monotonically to a value of 0.118 after 23 h at 550°C. Inserting the $f_v(0)$ values in Eq. 4, the maximum average fractional density of the boundary regions is found to vary between 0.617 and 0.656. Using the value for $f_v(0) = 0.133$ for the as-consolidated sample, and its known grain size (diameter = 12 nm) from electron microscopy observations [4], one can readily estimate the average width of the grain boundaries to be 0.5 nm (about two nearest-neighbor distances).

On the basis of positron annihilation experiments [4] on similar TiO$_2$ samples, it appears likely that a significant part of the first peak in the MEM distributions is due to the presence of voids (or pores). A model representation of the microstructure thought to exist in these nanophase TiO$_2$ compacts thus consists of theoretically dense grains of TiO$_2$, voids (or pores) between grains, and grain boundary regions having lower density than that of the grains. To illustrate the effect that including voids in this model for the microstructure would have on the remaining boundary density, the first peak has been eliminated for the as-compacted TiO$_2$ sample shown in Fig. 2(a) and δ_b(max) recalculated. One finds δ_b(max) increases from 0.635 to 0.718. It must be emphasized that the δ_b(max) values discussed here are absolute upper limits of the fractional density of TiO$_2$ in the boundary regions, since this parameter results from our setting f_v to unity in Eq. 3. In reality, the fractional density in the boundary regions must be somewhat lower; however, the present experiment does not permit us to extract this information.

In conclusion, we have shown that the maximum entropy method of analysis yields much more information than would a more conventional SANS analysis. In particular, the distribution of particle sizes giving rise to the observed scattering is readily obtained. We have been able to establish an upper limit of the fractional density of TiO$_2$ in the boundary regions between adjacent grains; for the as-consolidated material, this value is 0.635 if the boundaries are assumed to be homogeneous and 0.718 if the first MEM peak can be assigned to voids. In addition, the width of the grain boundaries could be estimated to be about 0.5 nm. The model proposed to represent the microstructure of the compacted and sintered nanocrystalline TiO$_2$ consists of the following: (1) theoretically dense, nanometer-sized grains of TiO$_2$, (2) voids (or pores) between these grains, and (3) grain boundary regions having lower ($\leq$ 60-70%) average density than that of the grains themselves and average widths of about 0.5 nm.

REFERENCES

1. R. Birringer, U. Herr and H. Gleiter, Suppl. Trans. Jpn. Inst. Met. 27, 43 (1986).
2. R. W. Siegel and H. Hahn, in **Current Trends in the Physics of Materials**, M. Yussouff, ed. (World Scientific Publ. Co., Singapore, 1987) p. 403.
3. R. W. Siegel and J. A. Eastman, these Proceedings.
4. R. W. Siegel, S. Ramasamy, H. Hahn, Z. Li, T. Lu, and R. Gronsky, J. Mater. Res. 3, 1367 (1988).
5. C. J. Borkowski and M. K. Kopp, J. Appl. Cryst. 11, 430 (1978).
6. F. S. Bates, private communication; F. S. Bates, J. Appl. Cryst. 21, 681 (1988).
7. J. A. Potton, G. J. Daniell, A. D. Eastop, M. Kitching, D. Melville, S. Poslad, B. D. Rainford, and H. Stanley, J. Magnetism and Magnetic Mater. 39, 95 (1983).
8. J. A. Potton, private communication, January 1988.
9. J. A. Eastman, J. E. Epperson, H. Hahn, T. E. Klippert, A. Narayanasamy, S. Ramasamy, R. W. Siegel, J. W. White, and F. Trouw, these Proceedings.

SMALL ANGLE NEUTRON SCATTERING FROM NANOPHASE TITANIUM AS A FUNCTION OF OXIDATION

J. A. EASTMAN, J. E. EPPERSON, H. HAHN,* T. E. KLIPPERT#, A. NARAYANASAMY[†], S. RAMASAMY[†], R. W. SIEGEL, J. W. WHITE,[††] and F. TROUW**
Materials Science Division, Argonne National Laboratory, Argonne, Illinois 60439 USA

ABSTRACT

Nanophase titanium, prepared by the gas-condensation method both as aggregated powder and in lightly compacted discs, has been studied by conventional small angle neutron scattering, and by use of contrast variation methods. The contrast has been changed (a), isotopically, by means of deuterated/protonated solvents distilled into the specimen and (b) by progressive incremental oxidation of the Ti particles using fixed doses of low-pressure oxygen. It was shown that some evolution of the small angle pattern for lightly compacted nanophase Ti occurred over a period of several months at 300 K. Contrast matching by external solvent works well and has allowed the scattering lengths of oxidized and unoxidized specimens to be followed. The results imply that the scattering from metal and oxide can be separated under suitable conditions. The partial oxidation experiments indicate that there is both a fast and slow oxidation at 300 K. Also, during slow oxidation, high scattering length density scattering centers were formed whose number increased, but whose size remained the same at about 2 nm; these centers are tentatively assumed to be TiO_2.

INTRODUCTION

Nanophase materials are consolidated under vacuum conditions from atomic clusters usually formed by condensation in an inert gas [1]. They are thus ultrafine-grained materials in which the individual grains are narrowly distributed about an average grain size, usually about 5 to 10 nm, and hence a large fraction of their volume is comprised of interfaces. When lightly consolidated, as in the present study, considerable open volume remains between grains, which is accessible to gaseous intrusion and subsequent reaction. In the present study of the oxidation of lightly compacted nanophase Ti, the evolution of the small angle neutron scattering (SANS) pattern as a function of oxidation was investigated. Both the effects of fast oxidation in gaseous oxygen, with

---

*Now at Materials Research Laboratory, University of Illinois, Urbana, Illinois 61801 USA
#Now at Advanced Photon Source, Argonne National Laboratory, Argonne, Illinois 60439 USA
[†]Permanent address: Department of Nuclear Physics, University of Madras, Madras, India
[††]Research School of Chemistry, Australian National University, Canberra, Australia and Argonne Fellow, Argonne National Laboratory, Argonne, Illinois 60439 USA
**Intense Pulsed Neutron Source, Argonne National Laboratory, Argonne, Illinois 60439 USA

Table 1. Coherent scattering lengths, b_{coh}, and scattering length densities, $\bar{\rho} = [N_A \rho / M] \Sigma_i b_i$, where N_A is Avogadro's number, ρ is the physical density, M is the molecular weight, and the b_i are the coherent scattering lengths per atom.

Material	b_{coh} (10^{-13} cm)	ρ (g cm^{-3})	M (g)	$\bar{\rho}$ (10^{10} cm^{-2})
Ti	-3.36	4.54	47.9	-1.92
TiO$_2$	8.26	4.23	80	2.63
TiO	2.45	4.23	64	0.97
C$_6$D$_6$	79.9	0.87	90	4.65
C$_6$H$_6$	17.5	0.87	78	1.18
CS$_2$	12.3	1.27	76	1.24

consequent formation of TiO$_2$ [1], and the nature of the oxidation produced by sequential small doses of oxygen at low partial pressures (slow oxidation) were investigated; however, only the latter results will be presented here because of space constraints. A complete report of our nanophase Ti oxidation experiments, including contrast variation studies, will appear elsewhere.

SANS is interesting for these oxidation studies for a number of reasons. As shown in Table 1, the negative scattering length density of titanium provides high contrast to both voids and the TiO$_2$ produced in the oxidation reaction. Furthermore, the $\bar{\rho}$ values indicate that a fortuitous null matrix may arise at the composition TiO$_{0.82}$ in the event of intimate mixing of Ti and TiO$_2$ in the compacts at a certain stage in the slow oxidation procedure. In addition, the possibilities of using various mixtures of the organic solvents shown can allow for suppression of the scattered intensity from Ti oxides. The study becomes further interesting in that the small angle diffractometer (SAD) at the Argonne National Laboratory-Intense Pulsed Neutron Source (IPNS) has the capability of monitoring the size scale for density discontinuity scattering from about 2 to 60 nm, without the necessity of altering the sample-detector distance.

SAMPLE PREPARATION AND EXPERIMENTAL PROCEDURE

Nanophase Ti particles of about 10 nm average diameter (determined by transmission electron microscopy) were produced by the gas-condensation method [2], deposited on a cold finger, scraped off and then lightly consolidated to form nanophase Ti pellets about 8 mm in diameter and 0.5 mm thick. The pellets were transferred under vacuum to high-purity silica cells and sealed off. These cells were provided with "break-off tips" which allowed O$_2$ to be subsequently introduced. Scattering measurements were made on the IPNS-SAD instrument, with normalization and conversion to absolute units being carried out as described in [3].

RESULTS AND DISCUSSION

There was essentially no change in the SANS pattern for a nanophase Ti sample between its measurements one and three days after production, and good reproducibility in the unoxidized state was demonstrated, as shown in Figure 1. However, samples investigated three and half months after fabrication had a lower Porod slope than did the fresher samples indicating some changes over this longer period. The effects produced on the small angle scattering by slow dosing of a nanophase Ti sample with oxygen are also summarized in Figure 1, in which it is shown that oxygen dosing produces a general increase in the scattered intensity S(q) at high q values (at ln q's above about -3.6) and a slight decrease at lower q's (momentum transfer $q = \{4\pi/\lambda\} \sin\theta$). The Porod slope before oxidation is nearly -4, as would be expected from smooth surfaces. During the slow oxidation, this parameter changes to a nearly constant value of -3, and interestingly it

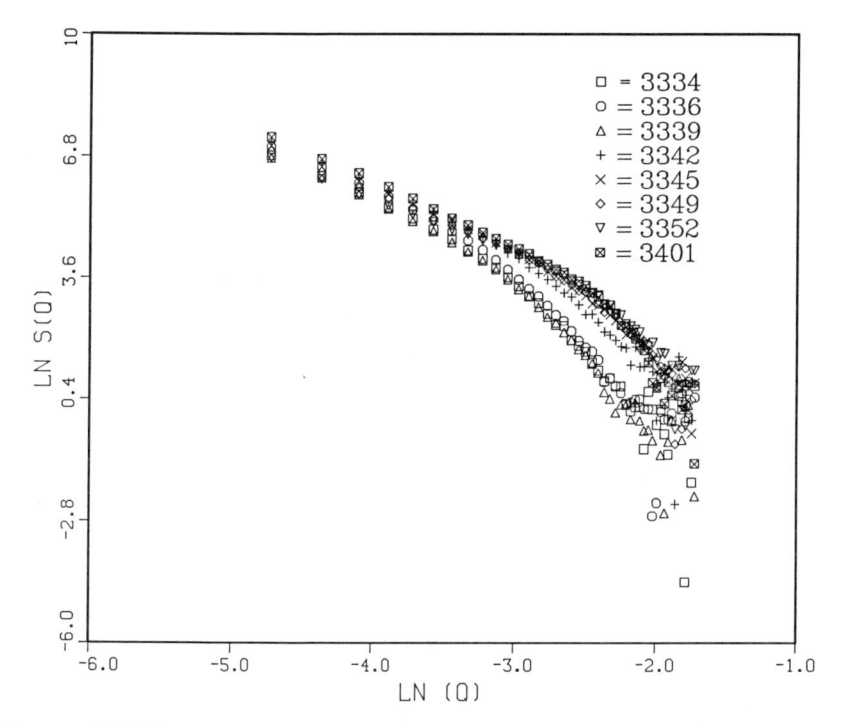

Figure 1. SANS data presented in ln S(q) vs ln q plots illustrating the reproducibility of scattering from two as-produced samples (3334 and 3336) and the effects of sequential dosing of a lightly compacted nanophase Ti sample (3334) with oxygen. The sequence of states of this latter sample investigated is as follows: 3334 = initial state (Ti); 3339 = 3334, 2 h later (Ti); 3342 = 3339 + dose 1 ($TiO_{0.115}$); 3345 = 3342 + dose 2 ($TiO_{0.37}$); 3349 = 3345 + dose 3 ($TiO_{0.44}$); 3352 = 3349 + dose 4 ($TiO_{0.83}$); 3401 = fully oxidized.

approaches -4 again after completion of the slow oxidation and exposure to excess oxygen. The behavior at intermediate dosing may be due to scattering from rough (or fractal type [4]) surfaces owing to only partial oxidation. The value of q at which ln S(q) vs ln q changes slope becomes progressively smaller as slow oxidation progresses. This can be represented as a characteristic length which, for example, decreases from a value of about 20 nm before dosing to about 7 nm after dosing to the composition $TiO_{0.44}$. Knowing the scattering length difference at the discontinuity which constitutes the interface, one can infer the surface area from the absolute difference in the Porod radii. Guinier analysis in selected q-regions shows that the radii of gyration (R_g) are relatively insensitive in the low q limit ($0.005 \leq q \leq 0.012$ Å^{-1}) to oxygen dosing. There is, however, a substantial linear region in the Guinier plots ($0.08 \leq q \leq 0.15$ Å^{-1}) which yields a relatively constant (actually slightly decreasing) R_g value of about 2 nm and a marked increase with oxygen doping of the intensity extrapolated to zero scattering angle S(q=0). Contrast variation techniques and an analysis of the radial scattering density distribution as a function of contrast may allow one to establish the factors contributing to the measured Guinier radii in such selected regions.

As an aid in understanding these observations, consider a very simple model for the progress of oxidation at the walls of a void. This model follows the discussion of the mechanism of oxidation of thin titanium films [5], where the growth of the oxide layer occurs at the oxide/metal interface by diffusion of oxygen through an oxide scale. The present model is clearly an oversimplification of the inherently complex nanophase Ti/TiO_2 system at hand, but preliminary useful information regarding the nature of the scattering centers can be obtained nevertheless. Consider, for simplicity, a spherical void of radius r_o formed by touching Ti metal clusters in a lightly compacted nanophase sample as illustrated schematically in Figure 2(a). Oxygen is considered to penetrate into this void and, as a result of the molar volume increase in formation of TiO_2 (+79.7%), there would be a reduction in the physical radius of the cavity as oxidation progresses, as illustrated in Figure 2(b). Because of the very different scattering length densities of Ti (metal), voids, and fully oxidized TiO_2, the scattering entity shown becomes more complex; the profiles of the scattering length density in a cut through the sample before and during oxidation are illustrated in the lower parts of Figures 2(a) and (b), respectively. Two extreme variants of this model may result: (1) inward oxide growth is allowed, which may result in some voids being closed off to oxidation and others being completely filled; and (2) oxide growth is allowed only into metal particles, without filling of the voids. This implicitly assumes formation of a TiO_2 layer which migrates into the Ti metal; however, it cannot be ruled out that a series of lower oxides of Ti may form as well. The different possibilities would have a dramatic consequence on the resulting SANS profile. Thus, such a simplified model will allow quantitative estimates of the radius of gyration and its variation with the degree of oxidation that can be compared with experimental observations.

For the slow oxidation experiment, Figure 3 shows the dependence upon oxidation of the radius of gyration and of the square root of the scattered intensity, extrapolated to zero scattering angle. As

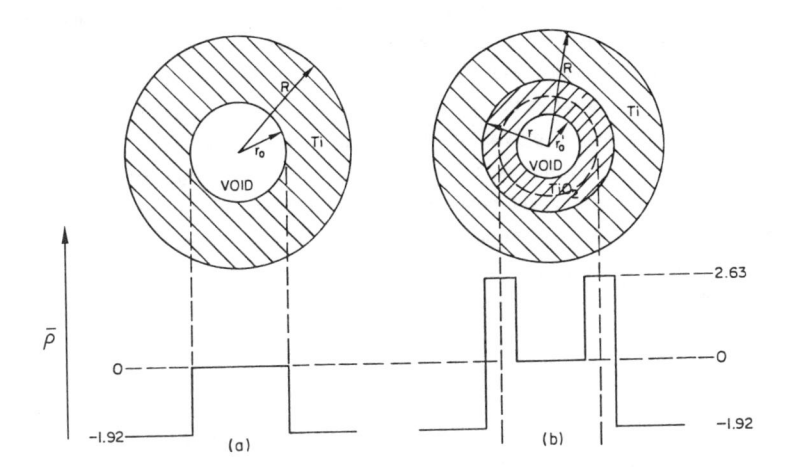

Figure 2. Model representation of the oxidation of nanophase Ti: (a) a spherical void in unoxidized Ti and (b) a partially oxidized sample. The profiles of the scattering length density $\bar{\rho}$ in a cut through each model sample are also illustrated.

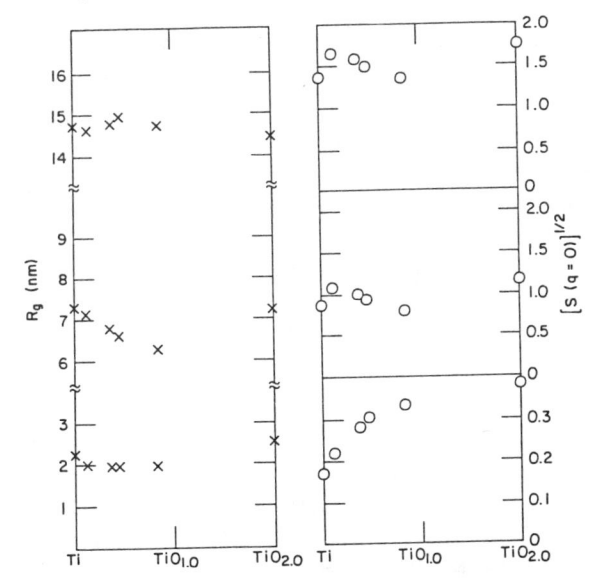

Figure 3. Oxidation of nanophase Ti. Left side: radii of gyration R_g determined from different regions in the scattering curve. Right side: corresponding extrapolated $[S(q=0)]^{1/2}$ values for the same q-regions. The upper portion of the figure corresponds to $0.005 \leq q \leq 0.012$ Å^{-1}, the middle portion to $0.012 \leq q \leq 0.033$ Å^{-1}, and the lower portion to $0.08 \leq q \leq 0.15$ Å^{-1}.

noted in the figure caption, these parameters were determined for the regions in q-space as follows: $0.005 \leq q \leq 0.012$ Å^{-1} (R_g~15 nm); $0.012 \leq q \leq 0.033$ Å^{-1} (R_g~7 nm); and $0.08 \leq q \leq 0.15$ Å^{-1} (R_g~2 nm). Unfortunately, insufficient data in the vicinity of the possible contrast match point are available presently to define the course of the curves on the right side of this figure. Further experiments are planned to provide the additional information needed. Some preliminary conclusions may be drawn, however. The scattering at low and intermediate q values may be assigned to voids whose size and contrast with respect to the surrounding matrix decrease as oxidation progresses. On the other hand, it seems reasonable to interpret the higher q scattering (into which intensity is displaced by oxidation) as coming from a high scattering length particle whose number (but not size) increases with oxidation; this is probably TiO_2.

CONCLUSIONS

There appear to be some differences detectable by SANS in the void scattering from freshly prepared and three month old titanium nanophase materials. The older sample, submitted to fast oxidation, gave SANS patterns consistent with growth of TiO_2 clusters of size about 10 nm at full oxidation (consistent with TEM observations [1]), but void scattering dominated the pattern from the unoxidized and lightly compacted material. By controlled, slow oxidation of a fresh sample, quite a different evolution of the size distribution dominating the scattering was produced. The most significant new feature is growth of a size distribution in the region of 2-3 nm radius, which is attributable to TiO_2 particles. This growth is associated with increased roughness of the void surfaces, but the final, fully oxidized material appears to be smooth on a distance scale of about 1 nm. Additional oxidation experiments, particularly in the vicinity of the contrast match point, are necessary to permit firm conclusions to be drawn. These experiments should provide detailed information about the internal structure of the partially oxidized particles.

This work was supported by the U. S. Department of Energy, BES-DMS, under Contract W-31-109-Eng-38.

REFERENCES

1. R. W. Siegel, S. Ramasamy, H. Hahn, Z. Li, T. Lu, and R. Gronsky, J. Mater. Res. 3, 1367 (1988).
2. R. W. Siegel and J. A. Eastman, these Proceedings.
3. J. E. Epperson, R. W. Siegel, J. W. White, T. E. Klippert, A. Narayanasamy, J. A. Eastman, and F. Trouw, these Proceedings
4. J. Teixeira, J. Appl. Cryst. 21, 781 (1988).
5. M. D. Bui, C. Jardin, J. P. Gauthier, G. Thollet, and P. Michel, in **Titanium Science and Technology '80** (Proc. 4th Intl. Conf on Titanium, Kyoto, 1980) H. Kimura and O. Izumi, eds.(The Metallurgical Society, Warrendale, 1980) p. 2819.

PHASE CONTROL IN NANOPHASE MATERIALS FORMED FROM ULTRAFINE Ti OR Pd POWDERS

J. A. Eastman
Materials Science Division, Argonne National Laboratory, Argonne, IL 60439

ABSTRACT

The effect of He gas pressure during evaporation and post-evaporation O_2 exposure on phase formation in nanophase materials has been examined. Ultrafine powders of Ti and Pd were prepared by the inert gas-condensation technique and bulk nanophase samples were formed by consolidation of these powders with or without prior exposure to O_2. Evaporation of Ti in He pressures greater than 500 Pa followed by exposure to O_2 results in the formation of ultrafine powders of crystalline rutile (TiO_2) which are compacted to form nanocrystalline TiO_2. Surprisingly, reducing the He pressure used during evaporation to less than approximately 500 Pa results in the formation of ultrafine powders of an amorphous phase. Room temperature consolidation of this powder under vacuum, however, results in nanocrystalline Ti being formed if the powder is not exposed to O_2 prior to consolidation, and a mixture of rutile and an unidentified crystalline phase if the powder has been previously exposed to O_2. Further reduction of the He pressure during evaporation of Ti to less than approximately 10 Pa results in the formation of crystalline Ti having a film-like morphology rather than than the desired ultrafine particles. Experiments on Pd evaporated in 10 to 6000 Pa of He have yielded ultrafine powders and consolidated samples of only the crystalline fcc phase, regardless of the He pressure.

INTRODUCTION

Ultrafine-grained (nanophase) materials formed by the solid-state consolidation of small particles have been found to possess many interesting and advantageous properties compared to materials having normal, larger grain sizes [1-3]. One reason which can be cited for interest in the study and development of nanophase materials is that the potential exists for using nanophase processing techniques to produce new materials, which can not be otherwise obtained. This potential derives from the small grain size of nanophase materials, which results in larger and more reactive interfacial areas (due to small radii of curvature), and also from the possibility of mixing materials on a very fine scale, which results in short diffusion distances during alloying. Metastable material production should be possible in some cases, since the relative energies of formation of phases can be altered by the increase in surface area which is obtained with small particles.

Most of the nanophase work which has been published to date has concentrated on producing and characterizing materials consisting of well-known phases, such as pure, single-component metals. Only one previously published study has noted the appearance of new, unreported phases as a result of nanophase processing [4]. In that study, Li et al. noted the formation of a previously unknown erbium oxide phase when ultrafine erbium powders were exposed to oxygen. While the effect of varying nanophase processing parameters on resulting particle size has been well characterized [5], the role of these parameters in controlling phase formation has not been investigated. The purpose of the present study is to examine the effects of processing parameters on phase formation in nanophase materials formed from Ti and Pd powders.

EXPERIMENTAL PROCEDURES

Ultrafine Ti and Pd powders were prepared by the gas-condensation process, which involves evaporating metals in an inert gas atmosphere. The powders are collected as fractal agglomerations on a liquid-N_2 cooled cold finger in a vacuum chamber. Bulk nanophase samples were formed by in situ consolidation of these powders with or without prior exposure to O_2. A schematic view of the system used is shown in Figure 1. The primary processing parameters controlled in producing ultrafine powders are the vapor pressure of the material being evaporated (the evaporation temperature), the composition of the inert gas and the inert gas pressure during evaporation. Increasing either the inert gas pressure or the evaporation temperature is known to increase the resulting particle size. Likewise, evaporating in a heavy inert gas such as Xe results in larger powders than are obtained using the same pressure of a lighter gas such as He [5].

In the current study, the inert gas used was He (99.999% purity) and the vapor pressure of the material being evaporated was maintained at approximately 10 Pa. This was achieved by evaporating both materials at just above their melting points (1552°C for Pd and 1668°C for Ti at atmospheric pressure). Pd was evaporated from ceramic boats, while tungsten boats were used for the evaporation of Ti. The time of evaporation varied from a few minutes to as long as one hour. The vacuum chamber was evacuated to a pressure of 10^{-5} Pa prior to backfilling with He. The role of He pressure was examined by producing samples using He pressures ranging from 10 - 6000 Pa. Bulk Pd and Ti nanophase samples were produced without any exposure

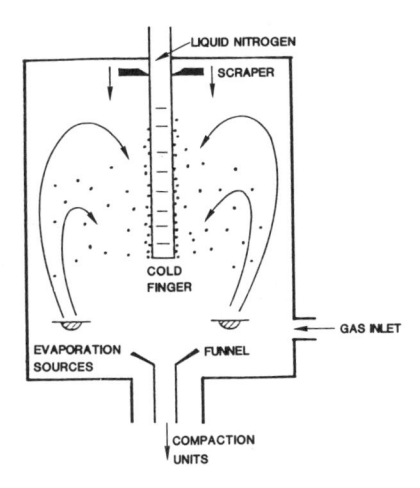

Figure 1. Schematic drawing of the main vacuum chamber of a gas-condensation system for the production of nanophase materials. Material evaporated from the sources condenses in an inert gas atmosphere and is transported via convection to a liquid-N_2 filled cold finger. The powders are scraped from the cold finger, collected via a funnel and compacted under vacuum in attached compaction devices. After reference [3].

of the powders to O_2, while nanophase TiO_2 was produced by oxidizing Ti powders prior to compaction by the rapid introduction to the vacuum chamber of approximately 3 kPa of O_2 at room temperature. Bulk samples were compacted under vacuum conditions using pressures of 1.4 GPa and the resulting samples were disc-shaped with diameters of 8 mm and a typical thickness of a few hundred microns.

Electron microscopy observations were carried out using either a Phillips 420 microscope operating at 120 kV or a JEOL 100CX operating at 100 kV. Both bright field (BF) and dark field (DF) diffraction-contrast imaging and selected area diffraction (SAD) techniques were used to characterize the materials produced. Powder samples required no special preparation in order to be examined by TEM, while bulk samples were thinned by jet-polishing. X-ray diffraction studies of bulk samples were done in reflection mode using a Phillips x-ray generator with a Cu target operated at 45 kV and 30 ma together with a θ-2θ diffractometer.

RESULTS AND DISCUSSION

TiO_2 powder which has been produced by evaporation of Ti in 500 Pa of He followed by exposure to O_2 by rapid introduction of the gas to the vacuum chamber is seen in Figure 2. The material is the crystalline rutile phase and has a nominal particle size of 12 nm. The powder is identical in appearance to TiO_2 powder produced in previous studies of nanophase TiO_2 under similar conditions [2]. Decreasing the He pressure during evaporation to less than 500 Pa, but more than approximately 10 Pa, and exposing the resulting powder to either air or O$_2$ produces material which is strikingly different than material produced at the higher gas pressures. In Figure 3 it is seen that material produced under these conditions yields a diffraction pattern characteristic of an amorphous material rather than the crystalline rutile pattern seen in Figure 2. Furthermore, the DF image seen in Figure 3 shows only a very low overall background contrast rather than strong contrast from individual crystallites which is clearly seen in Figure 2. Close inspection reveals only a few extremely small crystallites (of size $\leq$ 2 nm) in the powder material produced using 60 Pa of He. Thus, reduction of the He pressure during evaporation, which is known to reduce the resulting particle size, has resulted in the production of a different phase in this case.

Further evidence that a phase change has occurred as a result of reducing the He pressure is seen by observing the color of the materials before and after exposure to O$_2$. Ti evaporated in $\geq$ 500 Pa He pressures always appears black on the cold finger prior to oxidation. When this material is exposed to oxygen a noticeable "flash"

Figure 2. TiO$_2$ powder prepared by evaporating Ti in 500 Pa of He and then subsequently exposing the powder to O$_2$. (a) BF image of powder collected in situ directly on a TEM grid. (b) DF image which shows individual crystallites of 12 nm nominal size. (c) SAD pattern identifying the material as the rutile phase of TiO$_2$.

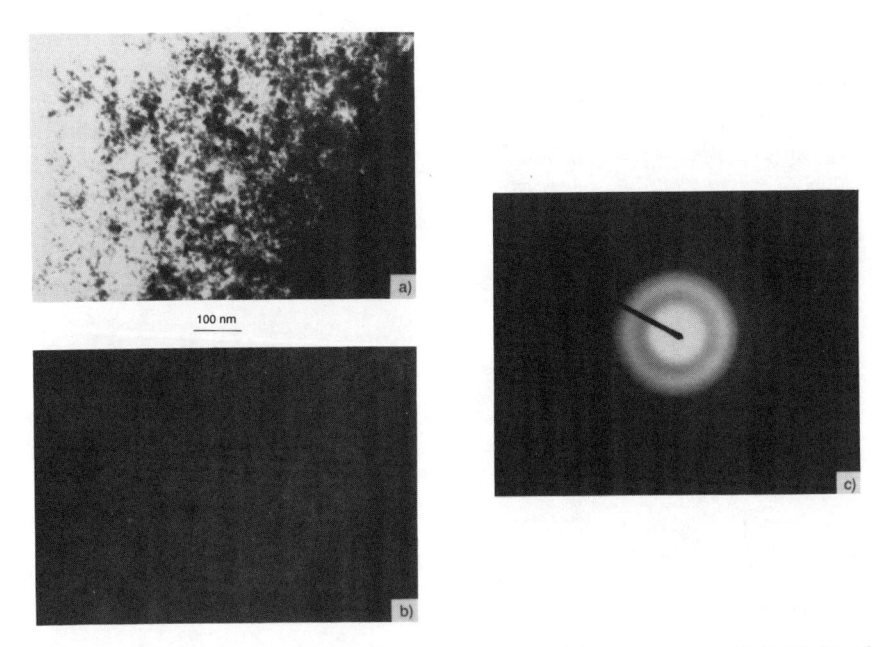

Figure 3. Ti powder of unknown O$_2$ content prepared by evaporating Ti in 60 Pa of He followed by exposure to O$_2$. (a) BF image. (b) DF image formed using a portion of the diffuse ring seen in the diffraction pattern showing uniform low background contrast. (c) SAD pattern revealing the powder to be amorphous.

is observed as the material spontaneously oxidizes, and the resulting TiO$_2$ always has a characteristic bluish-white color. In contrast to this behavior, Ti evaporated in lower pressures of He is grey-colored both before and after exposure to O$_2$. No visible evidence for oxidation of this powder is seen, although preliminary electron energy loss spectroscopy results indicate that material which has been exposed to air does contain oxygen (but less than TiO$_2$).

Reduction of the He pressure used during the evaporation process to less than about 10 Pa results in another change in the resulting material. In this case large-grained crystalline Ti having the appearance of a metallic film rather than that of ultrafine particles is obtained. This is perhaps not surprising because the mean free path between collisions with He atoms increases with decreasing He pressure and, eventually, is larger than the distance between the evaporation source and the cold finger. Pd powders have also been produced using a wide range of He pressures

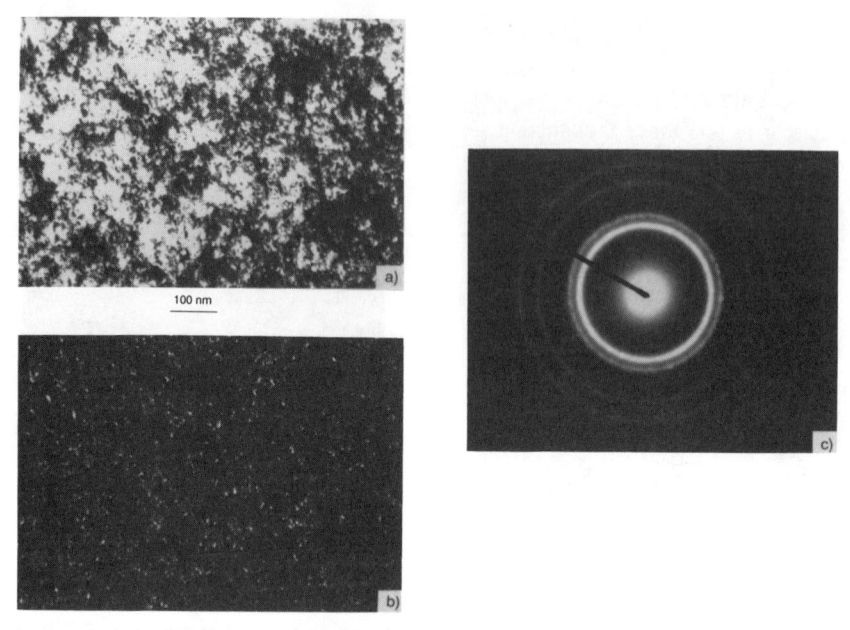

Figure 4. Pd powder prepared by evaporating Pd in 500 Pa of He. The powder was exposed to air prior to TEM observation. (a) BF image. (b) DF image showing the characteristic 5 nm diameter size of this powder. (c) Corresponding SAD pattern identifying the material as being the normal fcc phase.

and, unlike the case of Ti, no noticeable effect of He pressure on phase formation is observed. Pd produced using 500 Pa of He (Figure 4) is essentially identical to that produced using 10 Pa. In both cases, normal crystalline fcc particles nominally 5 nm in size are obtained.

Compacted Ti-based samples also show differences in phase depending upon the He pressure during powder production and on whether or not the powders are exposed to O_2 prior to compaction. These results are seen in the x-ray spectra of Figure 5. If the original powders are produced using He pressures $\geq$ 500 Pa, the bulk nanophase material is of the same phase (that is, rutile if the powders have been exposed to O_2 and Ti metal if no O_2 exposure has occurred). However, powders which are amorphous after air or O_2 exposure (< 500 Pa He pressures) yield bulk samples which are always crystalline, but the crystalline phase or phases present depend on whether or not the powder was exposed to O_2 prior to compaction. Samples compacted without prior O_2 exposure contain only crystalline Ti metal typically having grain sizes of 20 - 50 nm (significantly larger than that obtained if a 500 Pa He pressure is used, even though a smaller pressure is expected to result in a smaller grain size). This may be an indication that either grain growth or a crystallization of amorphous material occurs in this case, although it has not been possible to test these theories since the existing experimental apparatus does not allow characterization of powder without exposure of the material to air. If,

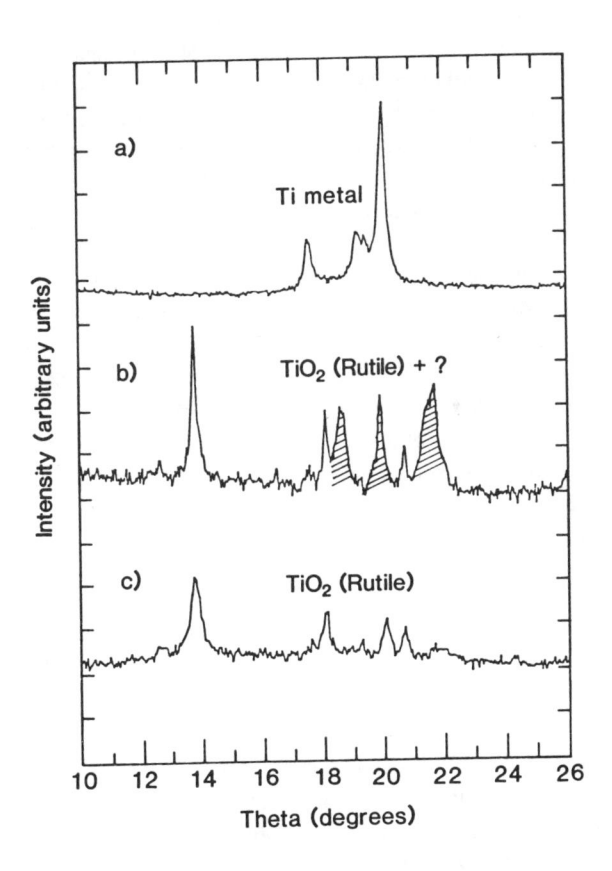

Figure 5. X-ray θ-2θ plots of nanophase material compacted using 1.4 GPa at ambient temperature. (a) Ti evaporated in 60 Pa of He and compacted without any exposure to O_2. (b) Ti evaporated in 60 Pa of He and exposed to oxygen prior to compaction. The shaded peaks are due to an unidentified phase or phases. (c) Ti evaporated in 500 Pa of He followed by exposure to O_2 to produce rutile.

however, the powder is exposed to O_2 and the O_2 is subsequently removed prior to scraping the powders from the cold finger, the compacted samples are observed to undergo what appears to be an uncontrolled oxidation reaction upon exposure to air. The resulting material is a mixture of rutile and an unidentified phase which is not Ti metal or any of the three common TiO_2 phases (rutile, anatase and brookite [6]). TEM characterization of the grain size and oxygen content of these samples has not yet been carried out.

Much work remains to be done in order to fully understand why changes in He pressure and O_2 exposure result in the formation of different phases for the case of Ti but not for Pd. It remains to be seen whether the types of phase changes observed in this study also occur in other materials. One can speculate that a possible explanation for the behavior seen in Ti might be that a critical particle size exists, below which the normal Ti hcp phase is not stable because of surface energy. The surface energy contribution increases both because the fraction of surface material increases with decreasing particle size and because the radii of curvature decrease. This new stable (or metastable) phase could be an amorphous phase or another crystalline phase which transforms to an amorphous phase upon exposure to O_2. Recently Kelly and co-workers have reported that a number of pure metals and metal alloys form amorphous phases when very small particles (less than about 30 nm diameter for pure Fe) are formed by electrohydrodynamic atomization and normal crystalline phases when larger particles are formed by the same technique [7]. Regardless of the interpretation of the phenomena, the present work has shown that phase control using nanophase processing is possible under certain conditions and has demonstrated that this direction in nanophase processing holds much promise for future materials engineering.

This work was supported by the U.S. Department of Energy, BES-DMS, under Contract W-31-109-Eng-38.

REFERENCES

1. R. Birringer, U. Herr and H. Gleiter, Suppl. Trans. Jpn. Inst. Met. 27, 43 (1986).
2. R. W. Siegel and H. Hahn, in **Current Trends in the Physics of Materials**, M. Yussouff, ed. (World Sci. Publ. Co., Singapore, 1987) p. 403.
3. R. W. Siegel and J. A. Eastman, these proceedings.
4. Z. Li, H. Hahn and R. W. Siegel, Mat. Lett. 6, 342 (1988).
5. C. G. Granqvist and R. A. Buhrman, J. Appl. Phys. 47, 2200 (1976).
6. J. L. Murray and H. A. Wriedt, Bull. Alloy Phase Diag. 8, 148 (1987).
7. Y.-W. Kim, H.-M. Lin and T. F. Kelly, Acta Met. 36, 2525 (1988).

MICROSTRUCTURE OF NANOCRYSTALLINE CERAMICS

H. HAHN[*], J. LOGAS[*], H.J. HÖFLER[*], Th. BIER[**] and R.S. AVERBACK[*]
[*] Materials Science and Engineering and Materials Research Laboratory,
[**] Center for Cement Composite Materials, University of Illinois, Urbana, Il. 61801.

ABSTRACT

The microstructure of nanocrystalline (n-) TiO_2 was studied as a function of sintering temperature up to 1273 K. Grain growth was monitored using x-ray diffraction and scanning electron microscopy. Measurements of density and permeability of He and Ar were also conducted. The specific surface area and the total pore volume were determined quantitatively using the nitrogen adsorption method. These measurements revealed that highly compacted n-TiO_2 had green body densities as high as 75 % of bulk density and that sintering occurred at much lower temperatures than in conventional powder. Densification proceeded by loss of the small pores first. The possibilities of achieving high densities with limited grain growth will be discussed.

INTRODUCTION

The development of advanced high density ceramic materials for use as structural or electrical components has been focussed towards finer grained starting materials. A novel technique to produce materials with grain sizes in the range of 5 - 50 nm, i.e. nanocrystalline materials, has been developed by Gleiter et al. [1]. Aside from their extremely small grain size, the advantage of this in situ processing technique which uses physical vapor deposition are clean particle surfaces.
The main motivation toward small grain sizes in ceramics processing is well illustrated by the following examples: 1) The densification rate of a green ceramic in its final stage of sintering depends on the inverse of the fourth power of the grain size. Indeed, a significant reduction of the sintering temperature of nanocrystalline (n-) TiO_2 with an average grain size of 12 nm by 600 degrees has been already observed [2]. 2) The deformation rate by diffusional creep via grain boundary diffusion, i.e. Coble creep, depends on the inverse of the third power of the grain size. Superplastic behavior has been observed in n-TiO_2 and n-CaF_2 at low temperatures [3].
Despite the potential importance of nanocrystals, their microstructure has not been investigated in any detail. This paper reports on our preliminary investigation of the microstructure of the nanocrystalline ceramic n-TiO_2. Grain size, density, internal surface area, permeability and porosity have been measured in n-TiO_2 as a function of annealing temperature and applied pressure.

EXPERIMENTAL

Nanocrystalline oxide ceramics for these studies were produced using a modification of the process originally described by Gleiter et al. for the preparation of nanocrystalline metals [1]. This process employs the method of inert gas condensation of metal vapors, post-oxidation of the nanocrystalline particles and in situ compaction at low temperatures. In the first step of the process a metal is evaporated from a Joule heated tungsten boat in an inert gas atmosphere. Small particles condense in the supersaturated metal vapor and are transported by the convective flow of the inert gas to a cold finger maintained at liquid nitrogen temperature. In order to prepare oxide nanocrystals, pure oxygen is then backfilled into the chamber, followed by the scraping of the powder from the cold finger. The particles are then collected in a compaction device where compacted at high pressures (up to 2 GPa) and temperatures as high as (to date) 800 K. The samples are disc-shaped with a diameter of 9 mm and thicknesses up to 300 µm. Cleanliness of the particle surfaces and subsequent interfaces is maintained by the in situ processing in a HV-chamber (residual gas pressure 1×10^{-8} mbar).
The crystal structure and grain sizes of the nanocrystalline samples were determined using

transmisson electron microscopy (TEM), high resolution scanning electron microscopy (SEM) and x-ray diffraction. In the latter case the grain size was determined from the line broadening according to the Scherrer-formula. The porosity was evaluated qualitatively by permeation measurements of He and Ar through the sample. A more quantitative study of the specific surface area, total porosity and pore size distribution was performed using nitrogen adsorption (BET). The density was measured using Archimedes principle by weighing in air, water and CCl_4 and, where possible, from the total porosity determined by BET. Annealing of the nanocrystalline samples was performed in an oxygen atmosphere in sealed quartz ampoules.

RESULTS AND DISCUSSION

A) Crystal structure, density and grain growth

X-ray diffraction studies showed that the stable rutile structure was the only phase present in the samples. Our as-compacted n-TiO_2 pellets are black in color whereas pure rutile is white. We suspect that defects caused by oxygen deficiency are present in the sample. Continued annealing at 573 - 823 K in an oxygen atmosphere gradually changes the color to white without grain growth, indicating microstructural changes. The grain size as a function of sintering temperature is shown in Table 1. Below sintering temperatures of 823 K, grain growth is negligible. Sintering at higher temperatures results in substantial grain growth with the grain size after the final annealing at 1273 K ~ 1 µm. We believe the small amount of grain growth at low temperatures, but at temperatures where sintering occurs, is due to the uniform (log normal) grain size distribution and possibly also to the pinning of grain boundaries by the small pores. To further suppress grain growth during sintering, some pellets were compacted under uniaxial load at high temperatures as pressure is known to enhance the driving force for sintering while inhibiting grain growth. Table 1 includes the results from two different samples annealed at 723 and 833 K in an oxygen atmosphere and with an applied pressure of ~ 1 GPa. Grain growth is seen to be suppressed in these samples. In addition a comparison of the density of samples sintered with and without pressure shows enhancement of sintering.

Fig. 1 shows high resolution SEM micrographs of fracture surfaces of n-TiO_2 after different sintering treatments. The individual grains of the as-prepared sample have an average size of ~ 30 nm (Fig. 1a). However, there is some discrepancy between the average grain size determined using x-ray and TEM with that determined by SEM being by a factor of two larger. We suspect that some of the individual particles seen in the SEM micrograph actually consist of a few grains. There is some indication of porosity in the as prepared sample, although SEM cannot be used to determine any density deficit in a quantitative manner. A SEM micrograph of the sample after sintering at 1173 K is shown in Fig. 1 b. A much denser structure than in the as-prepared sample with more faceted grain shapes is observed. The fracture mode in all cases is intergranular. After the highest annealing temperatures, no indication of porosity can be detected by SEM. Density measurements of as prepared n-TiO_2 using Archimedes principle show a density of about 75 % of the rutile density. This value increases to greater than 90 % of the theoretical value in the sample annealed at 1073 K.

B) Permeability measurements

In order to evaluate the nature of the open porosity of n-TiO_2, He and Ar permeation through the sample was measured at room temperature. A n-TiO_2 sample with a thickness of 150 µm was mounted on a solid Cu-disc with a 1mm hole. On the reservoir side of the sample, the gas pressure was varied from 10^{-2} to 30 mbar. The flux of the gas through the specimen was monitored using a residual gas analyser (RGA) with continuous pumping which resulted in total pressures in the 10^{-8} mbar range. A control measurement using a nonpermeable metal foil showed no permeation at all.

In both the as-prepared sample and the one annealed at 623 K, the He (Ar) flux increased on two distinct time scales. Upon applying gas pressure on the reservoir side, an immediate response was observed on the RGA side. This quick response was followed by an increase of the RGA signal over several hours before reaching a stable condition. Fig. 2 shows the pressure on the RGA side as a function of the He (Ar) pressure on the reservoir side for the as prepared and annealed

TABLE 1:

sintering treatment	grain size (nm)	density (%theoretical) Archimedes	BET	spec. surface area (m^2/g)	total porosity (cm^3/g)
as prepared	13.5	75	74.9	22.0	0.0787
2 hr @ 623 K	18.0	*	76.0	12.0	0.0750
22 hr @ 823 K	19.0	*	76.7	9.0	0.0715
15 hr @ 973 K	27.5	83-87	83.1	5.0	0.0478
20 hr @ 1073 K	30.5	90	**	**	**
14 hr @ 1173 K	51.0	91	**	**	**
18 hr @ 1273 K	~ 1000[1]	96-99	**	**	**

pressure assisted sintering (applied pressure 1 GPa):

| 21 hr @ 723 K | 15.0 | 87-90 | 86.6 | 5.0 | 0.0430 |
| 18 hr @ 833 K | 13.7 | 95 | ** | ** | ** |

* not measured; ** closed porosity; [1] grain size determined by SEM

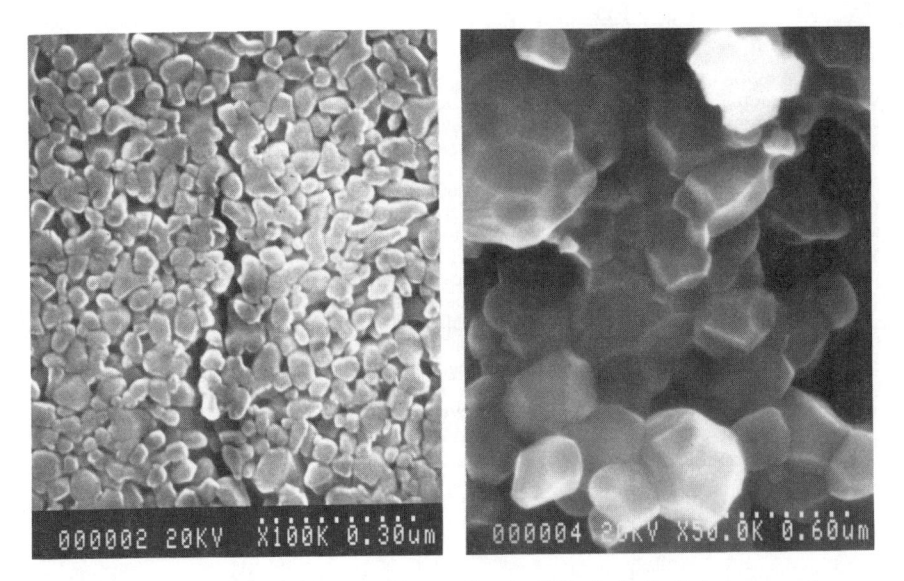

Fig. 1: High resolution scanning electron micrographs of n-TiO$_2$. Left side (1a): as prepared; right side(1b): annealed for 14 hours at 1173 K in an oxygen atmosphere.

samples. Only the immediate pressure increase is shown. Annealing strongly reduces the permeability of the sample. These results show that there is interconnected porosity in the sample with 75 - 80 % density. The increase of the He (Ar) flux on the different time scales indicates that two channels for gas flow are present: 1) the immediate response is caused by large interconnected pores; 2) the slower increase (not included in Fig. 2) is a consequence of a finer pore system that can only be filled much slower. After a sintering treatment at 623 K the porosity is reduced, although still interconnected. As expected from the delayed increase of the flux, He (Ar) gas is released by the sample over a period of several hours after the gas is pumped out of the reservoir.

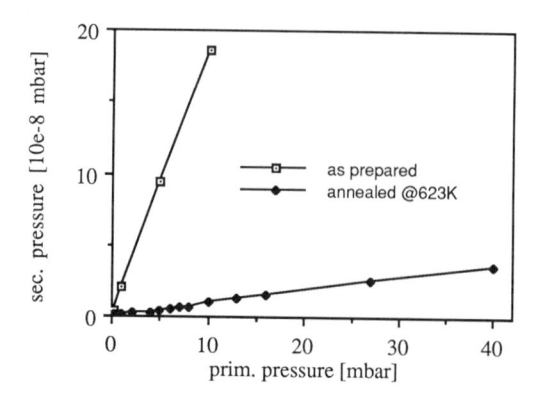

Fig. 2: He-Permeability in n-TiO$_2$ measured at room temperature before and after annealing at 623 K.

C) Nitrogen adsorption (BET)

The adsorption of a gas by a porous solid can be used to yield quantitative information on its surface area and pore structure. The build up of multilayers of the adsorbing gas and the capillary condensation in pores is theoretically described by the BET model [4]. In the current experiments, nitrogen gas at its boiling point, 77 K, was used as the adsorbing gas. Fig. 3 shows typical adsorption isotherms of nitrogen in n-TiO$_2$ after different sintering treatments. Consistent with the density and permeation experiments, a type IV-isotherm, typical of porous systems is observed [4]. From this isotherm both the specific surface area and the porosity can be determined. The isotherms at low relative pressures $p/p^0 < 0.1$ (p^0 being the saturation vapor pressure) yield the specific surface area via the BET equation. As a consequence of capillary condensation at larger values p/p^0 the adsorbed volume increases faster than expected from a simple build up of monolayers. This effect can be seen most clearly in the isotherm of the as-prepared sample at a relative pressure of ~0.5. Since capillary condensation occurs for small pores at lower p/p^0 values than for large pores, the pore size distribution can be calculated from a type IV isotherm. We have used the Kelvin equation which gives a linear relationship between the logarithm of the relative pressure p/p^0 and the inverse radius of curvature r_m of a meniscus in a capillary. The model assumes that the initial part of the isotherm results from multilayer coverage and at higher relative pressures capillary

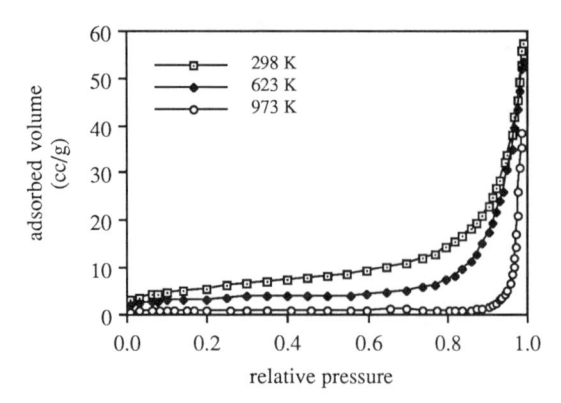

Fig. 3: Adsorption isotherms of nitrogen in n-TiO$_2$ after different sintering temperatures.

condensation commences in the smallest pores. At increasing pressures larger pores are filled until the entire system is filled with condensate at the saturation pressure. Pores in the size range of 1 - 100 nm can be detected by this method and thus BET is ideal for the examination of porosity in n-TiO$_2$.

The values of the specific surface area and the total pore volume are shown in Table 1. The specific surface area, as determined from the isotherms at low p/p^0 values (<0.1) decreases with increasing sintering temperature, indicating sintering processes even at the lowest temperatures. The initial value of 22 m^2/g for the as-prepared sample is much smaller than the value calculated for isolated 15 nm size particles, 100 m^2/g, suggesting extensive sintering during compaction. The onset of the capillary condensation, shifts to higher p/p^0 values in the annealed samples. A detailed calculation of the pore size distribution on the base of the Kelvin equation shows that the smallest pores in the range of 1 - 20 nm disappear during low-temperature sintering whereas the number and size of large pores remain unchanged. A limitation of the BET technique is that closed pores cannot be reached by the adsorbing gas causing an underestimated total porosity. A comparison of the density determined from the total porosity measured by BET with that measured by Archimedes principle, shown in Table 1, shows reasonable agreement. Thus, the entire pore system in n-TiO$_2$ up to annealing temperatures of 973 K is revealed by the BET method. Annealing of n-TiO$_2$ above 973 K, however, results in extensive closed porosity precluding further BET measurements at higher temperatures. The improvement of sintering as indicated already by the density measurements is clearly seen in the values of surface area and total pore volume.

ACKNOWLEDGEMENTS

This work was supported by the U.S. Department of Energy, Basic Energy Sciences, under contract DE-AC02-76ER01198 and by the U.S. Army Research Office, under contract DAAL03-88-K-0094. Discussions with Prof. H. Gleiter are gratefully acknowledged.

REFERENCES

[1] H. Gleiter, in Deformation of Polycrystals: Mechanisms and Microstructures, Proc. of 2nd RISØ Internat. Symp. (1981), p. 15.
[2] H. Hahn, J.A. Eastman, R.W. Siegel, Proceedings of the First Internat. Conf. on Ceramic Powder Process. Sci. (American Ceramic Society, Westerville, in press).
[3] H. Karch, R. Birringer, H. Gleiter, Nature 330, 556 (1987).
[4] S.J. Gregg, K.S.W. Sing, Adsorption, Surface Area and Porosity (1982), Academic Press, London, New York.

PHYSICAL-CHEMICAL PROPERTIES OF TiO$_2$ MEMBRANES CONTROLLED BY SOL-GEL PROCESSING

QUNYIN XU AND MARC A. ANDERSON
Water Chemistry Program, University of Wisconsin, 660 N. Park St., Madison, WI 53706

ABSTRACT

Particulate TiO$_2$ membranes have been prepared by sol-gel techniques from alkoxide precursors. The degree of aggregation, which is controlled by physical chemical processes during sol preparation influences the gelling volume and gel structure. While three types of gel structures have been proposed, the porosity of the final membrane seems to be little affected by these hydrogel structures. Firing temperature is a much more critical parameter. It is concluded that the primary particle size determines the final membrane porosity in this TiO$_2$ system.

INTRODUCTION

Ceramic membranes are playing a growing role in membrane technology since sol-gel techniques allow the fabrication of membranes with controllable microstructure from a variety of multicomponent inorganic materials. Gel derived oxide membranes have been synthesized in several laboratories around the world including ours. These membranes are being applied in industrial gas and liquid separation processing, as well as catalytic reactors. A membrane produced in our laboratory shows promise in photocatalytically degrading such environmental contaminants as PCBs.

In this paper, we focus our attention on the sol-gel process for synthesizing membranes with controlled structure and properties. For convenient characterization, unsupported TiO$_2$ membranes will be used and discussed in this paper.

EXPERIMENTAL

Particulate TiO$_2$ membranes were prepared by the sol-gel technique, starting with titanium tetraisopropoxide (a product of Aldrich Chemical Co. Inc.). The procedure which has been reported previously [1] includes: (1) hydrolysis of Ti(i-C$_3$H$_7$O)$_4$ in aqueous medium; (2) preparation of a stable colloidal suspension through peptization with HNO$_3$; (3) gel formation through evaporation of H$_2$O; (4) drying of the hydrogel to the crack-free xerogel under constant relative humidity at room temperature; (5) production of unsupported TiO$_2$ membranes by firing xerogels at 250°C to 500°C. All chemicals were used as received without further purification.

Particle sizes in various sols were measured by quasi-elastic light scattering (Brookhaven Instruments Inc.) operating at an angle of 150°. The primary particle size in the TiO$_2$ sols was measured by TEM. TG and DTA measurements were conducted on a Netzsch Thermal Analysis System in air using 100 mg ground xerogel samples. Degrees of crystallinity and phase changes at different temperatures were detected by powder x-ray diffraction (Scintag Inc.) using monochromic Cu K$_\alpha$ radiation. Percentages of crystallinity were calculated from the width at half height of the anatase {101} peak by setting the value at 500°C as 100%. The percentages of anatase to rutile transformation were determined by comparing the intensities of the anatase {101} and rutile {111} peaks. Specific surface area, porosity and pore size distribution were estimated by N_2 adsorption, using the BET model and capillary condensation theory of Zsigmondy [2] and assuming cylindric pores.

RESULTS AND DISCUSSION

Sol Preparation And Its Effect On Gel Structure

Preparation of TiO_2 lyophobic sols was conducted in two stages: 1) hydrolyzing titanium alkoxide in an aqueous system; 2) peptizing the resulting TiO_2 precipitates with appropriate amounts of acid. A large excess of water was used in hydrolysis to assure fast, although uncontrolled, precipitation. Peptization of precipitates with acid was carried out at 80°C in an ultrasonic field. This so-called peptization, a common method for preparing stable colloidal suspensions from bulk matter, involves three possible processes which could occur simultaneously: 1) aggregate break-up into particles of colloidal dimension (either primary particles or smaller aggregates) by providing thermal or other energies; 2) particle charging by proton adsorption which, in turn, stabilizes the suspension through electrostatic repulsion; 3) particle reaggregation caused by collision to give larger but more weakly bound aggregates (called secondary particles [3]) whose size remains in the colloidal range. The overall process of peptization is complex and the average particle size in the end product will depend on the relative rates of breakdown and aggregation. In this process, a number of factors must be considered: 1) pH of solution -- this determines the surface potential of particles; 2) ionic strength -- this variable controls the thickness of the double layer surrounding the particles; 3) particle concentration -- the greater the particle number, the higher the probability of collision; 4) temperature -- this variable not only provides energy for breaking apart precipitates but also for reaggregating particles.

The sol obtained after peptization usually contains various sizes of aggregates rather than primary particles. Figure 1 shows the average particle size, measured by quasi-elastic light scattering, varying with the mole ratio of acid to titanium for three sol concentrations. All three curves drop off rapidly with initially increasing H^+/Ti mole ratio, then pass almost the same minimal point (where $H^+/Ti=0.2$, diameter=80nm), and finally increase with increasing acid concentration. DLVO theory [4] of colloidal stability using a simple two-body model provides a good explanation of the above experimental results. Figure 2 illustrates the total potential energy of interaction between approaching spherical particles. Until recently, only attractive Van der Walls and repulsive electrostatic double-layer forces were considered to be important in this system. The attractive energy has an inverse power-law distance dependence and is only related to the nature of the particle material, while the repulsive energy increases exponentially as particles approach each other and is a function of surface potential and the thickness of double layer which, in turn, is related to ionic strength (IS). Curve (a) in Figure 2 represents non-charged or weakly-charged particles where

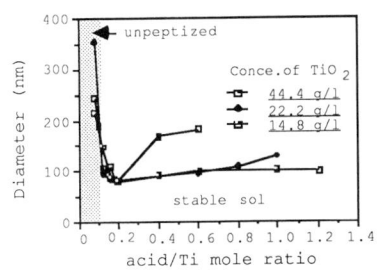

Figure 1. Average particle size in various sols

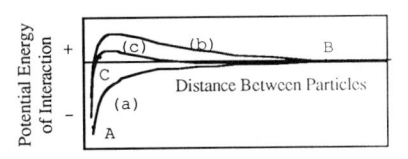

Figure 2. Schematic diagram of total potential energy of interaction (from [4]) between uncharged particles(a); charged particles in the lower IS solution(b) and in the higher IS solution(c). A,B and C are positions the particles present in precipitates, stable sols, and reaggregates respectively

attractive forces dominate resulting in strong adhesion at contact (position A in Fig.2). This situation occurs in hydroxide precipitation. During peptization, acid addition causes surface charge on the particles to become more positive and the surface potential to become higher (since the isoelectric point in TiO_2 sols was measured as 6.8 [1] and pH in these systems is less than 2); correspondingly, the potential energy curve shifts from (a) to (b) in which a repulsive energy barrier has built up and stabilizes the particles. Position B in Figure 2 likely represents the distance between two particles which exist in stable sols. By continuing to increase acid concentration, we also increase the counterion concentration (NO_3^- in our particular sols) and the overall IS. This results in double layer compression and reduces the repulsive barrier on the potential energy curve (Fig.2 (c)). Aggregation takes place if the kinetic energy of particle motion overcomes the small potential barrier. Fortunately, aggregated particles have a weaker structure (position C) than unpeptized precipitates (position A) and can be more easily rebroken by sonication at room temperature. This reduces aggregate size to 19nm according to our experiments, and these highly dispersed sols are stable for several days.

The difference of acid and particle concentration in sols leads to a distribution of gelling volume which is defined as the volume of the suspension at gelling point. As illustrated in Figure 3, the sols containing higher concentration of acid or particles and, in turn, larger aggregates (see Fig.1) preferred to form looser gels. Sonicated sols tend to form denser gels due to smaller aggregates being present and the gelling volume decreases by about 22% relative to those without sonication.

Sol-Gel Transformation And Its Effect On Membrane Structure

Sol to gel transformation takes place by water evaporation from the sol. However, evaporation does not always produce gels, as precipitates which yield powders upon drying are obtained in many cases. Although both the gel and precipitates are products of sol destabilization, a gel is defined as a semi-solid system with a continuous particle network throughout the whole dispersion medium. Any failure to produce such a network during aggregation will lead to precipitation. Comparing the structure between gel and precipitate, we easily notice that the average coordination number around each particle in the precipitate is larger than that

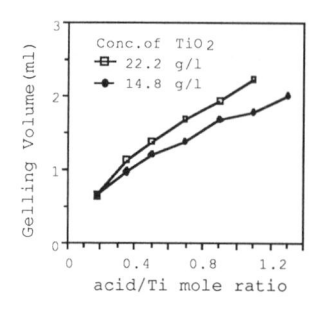

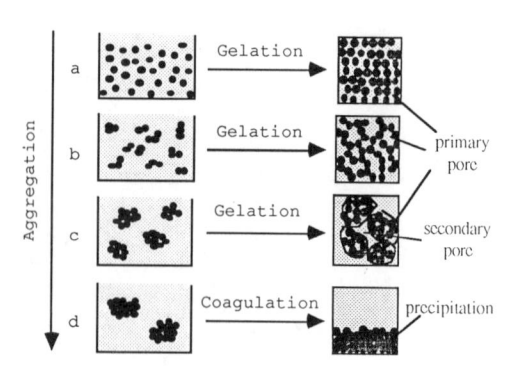

Figure 3 Gelling volume of various sols starting with 7 ml

Figure 4. Increasing degrees of aggregation in the sols from (a) to (d) lead to three types of gel and precipitation

in the gel. The former is normally greater than four, while the latter is two to three. Apparently, in order to obtain a gel, the action of pH and IS during volume reduction must stabilize the particles until the volume reaches a point where interaction of electric field around particles is so strong that particles are prevented from moving together to coagulate.

As of this writing, few publications report on the mechanism of gelation from colloidal TiO_2 sols. According to our observations, TiO_2 sols undergo two stages to form gels. During the first stage, more than 50% of the water is removed. The volume of sol is reduced without coagulation until the gelling point has been reached. At this stage, a non-fluid white rigid mass exists throughout the whole volume. During the second stage, water continues to be removed, the gel shrinks and turns transparent due to elimination of the boundary between particles which scatters light. This phenomenon might be due to chemical condensation between particles:

(Part.)Ti-OH + HO-Ti(Part.) –> (Part.)Ti-O-Ti(Part.) + H_2O

in which H_2O evaporation shifts this reaction to the right side. Zarzycki[3] reported that the aging mechanism of silica gels involves formation of Si-O-Si bonds between particles and strengthens particulate chains by depositing soluble silica at the particle necks. However, no experimental evidence has been found to substantiate this mechanism. In the case of our TiO_2 gel, we have noted that the gel was much easier to disperse in the first white stage than in the second transparent stage. Direct investigation of chemical bonding is currently being performed using in situ FTIR and UV spectroscopy.

Figure 4 proposes three types of gel structures derived from sols containing different degrees of aggregation. As mentioned before, particles can exist in sols as monodispersed primary particles, as small aggregates made by two or three primary particles or as large aggregates in which more than four primary particles combine together, forming primary pores between them. The gelation of these sol types would produce (a) denser packing of primary particles, producing micro-primary pores between them; (b) looser packing of small aggregates, producing meso-primary pores between them; (c) looser packing of large aggregates, producing macro-secondary pores or open channels between aggregates, with remaining meso-primary pores inside those aggregates. The primary particle size in our TiO_2 sols is 7nm as measured by TEM. A sonicated sol which contains an average particle size about 19nm and illustrates case (b) in the above models, while all sols made without sonication fall into category (c) since the minimal particle size in those sols is 80nm (Fig.1).

Does sol chemistry and gel structure determine the pore size of the ultimate membrane? Our experimental results (Fig.5 a and f) show that the pore size distribution of samples either with or without sonication give the same median radius of 1.2 to 1.5 nm after firing at 250°C for 30 mins. The membrane pore radius calculated from a close packing model of primary particles having a diameter of 7nm is 1.5 nm. This agreement of measured and calculated values in pore size seems to indicate that particles tend to form close packing during drying rather than by maintaining the packing geometry of the wet gel.

While we were not able to alter pore size of our membranes by using sonicated sols containing small aggregates, crack-free gels could be obtained from these sols because smaller gelling volumes produced less shrinkage, hence less stress developed during the drying process.

Effect Of Thermal Treatment On Membrane Structure

TG and DTA data (Fig.6) show that xerogel to oxide membrane conversion undergoes four transitions: (a) physically absorbed water and organic solvents are removed at 100°C; (b) bonded organic groups and nitrate ligands are burned off at 200-350°C; (c) crystallization of TiO_2 takes place at 350-450°C; (d) anatase to rutile phase change occurs at 450-600°C. Parallel studies of crystallization (Fig.7) and specific surface area (SSA) as well as porosity (Fig.8) reveal that by raising temperature, the membrane densifies with growing anatase crystal size. SSA and porosity drop down sharply to 5 m^2/g and 3% at 500°C where x-ray diffraction data indicates completion of crystallization and beginning of the transition to rutile. Further, pore size distribution diagrams (Fig.5) show the median pore radius shifts from 1.1 nm at 250°C to

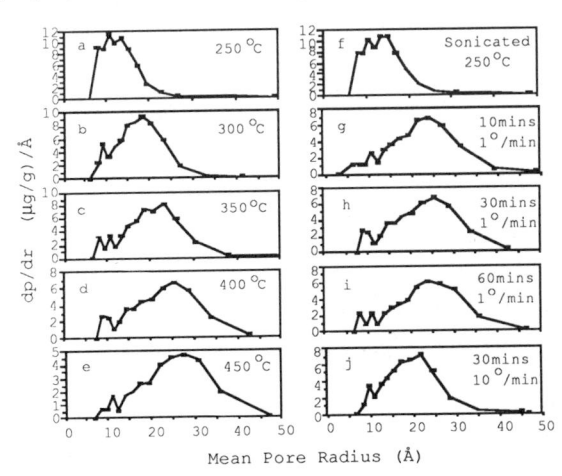

Figure 5. Effect of temperature, heating rate and dwell time on pore size distribution which was derived from adsorption branch of isotherms

Sample a-f have a dwell time of 30mins and a heating rate of 1°/min.
Sample g-j were fired at 400°C

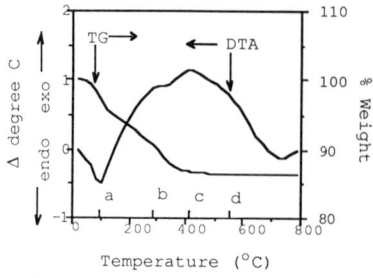

Figure 6. Thermal analysis data, a,b,c,d represent four transition temperature

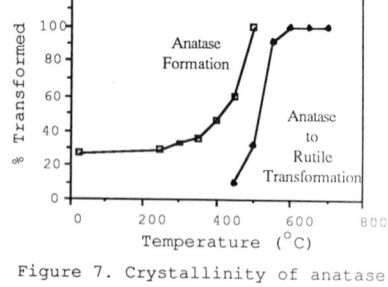

Figure 7. Crystallinity of anatase and phase change of TiO$_2$ membrane during thermal treatment

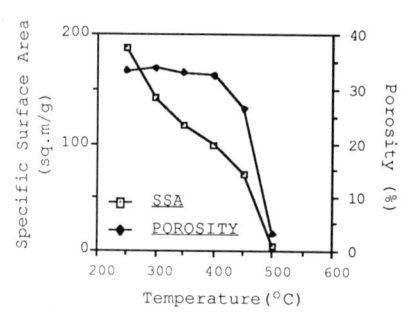

Figure 8. Specific surface area and porosity vary with firing temperature (dwell time: 30mins; heating rate: 1°C/min)

Table I. Specific surface area and porosity vary with dwell time and heating rate*

Time (mins)	SSA (sq.m/g)	Porosity (%)
10	99.7	31.7
30	98.7	32.7
60	91.2	31.1

Heating rate (°C/min)		
1	98.7	32.7
10	105.0	29.6

* All samples were fired at 400°C

2.7 nm at 450°C. Finally, the effects of heating rate and dwell time of firing upon either SSA and porosity or pore size distribution, as given in Table I and Figure 5, are much less than that of firing temperature.

CONCLUSION

The conditions of sol preparation determine the nature of the sol and lead to different packing geometries in the hydrogel. The pore structure of membranes is mainly determined by the size of close packed primary particles.

REFERENCES

1. M.A. Anderson, M. Gieselmann and Q. Xu, J. Membrane Sci., 39, 243 (1988).
2. A. Zsigmondy, J. Inorg. Chem., 71, 356 (1911).
3. J. Zarzycki, M. Prassas and J. Phalippou, J. Mate. Sci., 17, 3371 (1982).
4. B.V. Derjaguin and L. Landau, Acta Physicochim. URSS 14, 633 (1941); E.J.W. Verwey and J.Th.G. Overbeek, Theory of the Stability of Lyophobic Colloids, (Elsevier, Amsterdam, 1948).

PART II

Composites

CHEMICAL VAPOR DEPOSITION OF ULTRAFINE CERAMIC STRUCTURES

B.M. GALLOIS, R. MATHUR, S.R. LEE AND J.Y. YOO
Department of Materials and Metallurgical Engineering
Stevens Institute of Technology, Hoboken, NJ 07030

ABSTRACT

Ultrafine ceramic structures based on the nitrides and carbides of titanium and silicon have been synthesized in a computer-controlled hot-wall CVD reactor. Layered deposits have been produced by pulsing the reactant gases judiciously under software control. The development of a columnar structure which is endemic to most CVD materials has been suppressed. Skeletal structures of filaments have also been grown with appropriate catalysts by the vapor-liquid-solid mechanism and immediately infiltrated in situ with different materials to produce filament-reinforced composite coatings.
Ultrafine-grained carbon films and filaments have been grown from methane-hydrogen mixtures by RF plasma-assisted CVD. The microstructural features of these materials are of the order of 20 to 100 nm. The subgrain structure determined by Raman spectroscopy varies from 2 to 3 nm.

INTRODUCTION

The growth of columnar grains in CVD is characteristic of many materials over a wide range of experimental conditions particularly when the thickness of the deposit exceeds a few microns [1]. In micro- electronic applications where the thickness of the film is kept below one micrometer, ultrafine structures are occasionally observed [2]. The facetted surface accompanying columnar growth in structural coatings is undesirable when a smooth surface finish is required as in tribological applications. The anisotropy in the mechanical properties and particularly the lack of toughness, is also detrimental to the wear behavior of coatings [3] and in most other applications. For maximum strength and smooth surface finish, deposition methodologies should be explored to give fine- or ultrafine-grained materials. Holman and Heugel [4] were able to eliminate columnar growth in tungsten-rhenium deposits by rubbing the surface with a tungsten brush during deposition. The perturbation of the boundary layer may have facilitated renucleation at the points of contact because of local increases in supersaturation. Recently, Sugiyama et al. [5] irradiated substrates with acoustic waves. They obtained uniform thick films consisting of a fine-grained rather than a columnar structure. The controlled nucleation thermochemical deposition method developed by Holzl [6] has produced grain sizes of the order of 100 nm in materials such as W-C alloys, Ta-C alloys, TiB_2 and SiC. These methods present some interest but they may not be widely applicable. They lead to microstructures exhibiting grain sizes of the order of micrometers, apart from the latter which is not a true CVD technique.
In contrast to monolithic coatings, the deposition of ultrafine layered, graded or composite materials has received scant attention. Lackey et al. [7] have recently reviewed the codeposition of dispersed phase ceramic composites by CVD. Among the investigators who have used this technique, only Hirai and Hayashi [8] report the formation of ultrafine dispersoids (3 nm) of TiN in Si_3N_4. There is a dearth of experimental information on the processing of CVD ceramic composite structures exhibiting fibrous or lamellar modulations of a fraction of a micron or less. While monolithic coarse-grained or columnar coatings require relatively

simple reactors, the formation of modulated and composite structures can only be achieved by controlling the reactant gases accurately and rapidly, a task which is best acomplished by a computerized system. Computer-aided CVD of ceramic materials offers the potential for the development of ultrafine materials with unique properties. For many applications, CVD is the most economical and practical means of coating materials and it may be especially suited for large work pieces because of its superior throwing power.

In the first part of this contribution, we report the synthesis of ultrafine layered and filament-reinforced structures based on titanium carbide, titanium nitride and silicon carbide in a hot-wall, reduced-pressure reactor. In the second part, the deposition of ultrafine carbon films and filaments by RF plasma-assisted CVD is described. In keeping with the spirit of this Symposium, we do not address the synthesis of ultrafine powders by CVD methods.

CERAMIC COMPOSITE STRUCTURES

Experimental Details

A schematic of the hot-wall reduced-pressure reactor appears in Fig. 1. The gas trains are equipped with seven mass flow controllers. Ahead of the controllers are placed air-actuated fast-acting valves. Two of the lines include teflon-coated stainless steel bubblers which contain the liquid precursors $TiCl_4$ and $SiCH_3Cl_3$. The bubblers are controlled through a 5-valve manifold used for purging and pressure balancing during processing. The remaining lines are for argon, hydrogen, methane, nitrogen and other reactants. The gases are pumped by a 550 l/min mechanical pump equipped with inlet and outlet filters and an oil-neutralizing recirculation unit. The exhaust gases are first scrubbed and then burned in a hood. The reaction chamber is heated by a tubular globar furnace which provide a hot zone of 10 cm (±1 K) at 1400 K. The wedge-shaped specimen holder is mounted on a silica rod which can be retracted rapidly into the cold zone by means of a magnetic clutch. The entire system is controlled by a microcomputer equipped with a 16-channel, 12 bit A/D converter, a 8-bit D/A converter and a 32-line digital interface to which the various process controllers and valves are interfaced. The operation of the system is fully automated. A processing file is first created, e.g. to produce a layered or graded material and then down-loaded into the central processor. All the process parameters can be recorded periodically as the desired structure is produced under software control.

The substrates used in this study consisted of an electronic-grade graphite, a pyrolytic graphite and nickel foil. They were polished by standard metallographic techniques to a mirror finish and ultrasonically cleaned in acetone and alcohol. Titanium carbide was deposited from commonly used mixtures of hydrogen, methane and titanium tetrachloride. Nitrogen was substituted for methane in the case of titanium nitride. Methyltrichlorosilane was pyrolyzed in the presence of hydrogen to produce β-silicon carbide. The depositions were conducted at pressures ranging from 30 to 300 torr and temperatures between 1223 K and 1473 K. Typical flowrates in the 40 mm diameter reaction chamber were of the order of 300 to 800 sccm corresponding to a residence time of the order of seconds.

Phase identification was conducted by means of standard X-ray techniques. The stoichiometry of the compounds was determined by the measurement of precision lattice constants. The microstructure of the deposits was examined in a JEOL 840 scanning electron microscope. The thickness of the coatings was also determined by means of this instrument or by a stylus profilometer after masking and etching of a portion of the deposit to create a step.

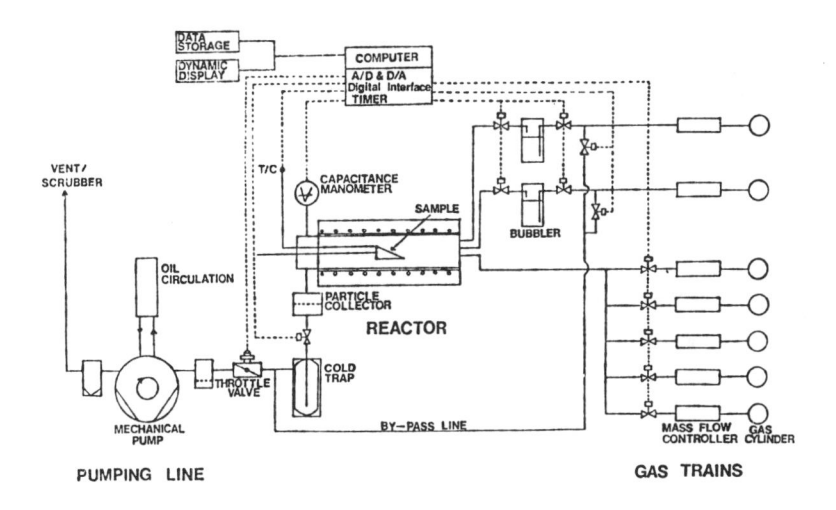

Fig. 1. Schematic of the computer-controlled hot wall CVD reactor.

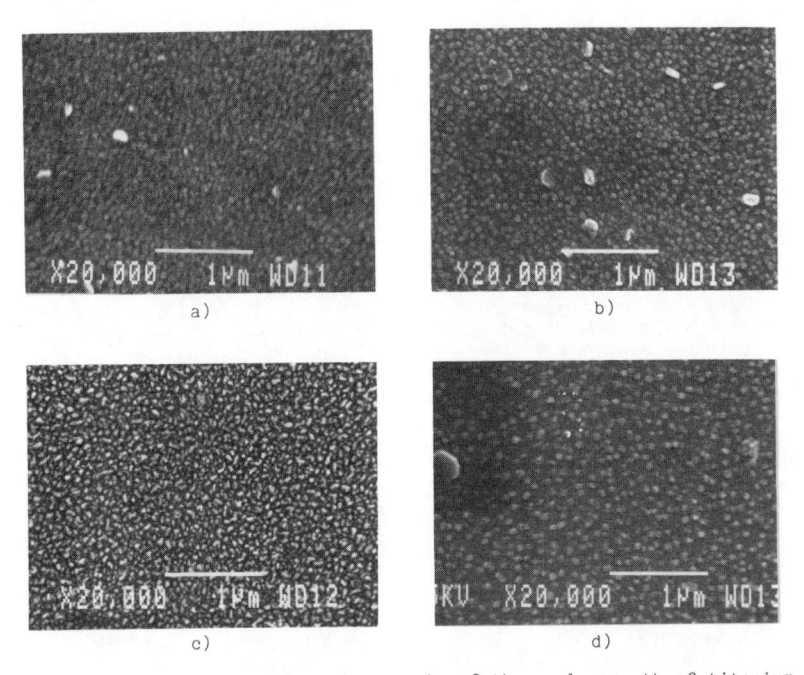

Fig. 2. Scanning electron micrographs of the early growth of titanium nitride on graphite as a function of time: a) t = 2 min, b) t = 4 min, c) t = 10 min, d) t = 20 min.

Layered Structures

The early growth morphology of CVD deposits often consists of fine equiaxed grains which are quickly replaced by a well-developed columnar structure. Fig. 2 shows several stages in the deposition of TiN on pyrolytic graphite as a function of time. The thickness of the deposit is plotted in Fig. 3 as a function of time. Initially the density of the ultrafine grains increases until the first layer is completed in less than 10 min. The substrate is now covered with a dense layer of ultrafine Ti_2N as determined by thin film X-ray diffractometry. The apparent thickness measured by the profilometer appears to be constant because the radius of curvature of the stylus is much greater than the size of the particles on which it slides. The second layer consists of substoichiometric TiN. It has the same grain size of about 50 nm and its growth is progressively completed in another 10 min. at which time larger grains appear which mark the onset of columnar growth. The early growth of TiC follows the same pattern, the only differences being that at the same temperature, the grain size is of the order of 100 nm and that a morphological transition from an ultrafine-grained structure to an acicular structure is observed.

Fig. 4 shows a top view of a five-layer TiN/TiC deposit. Each layer was deposited for a period of 10 min. which corresponds approximately to the time needed to complete a dense layer of ultrafine equiaxed grains for each material. The total thickness of the deposit is around 500 nm. The average grain size after growing five layers is of the order of 300 nm. There is no evidence of a transition to a columnar or to an acicular structure. X-ray diffraction shows peaks corresponding to discrete phases of TiN and TiC and not the carbonitride. TiN and TiC have the same cubic structure and they form a continuous series of solid solution. The lattice mismatch is of the order of 2 % which may lead to heteroepitaxial growth of one compound on the other. Pulsing of the reactants may have induced the renucleation of one phase on the other thereby suppressing the transition to a different growth mode.

The same type of experiment was conducted on nickel substrates. Nickel catalyzes the growth of both TiN and TiC. The deposition rate is increased by an order of magnitude as compared to graphite [9]. The mechanism involves the diffusion of nickel through the growing deposit to its surface where it may catalyze surface reactions. The deposition of TiC in the kinetic regime, for example, is known to be limited by the decomposition of methane [10]. The reactants were pulsed alternatively in the reaction chamber for different times. Fig. 5 shows typical SEM micrographs obtained on fracture surfaces. The grain size decreases from an average value of 400 nm for a cycle of 6 min. TiN/ 4 min. TiC (20 layers) to about 200 nm for a 45s TiN/ 30s TiC (120 layers). The X-ray diffractogram indicated that the deposit consisted mainly of TiN. Further analysis appears to be warranted to elucidate the mechanisms which lead to the elimination of the columnar structure and considerable grain refinement. The method appears to be particularly attractive since there is no limit, apart from time, to the thickness which can be achieved. The possibility of using nucleating agents from the gas phase is another attractive alternative. Blocher [11] recognized early that "the potential of inhibiting grain growth by co-deposition of second phase materials in CVD has been explored only superficially and constitutes a potentially highly productive area of research."

Growth of Filaments

The growth of filaments and whiskers by CVD and by means of the vapor-liquid-solid mechanism has been the subject of numerous investigations. The basic mechanisms are well understood and have been reviewed by Wagner [12]. The method consists in seeding a substrate with a suitable catalytic particle which can form a liquid alloy with the growing phase. The liquid

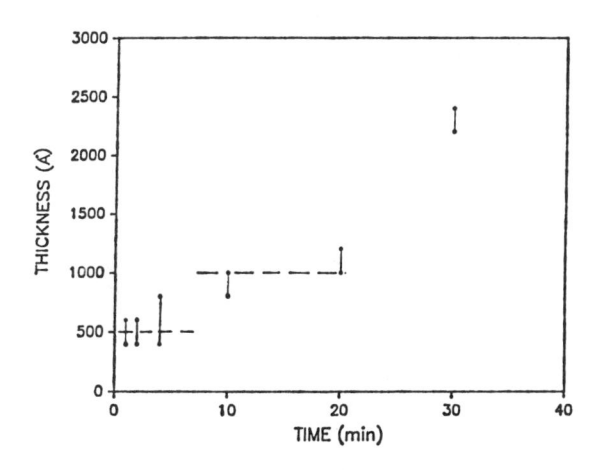

Fig. 3. Thickness of titanium nitride deposits on graphite as a function of time.

Fig. 4. Scanning electron micrograph of a five-layer TiN/TiC deposit.

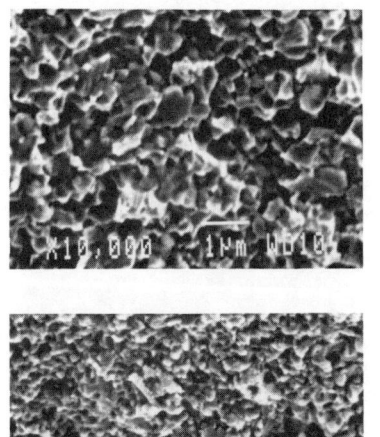

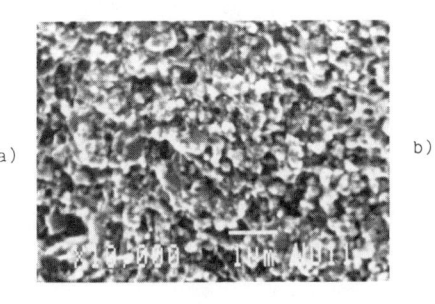

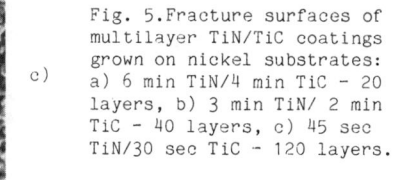

Fig. 5. Fracture surfaces of multilayer TiN/TiC coatings grown on nickel substrates: a) 6 min TiN/4 min TiC - 20 layers, b) 3 min TiN/ 2 min TiC - 40 layers, c) 45 sec TiN/30 sec TiC - 120 layers.

becomes supersaturated with material supplied from the vapor and crystal growth occurs by precipitation at the solid/liquid interface. The attention of most investigators has been focussed on the controlled growth of large whiskers and occasional reports have mentioned the occurrence of filamentary or wool-like structures [13].

In the present work, thin films of nickel were evaporated or pulse-plated on polished graphite substrates. They were subsequently annealed in an argon atmosphere at temperatures of 1250 K to 1420 K, for times of the order of minutes, to produce a homogeneous dispersion of particles of near uniform size. The average particle size can be varied from 10 nm to several µm. Typical filamentary growths of TiC appear in Fig. 6. Individual filaments can grow from a single nickel particle (a) or a mass of filaments can nucleate on larger particles (b). By suitable dispersion of nickel on the substrate, it is possible to produce skeletal structures of intertwined whiskers or filaments. The filaments grow initially in length very rapidly and then thicken with further processing as shown in Fig. 7. The time scale of the process is such that it is experimentally possible to select the final size of the filaments for further processing. The use of drop towers or fluidized bed technology offers attractive possibilities for the extension of these techniques to the mass production of ultrafine filaments and whiskers for use in short fiber composites.

Filament-reinforced Structures

Chemical vapor infiltration has been used to fabricate composites from fiber preforms. It is well established that the process should be conducted at low temperature and low pressure to optimize the densification of the composite. Several of the filamentary structures which were described in the preceding section were grown in the reactor and immediately infiltrated in situ with a different material. Fig. 8(a) shows a structure consisting of fine filaments of β-silicon carbide grown on graphite. A top view of this structure partially infiltrated by titanium nitride appears in Fig. 8(b). This phase nucleates and grows preferential- ly on the filaments. In Fig. 8(c), the resultant composite coating exhibits a grain size of the order of 100 to 200 nm. Some porosity can be seen. Another example is shown in Fig. 9. A dense mat of titanium carbide filaments about 100 nm in diameter (a) has been infiltrated with titanium nitride to yield a composite coating with a grain size of the order of 200 nm (b). These structures should find applications in tribology. It has been shown that the wear behavior of toughened ceramics depend strongly on the toughness and not the hardness of the materials [14]. The techniques we have developed would also permit the production of filament-reinforced coatings in which the scale of the microstructure i.e. the size and density of the filaments can be varied. In the machining of metallic materials, and particularly steel, the surface temperature may reach elevated values. These coatings could also provide high toughness in such applications and lead to improve performance.

ULTRAFINE CARBON STRUCTURES

Experimental Details

The depositions were carried out in a planar-diode reactor. The design and the operation of the gas trains are similar to those of the hot wall reactor. The reaction chamber is assembled from ultrahigh vacuum components to maintain a clean leak-tight environment. It is equipped with several windows for optical diagnostics. The bottom molybdenum electrode, the anode, is 8 cm in diameter and 1.8 mm in thickness. A 75 µm thermocouple is spotwelded to the underside of the anode to measure temperature. The thermocouple wires are connected to a RF choke to avoid interference

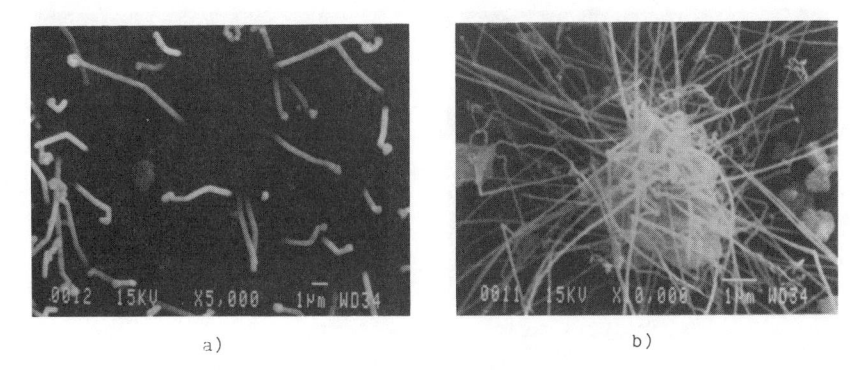

Fig. 6. Filamentary growth of titanium carbide on nickel particles dispersed on a graphite substrate: a) individual filaments grown from single particles, b) multiple growth on a 4µ particle.

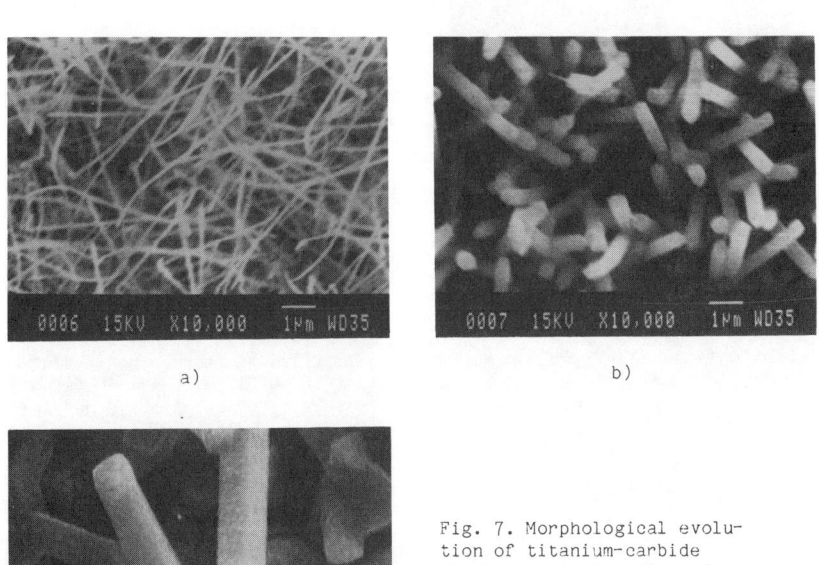

Fig. 7. Morphological evolution of titanium-carbide whiskers as a function of processing time: a) t = 2 min, b) t = 5 min, c) t = 10 min.

Fig. 8. Scanning electron micrographs of whisker-reinforced composite coating: a) as-growth β-SiC whiskers, b) top view of whisker structure partially infiltrated by titanium nitride, c) fracture surface of composite structure.

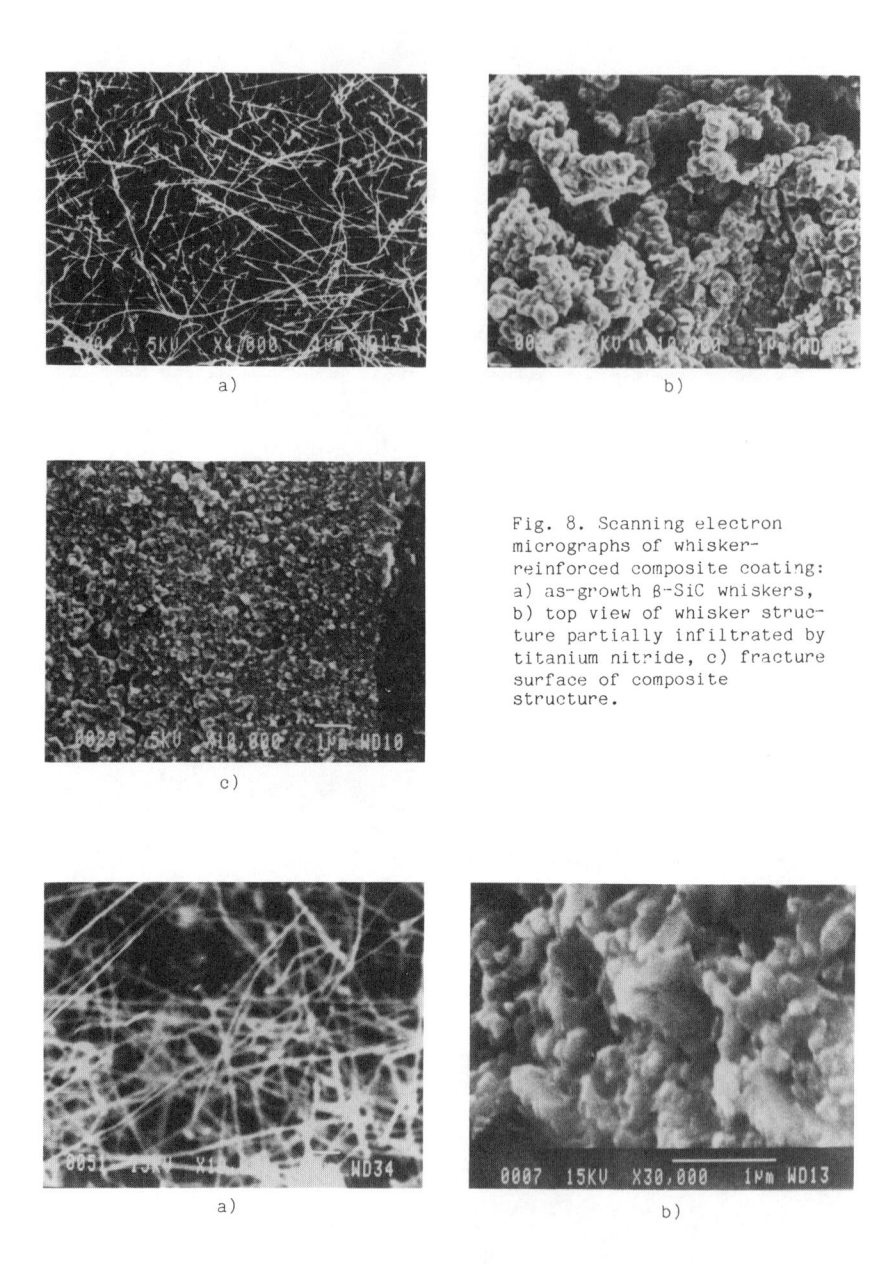

Fig. 9. Scanning electron micrographs of whisker-reinforced composite coating: a) as-grown titanium carbide whiskers, b) fracture surface of whisker structure infiltrated by titanium nitride.

from the glow discharge. The grounded anode can be heated radiantly up to 1300 K by a DC-powered graphite susceptor. The molybdenum cathode, 3 cm in diameter, is powered by a computer-controlled 500 W RF generator operating at 13.56 MHz which is connected to an automatic impedance matching network. The DC bias between the anode and the cathode was measured to be proportional to the RF power and inversely proportional to the pressure and the inter-electrode volume.

The deposits were prepared at pressures of 10 to 30 torr, temperatures of 300 to 1300 K and total flowrates of 20 to 200 sccm. The mole fraction of methane in the hydrogen carrier gas was varied from 1 to 100 %. The silicon and silicon carbide substrates were ultrasonically cleaned in acetone and alcohol. Prior to the deposition, they were exposed to a pure hydrogen discharge at the process temperature for five minutes. The chamber was then flushed and purged prior to the introduction of the reactant gases.

The microstructures of the deposits were examined in the SEM. Energy dispersive analysis and Auger spectroscopy revealed the absence of both metallic contaminants and oxygen. The domain size of the carbon deposits was measured by Raman spectroscopy. Several specimens were examined in a dual beam IR spectrometer to determine the presence and the bonding of hydrogen to carbon as well as the carbon-carbon bonding.

Carbon Films

Continuous carbon films were produced and the results are summarized in a portion of a processing map in Fig. 10. A top view of the coarse structure is illustrated in Fig. 11(a). It exhibits a fractal-like microstructure which has been observed in sputtered carbon films [15]. A micrograph of a fracture surface at higher magnification (b) indicates the presence of ultrafine grains. The top view of the fine structure (c) shows a smooth surface whereas the cross section (d) exhibits morphological features similar to the coarse film. Preliminary measurements by IR spectroscopy show the presence of symmetric and asymmetric CH and CH_2 stretch modes in the range of 2830 to 2950 cm^{-1}. The hydrogen content was found to be of the order of 1 % by a nuclear profiling technique. The subgrain size as determined by the method of Tuinstra and Koenig [16] is indicated in Fig. 10. It is typically of the order of a few nanometers.

Carbon Filaments

When the methane concentration is reduced to 1 %, dense films are still produced but their surface morphology now consists of ultrafine cones and filaments. A partial section of the processing map appear in Fig. 12. Filamentary growth is prevalent at the lower flowrates (up to 100 sccm) while the cone structure is typically observed at 200 sccm. The cones (Fig. 13(a)) and the filaments (b) appear to be made up of fine globules about 20 nm in size. Raman spectroscopy indicates that the subgrain size of these carbon structures is again of the order of several nanometers as indicated in Fig. 12. Both symmetric and asymmetric CH_2 and CH stretch modes are visible in the IR spectrum indicating that small amounts of hydrogen have been incorporated in the growing film. It should be noted that both types of structure would constitute ideal solar absorbers.

CONCLUSION

We have demonstrated that a variety of ultrafine structures can be produced by computer-aided CVD methods.

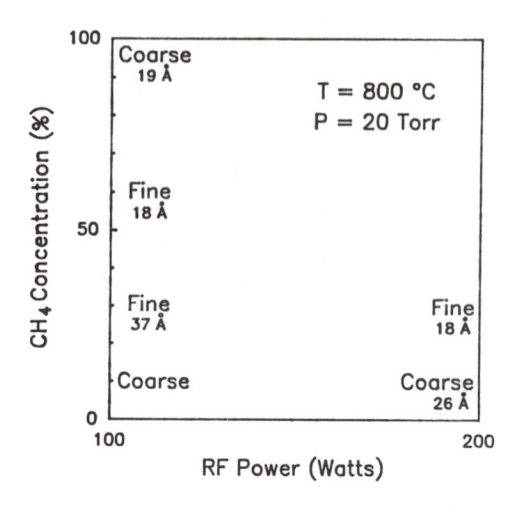

Fig. 10. Partial processing map of carbon films. The subgrain size is measured by Raman spectroscopy.

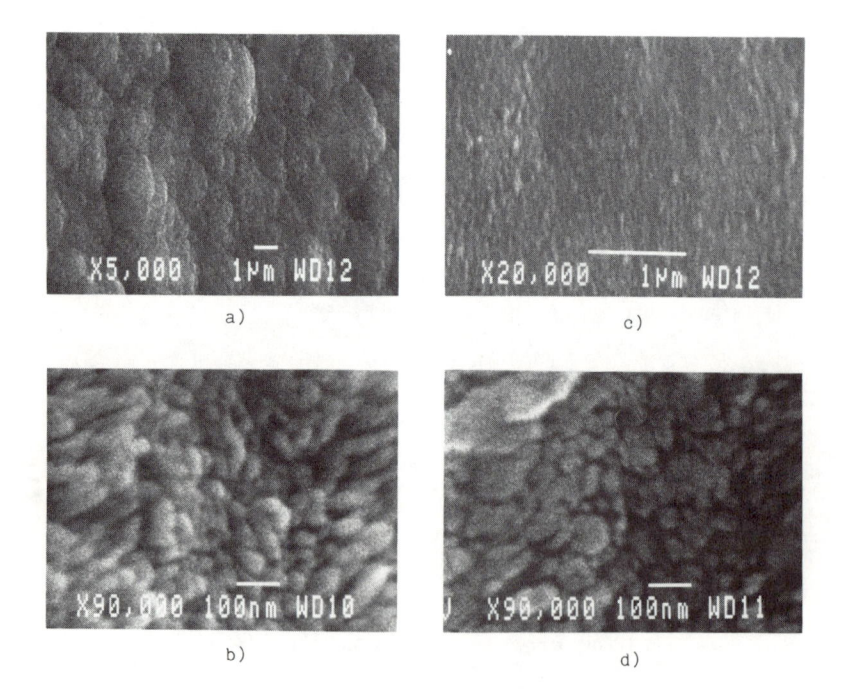

Fig. 11. Scanning electron micrographs of carbon films: a) fractal-like growth, b) the corresponding fracture surface, c) smooth growth, d) the corresponding fracture surface.

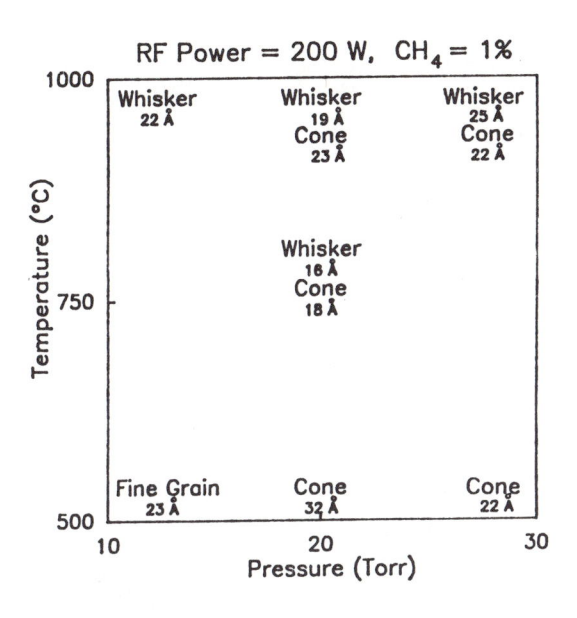

Fig. 12. Partial processing map of carbon filaments. The subgrain size is determined by Raman spectroscopy.

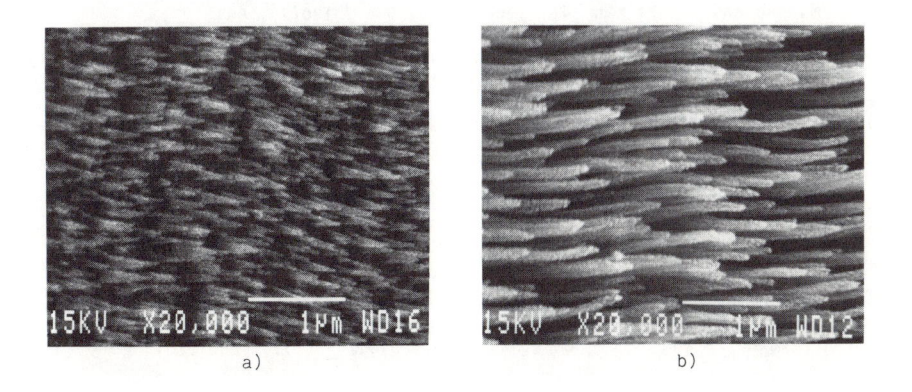

Fig. 13. Scanning electron micrographs of carbon deposits: a) cone-like growth, b) filamentary growth.

ACKNOWLEDGMENTS

The research work reported herein was sponsored by the Army Research Office, Division of Materials Science, under contract DAAG29-85-K-0214 and by the New Jersey Commission on Science and Technology.

REFERENCES

1. W. Schintlmeister, O. Packer, K. Pfaffinger and T. Raine, Proc. Fifth Int. Conf. CVD, p.523, ed. J.M. Blocher Jr., Electrochem. Soc., (1975).

2. M.L. Green and R.A. Levy, J. Electrochem. Soc., 132, 1243(1985).

3. J.R. Peterson, J. Vac. Sci. Technol., 11, 715(1974).

4. W.R. Holman and F.J. Heugel, Proc. Conf. CVD Refractory Metals, Alloys, Compounds, p. 127 and p. 427, ed. A.C. Schaffhauser, Am. Nuc. Soc. (1957).

5. K. Sugiyama, K. Kinbara and H. Itoh, Thin Solid Films, 112, 257(1984).

6. R.A. Holzl, U.S Patent 4040870, Aug. 9, 1977; U.S Patent 4147820, Apr. 3, 1979; U.S Patent 4193230, May 8, 1979.

7. W.J. Lackey, A.W. Smith, D.M. Dillard and D.J. Twait, Proc. Tenth Int. Conf. CVD, p. 1008, eds. G.W. Cullen, Electrochem. Soc., (1987).

8. T. Hirai and S. Hayashi, Proc. Eighth Int. Conf. CVD, eds. J.M. Blocher Jr. and G.E. Vuillard, p.790, Electrochem. Soc., (1981).

9. J.S. Paik and B.M. Gallois, unpublished results, Stevens Institute of Technology.

10. M. Lee and H. Richman, J. Electrochem. Soc., 120(7), 993(1973).

11. J.M. Blocher in "Deposition Technologies for Films and Coatings," p. 335, ed. R. Bunshah, (1983), Noyes Data Corp. Parkridge, NJ.

12. R. Wagner in "Whisker Technology", p.47, ed. A.P. Levitt, Wiley Interscience, NY, 1970.

13. S. Motojima and M. Hasegewa, J. Cryst. Growth, 87, 311(1988).

14. T.E. Fischer, M.P. Anderson, S. Jahanmir and R. Salher, J. Am. Cer. Soc., in press.

15. R. Messier and J.E. Yehoda, J. Appl. Phys. 58(10), 3739(1985).

16. F. Tuinstra and J.L. Koenig, J. Chem. Phys., 53(3), 1126(1970).

ULTRAFINE COMPOSITE SYNTHESIS BY LASER-REACTIONS AND RAPID CONDENSATION

GAN-MOOG CHOW and PETER R. STRUTT.
The University of Connecticut, Institute of Materials Science, Box U-136, Storrs, CT 06268.

Abstract

A new approach in ultrafine composite synthesis involves rapid condensation of metallic and non-metallic species, produced by laser-induced evaporation. A heated tungsten filament is simultaneously employed for codeposition of W via chemical transport mechanisms. This process occurs in a reducing environment of hydrogen gas, where evaporated species were produced by a laser-induced plume. Composite layers were formed on a Ni alloy substrate surface, at a rate of about $1 \mu m$/sec. The matrix of the composite films was either Al or W, and the dispersed phase was amorphous silica fibers. The diameter of the fibers was between 25nm and 120nm, depending on the interaction times. Various analytical techniques have been employed to characterize the as-synthesized layers. Experimental evidences do not support the Vapor-Liquid-Solid model for fiber growth. An alternative fiber growth mechanism is proposed.

Introduction

The intent in synthesizing composite materials is to produce unique properties by the selection of appropriate constituent phases, and then by achieving control of morphology and microstructural scale. From a practical standpoint, the latter is critically important, since, for example, methods leading to a drastic reduction in interfiber spacing, will result in increased strength and fracture toughness. To date, methods for drastically reducing microstructural scale have been an equiaxed structures as in the pioneering work of Gleiter [1]. Here, the synthesis of 1-100 nm particle materials resulted in the discovery of unique physical and chemical properties [1, 2, 3]. Gleiter [1] synthesized nanoscale powder materials by evaporation, followed by rapid condensation. Conversion into bulk form was then achieved by removal from the processing chamber, and powder compaction.

This paper is a first report of a new technique showing how the rapid condensation approach of Gleiter [1] can be extended to the synthesis of nanoscale fibers. Furthermore, it demonstrates that nanocomposites can be synthesized in a single process.

Experimental Methods and Results

The processing set-up used for co-deposition of metal and nanoscale ceramic material is shown schematically in Fig. 1. In this arrangement, a nickel alloy disc of composition (wt %) 58% Ni, 4% Al, 9% Cr, balance of V, Ti, Mn, Co, C is mounted above a ceramic target. The composition of the latter (wt %) is 55% SiO_2, 42% Al_2O_3, balance TiO_2, MgO, CaO. Prior to laser-processing, the chamber is evacuated and flushed with nitrogen. Following this, and during laser irradiation a continuous flow of 98.5% hydrogen/1.5% methane is maintained at a flow rate of 150 sccm. The 99.9% purity tungsten filament, in proximity to the laser beam-material interaction zone, is switched on 15 minutes before each experiment, and is maintained at 1700°C, it remains on during actual laser irradiation.

The processing procedure involves irradiating the nickel-alloy disc with a cw carbon dioxide laser, using a power density of $2 \times 10^4 W/cm^2$. This intensity is not only sufficient for rapid evaporation, but for complete beam penetration through the disc at the focal spot. This immediately results in interaction with the underlying ceramic target to produce a plume, containing predominantly silica species. Subsequent condensation of these species, together with simultaneous:
(i) condensation of aluminum, from the aluminum containing nickel-alloy disc.
(ii) deposition of tungsten, via chemical transport from the heated filament.
results in deposition of a composite layer, on cooler region of the nickel-alloy disc. Experimentally it was found that (i) occured for shorter (2-3 sec), and (ii) occured for longer (8-10 sec) interaction times. Microstructural examination, using electron microscopy, revealed that silica incorporated in deposited films, frequently existed in the form of fibers, with dimensions in the 25-120 nm range. These fibers were also analysed by Auger, and Fourier transform infrared spectroscopy, and by X-ray and electron diffraction. In all cases they were amorphous. Observation of fiber-extraction replicas in the electron microscope was praticuarly valuable in studying initial fiber growth, and development of randomly distributed fiber networks. The effect of increasing the ceramic target-laser interaction time is clearly apparent by comparing fiber-structures in Figs. 2 and 3. With the short interaction time of about 2 sec (Fig. 2), 25 nm diameter fibers are seen to be forming by apparent coagulation of silica particles. However, a longer interaction-time (about 7-8sec), results in well-developed 100 nm diameter fibers, which form relatively dense networks. The rapidity of fiber-growth is particularly intriguing, and it has been established that composite-films can grow at rates of $\sim 1 \mu m$ /sec.

In the experiments, a critical factor was found to be the position of the filament relative to the nickel-alloy disc (see Fig. 1). With the filament in close proximity to the disc, for instance, nanoscale silica particles no longer formed fibers, but aggregated into large crystalline particles. These followed chain-like structures as shown in Fig. 4.

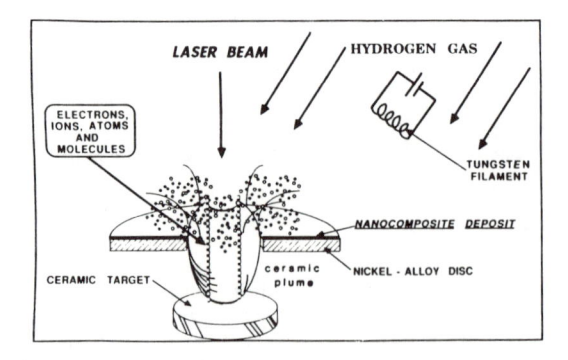

Fig. 1 A schematic showing the processing reactions

Fig. 2   100 nm   Fig. 3   25 nm

TEM extraction replica micrograph of silica fibers

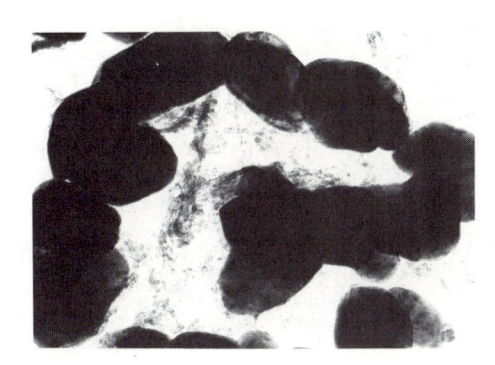

Fig. 4

TEM extraction replica micrograph of a silica linear aggregate

25nm

○ silica
ⵔⵔ silica linear aggregate
▭ silica fiber

CONVECTIVE FLOW PATH

composite film
Ni alloy substrate

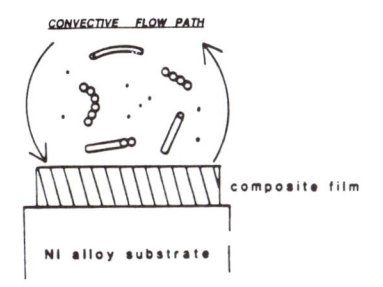

Fig. 5   A schematic showing the convection circuit

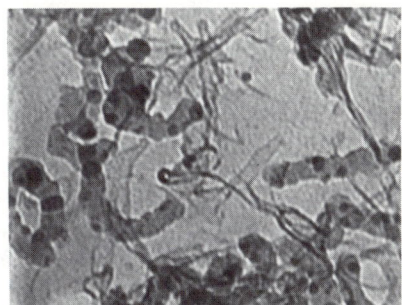

Discussion

The apparent coagulation of 25 nm, and smaller diameter, particles to form fibers (see Fig. 2) is particularly intriguing. So too is the rapidity with which this process occurs. In transmission electron microscope studies there was no discernable evidence of catalyst particles at the fiber tips, neither were the fiber tips hemispherical. These studies, it should be noted, included chemical analysis using the Philips 420 system. Thus fiber nucleation and growth does not appear to occur via the vapor-liquid-solid (VLS) mechanism [4].

An alternative fiber growth mechanism is proposed. SiO vapor species leave the ceramic plume and some arrive at the region above the Ni alloy substrate. They chemically react to form SiO_2. In the laser optical path, stable nuclei of silica can not form due to their revaporization as a result of absorption of the laser beam in the infrared. For those silica vapor molecules outside the optical path, they can form stable nuclei and subsequent condensation of the vapor allow the nuclei to grow into spherical droplets. Using the model by Homer and Prothero [5], it is calculated that a droplet of 50nm radius would form in 5×10^{-8} sec. These silica droplets and the mixture gas form a aerodisperse (aerosols) system.

Coagulation of aerosol liquid or solid particles occur when they come into contact and coalesce or adhere to one another. Aggregates of particles can assume different shapes upon coagulation, depending on the nature of applied forces. Usually, aerosol droplets are spherical and upon collision they may fuse together to produce a single spherical particle [6]. This of course is a natural consequence of favoring the lowest energy of the system. However, aeroslos such as Fe_2O_3 [7], MgO [8] and Al [9] can coagulate to form aggregates of thread- or chain-like shapes. Induced or permanent dipole moment possessed by these aerosols in a strong electric field, contact polarization or free surface charges of these particles are the mechanisms for this type of coagulation. In this work, it appears that silica droplets possess permanent dipole moment which may arise from the fact that they are small, and the net fluctuation in charge on one side of a droplet can be relatively large. As a result, linear aggregates of silica droplets are formed. Experimental evidence also supported the formation of linear chain aggregates (Fig. 4).

In thermal coagulation of these droplets, diffusion equation can be applied to Brownian motion since the experimental time intervals are large compared with the relaxation time. Values of diffusion coefficients increase as the particle size decreases. For particle of radius about 10^{-6} cm, diffusion coefficient D is about 10^{-6} cm^2/sec. In thermal coagulation of droplets, the droplet particles follow Brownian trajectory. Let L be the average diffusion distance of the droplets contained in the interaction volume:

$$L = N^{-1/3} \qquad (1)$$

where N is the droplet concentration in the interaction zone. Since coagulation only involves droplets which are close to each other, it is reasonable to subdivide the interaction volume into independent cells in which localized coagulation occurs. Consider an individual cell in which a number of droplets diffuse towards each other to coagulate and then transform into a fiber. For a fiber of 25 nm in diameter and 170 nm long, it is consisted of 10 droplets of individual diameter of 25 nm. The total diffusion distance that these 10 droplets will travel in order to coagulate to form a linear chain is L':

$$L' = n^{1/3} L \qquad (2)$$

where n is the number of droplets in the fiber. Therefore, L' is calculated to be 9.8×10^{-7} m. The total time of coagualtion of these 10 droplets to form a linear aggregate is $t_{coagulation}$:

$$t_{coagulation} = \frac{L'^2}{2D} \tag{3}$$

using $D = 10^{-6}$ cm^2/sec, $t_{coagualtion}$ is 0.005 sec. In other words, it takes 0.005 sec for 10 droplets of individual diameter of 25 nm to coagulate to form a linear aggregate, following by the transformation into a fiber. It should be mentioned that when these droplets are brought together by thermal diffusion, at a critical seperation distance where the droplets are very close to each other, there will be a sudden acceleration in coagulation time because of the effect of polarization coagulation.

Consider a single linear aggregate. This aggregate is metastable because it will transform to a more energetically favorable cylindrical shape of equal mass, or even to the most favorable spherical shape, provided appropriate conditions are available. At the initial stage of aggregation, the linear chain can have its shape preserved if its surface energy is lowered. It has been known that addition of small concentration of W to form WO_3 lowers the surface energy of many ceramic material systems. In this work, experimental evidence indicates that WO_3 from chemical transport from the heated W filament chemically adsorbs on the linear chain in its course of condensation. As a result, the surface energy of the metastable linear-chain aggregate is lowered and its shape is arrested. The linear aggregate grows lengthwise as each droplet is added to its end, and the end of the aggregate is hotter than the rest of its body. Similar to the VLS model, this liquid-like end has a larger accomodation coefficient for individual silica vapor molecules. Together with the polarization and also perhaps the surface charge effects, this causes lengthwise growth of the aggregate. If the degree of perservation of this chain is not high due to low concentration of WO_3 species in the interaction zone, the linear chain can transform to a cylindrical fiber to lower the system energy while the aggregation simultaneously continues at its tip.

The fibers were deposited simultaneously with either Al or W matrix. The fibers adhere to the growing composite film via three possible mechanisms, (i) settling due to gravity, (ii) brownian motion closely above the growing film, and (iii) natural convection circuit above the substrate. The settling velocity of a spherical particle of 50 nm radius in the experimental conditions is calculated to be 1×10^{-4} cm/sec. The settling velocity of a nonspherical particle is even smaller. Therefore, the settling due to gravity is not important for fine particles. Brownian motions of the fibers closely above the growing film surface can result in their adherence to the film. A convective circuit is also established above the Ni alloy substrate which is heated by the ceramic plume radiation. The silica fibers are carried by the convective current of the mixture gas which circulates above the substrate. Upon each pass of this gas parcel close to the surface of the growing film, there is a finite probability that a fiber can become adhered to the surface via electrostatic interaction or chemical adsorption. It is calculated that the convective circulation accounts for 50% of the measured fiber density. Fig. 5 depicts the schematic of this mechanism.

In summary, the new technique involves the rapid condensation of nanoscale particles (in this case silica), from the plume formed by the interaction of a CO_2 laser beam with a ceramic target. As the process occurs in a hydrogen/methane mixture gaseous environment, the nanoscale particles are sustained in Brownian motion. Coagulation of these particles by electrostatic interaction finally results in the formation of nanoscale fibers. These become incorporated by codeposition of tungsten. The tungsten is transported from a heated filament by a chemical transport mechanism.

Acknowledgements

The authors wish to thank the Office of Naval Research for its support: Contract N00014-78-C-0580. In addition, they wish to thank Professor Paul G. Klemens (Department of Physics, University of Connecticut) for helpful discussions, and F.P. Massicotte (Department of Metallurgy, University of Connecticut) for computer graphics. They thank C. Burila, G. Lang, B. Laube, D. Delong of United Technologies Research Center for assistance with the microprobe and Auger analyses. They also thank A. Strauss and M. Finn of Lincoln Lab for Auger analyses.

References

1. R. Birringer, V. Herr, H. Gleiter, presented at the 1986 MRS Fall Meeting, Symposium: 'Chemical Vapor Deposition of Ceramic Nanocomposites'.
2. C. Hayashi, J. Vac. Sci. Technol., 20 (3), (1987).
3. E.J. Pope, J.D. Mackenzie, MRS Bulletin, 20 (March, 1980).
4. R.S. Wagner, W.C. Ellis, Trans. Met. Soc. AIME, 233, 1053 (1965).
5. J.B. Homer, A.Prothero, J. Chem. Soc. Faraday Trans. I, 69, 673 (1973).
6. N.A. Fuchs,(translated by R.E. Daisley, M. Fuchs), The Mechanics of Aerosols, (Pergamon Press, New York, 1964).
7. O. Schweckendiek, Z. Naturforsch, 5a, 397 (1950).
8. D. Beischer, Z. Elektrochem, 44, 375 (1938).
9. R. Whytlaw-Gray, H.S. Patterson, Smoke, (London, 1932).

METASTABLE NANOCRYSTALLINE CARBIDES IN CHEMICALLY SYNTHESIZED W-Co-C TERNARY ALLOYS

L.E. McCANDLISH, B.H. KEAR, B.K. KIM and L.W. WU
Center for Materials Synthesis, Rutgers University, Piscataway, New Jersey 08855-0909

ABSTRACT

Nanophase materials can be prepared either by physical methods or chemical methods. Physical methods include thermal evaporation, sputtering and melt quenching, whereas chemical methods include glow-discharge decomposition, chemical vapor deposition, sol-gel dehydration and gas-solid reaction. Recently, we have used controlled activity gas-solid reactions to prepare nanophase WC-Co cermet powders at different WC loadings. In the process we have discovered some new metastable phases in the W-Co-C ternary system at temperatures below 1000 °C.

INTRODUCTION

The basic chemical synthesis method for preparing WC-Co nanophase composite powders consists of three steps [1,2]:

1) A precursor powder is prepared by solution chemistry methods. The cobalt and tungsten are mixed at the atomic level in the form of a crystalline compound, an amorphous solid, or a mixed microcrystalline/amorphous solid. Such solids may be molecular, ionic, or 3-dimensional bonded networks. The Co/W atom ratio is fixed by the composition of the precursor material.

2) A high-surface-area intermediate is formed by reductive decomposition of the precursor powder. Depending on the decomposition conditions, metals, carbides, or oxides are formed. The rapid removal of organic ligands from the precursor material during the thermal decomposition contributes greatly to the porosity and, therefore, to the high surface area and reactivity of the intermediate.

3) The reactive intermediate powder is carburized in a flowing mixture of CO_2/CO at a specific gas ratio, temperature and pressure. The gas phase acts as a buffered source or sink for carbon and oxygen. Under equilibrium conditions, the carbon and oxygen activities of the gas phase determine the composition and structure of the final powder product.

The optimum reaction temperature is determined, in a complex way, by the competition between reaction rate and the rate at which the microstructure coarsens. The situation is further complicated at low temperature by the tendency of the product composition to be reaction rate controlled, rather than thermodynamically controlled. Success in producing nanophase WC-Co powders is enhanced by optimizing reaction time and temperature. Generally, shortening the reaction time requires raising the reaction temperature and vice versa.

EXPERIMENTAL PROCEDURE

Precursor Powder Synthesis

Precursor powders were made from tris(ethylenediamine)-cobalt(II) tungstate, tungstic acid, and ammonium hydroxide using standard wet chemistry and spray drying techniques.

$Co(en)_3WO_4$ was prepared by mixing an aqueous solution of $CoCl_2 \cdot 6H_2O$ with a solution of H_2WO_4 in ethylenediamine (en) and H_2O. Crystals, in the form of hexagonal prisms, precipitate immediately. These crystals, in which the tungsten/cobalt ratio is equal to 1, are suitable precursors for making WC-Co having 23 weight per cent Co binder phase.

Commercially important grades of WC-Co have higher loadings of WC, e.g. typical hard grades of WC-Co contain 6-12% Co [3]. A different approach to precursor synthesis is necessary to adjust the W/Co atom ratio in this compositional range. Again, the primary requirement is to produce a solid material in which tungsten and cobalt are homogeneously mixed at the atomic level. Such a mixture can be obtained by <u>rapid</u> spray drying of a suitable solution of tungsten and cobalt compounds. A successful test was performed on a mixture of $Co(en)_3WO_4$ and H_2WO_4 in aqueous ammonium hydroxide. This produced spherical, amorphous particles (~15 μm diameter) having a W/Co ratio of 4.7, which corresponds to 6% Co in the final WC-Co product.

Spray drying was performed in a Bowen Engineering Inc. Laboratory Spray-Aire drier fitted with a 2" DH rotary atomizer spinning at 35000 rpm. The inlet and outlet temperatures were maintained, respectively, at a nominal 205 and 115 C. The feed solution was pumped into the drier at 156 ml-min^{-1}.

Reductive Pyrolysis and Carburization Reactions

Reductive pyrolysis and carburization of the precursor powder was carried out in a controlled atmosphere thermal gravimetric analyzer (TGA). Sample weight changes were recorded using a Cahn 1000 microbalance. Reactive gases were supplied from a gas manifold, via Brooks mass flow controllers, and the total pressure in the system was regulated with an MKS pressure controller. The sample was contained in a platinum basket, and suspended in a temperature controlled vertical tube furnace. The sample temperature was estimated using a type K thermocouple located several millimeters below the sample basket. Approximately 300 mg of precursor powder was used in each run.

The reductions were carried out in a 50/50 H_2/Ar gas mixture flowing at a total volumetric flow rate of 45 cm^3-min^{-1}. The reducing atmosphere having been established, the temperature of the sample was ramped up to the carburization temperature in a linear fashion over a 2.5 hr period.

The carburization reactions were carried out isothermally at three temperatures: 750, 850, and 950 °C. In each case the carbon activity of the CO_2/CO mixture was maintained at $a_c = 0.9$. The volumetric flow rate was slightly different in each run but in the range 72-82 $cm^3 \cdot min^{-1}$.

The carbon activity of the gas phase depends on the total reactive gas pressure, $P = P_{CO} + P_{CO_2}$, as well as the gas ratio, $r = P_{CO_2}/P_{CO}$, according to equation (1):

$$a_c = \{P/[r(r+1)]\}\exp(-\Delta G/RT) \qquad (1)$$

where $\Delta G = -39700 + 40.6T$ calories [4] is the standard free energy of formation of 1 mole of carbon in the reaction

$$2\ CO(g) = CO_2(g) + C(s) \qquad (2)$$

At these temperatures and at $a_c = 0.9$, the oxygen partial pressure in the system due to the disproportionation of $CO_2(g)$ to form $CO(g)$ and $O_2(g)$ is below the $O_2(g)$ pressure at which WO_2 could be expected to form [2].

RESULTS AND DISCUSSION

X-ray diffraction (XRD) and scanning electron microscopy (SEM) showed the precursor powder to be hollow, spherical particles about 5-30 microns in diameter. The powder prepared from $Co(en)_3WO_4$, with a W/Co ratio of 1, had an homogeneous microcrystalline structure, whereas the powder having a W/Co ratio of 4.7 had an homogeneous amorphous structure. The transformation of $Co(en)_3WO_4$ has been reported elsewhere [2]. In what follows we will describe the new findings obtained on the transformation of the tungsten rich powder.

In the as-spray-dried condition, the tungsten rich powder had an approximate composition: $[Co(en)_3WO_4]_{.175}[H_2WO_4]_{.65}$ which was consistent with the observed weight loss on reductive pyrolysis. To determine the composition of the reactive intermediate, one experiment was interrupted after the 850 C reduction and the sample cooled to 75 C under flowing H_2/Ar gas. The sample was passivated by further cooling to room temperature in flowing CO_2 gas. XRD analysis of the reduced sample showed it to be comprised mainly of W and W_2C, with trace amounts of $M_{12}C$ and M_6C. No diffraction peaks were observed that could be attributed to free cobalt. It appears that the small amount of carbon (and nitrogen?) released by the decomposition of ethylenediamine during the pyrolysis reaction reacted to form carbide or carbonitride phases.

Figure 1 shows the TGA weight-gain curves for the carburization reactions in CO_2/CO gas at $a_c = 0.9$ and at the various reaction temperatures. Carburization in all of the

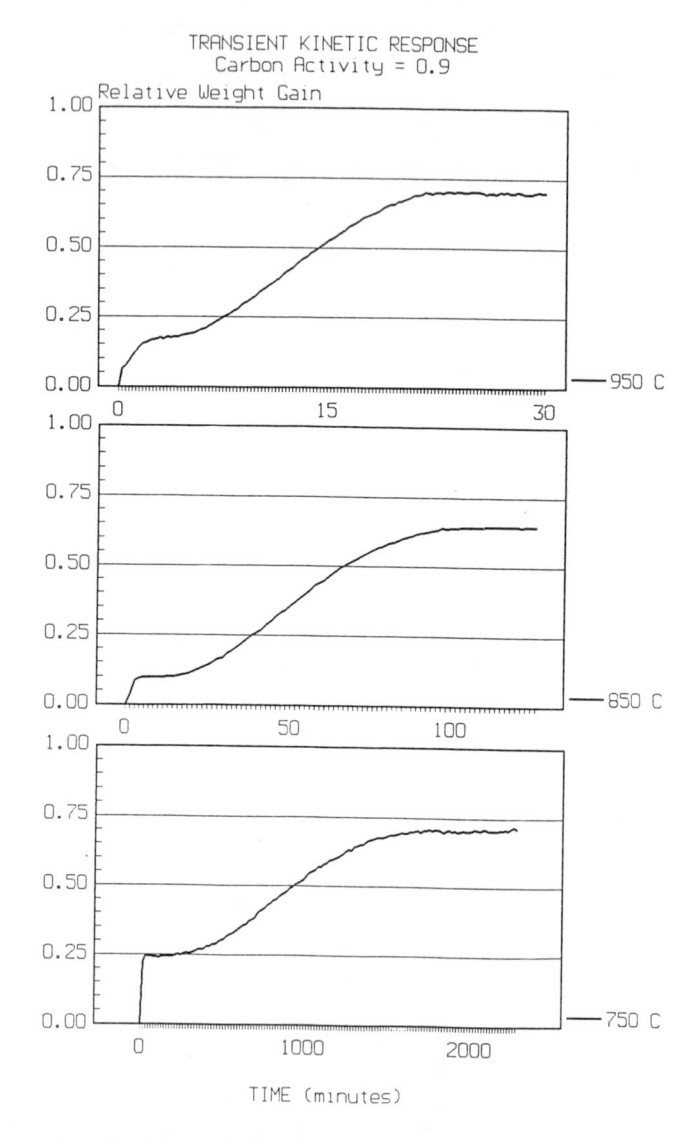

Figure 1. Carbon uptake at 950, 850, and 750 C in CO_2/CO gas at a_c = 0.9. A weight gain of 1.0 on this scale corresponds to W-Co going to WC-Co.

runs proceeds through two stages. There is a rapid carbon uptake culminating in a first plateau in less than 3 minutes. After the initial carbon uptake, and after an incubation period (which is longest for the lowest reaction temperature), additional carbon is consumed at a much slower rate, terminating in a second plateau. The total amount of carbon taken up at each stage depends on the reduction and carburization temperatures.

XRD patterns of material collected at the first plateau of the carburization curves showed major diffraction peaks at angles expected for W_2C and very minor peaks corresponding to M_6C. Similarly, the material collected at the second plateau showed major diffraction peaks at angles expected for WC and very minor peaks due to Co. The intensities of the Co peaks are smallest for lower reaction temperatures and shorter reaction times. At 750 C after 38 hours, and at 950 C after 1.25 hours, the most intense Co peak is just detectable. The relative intensity of the most intense Co peak was highest for a run at 850 C for 8 hours.

SEM micrographs showed the product particles to be pseudomorphic with the precursor powder particles. The particles were nanoporous and homogeneous in appearance and showed no evidence of phase segregation.

A typical isothermal (~1150 C) section of the Co-W-C phase diagram [5] is shown in Figure 2. At slightly higher temperature (~1200 C) the stability range of M_6C expands to the lower W/Co ratios (i.e. from Co_2W_4C to Co_3W_3C) and W_2C becomes thermodynamically stable. Figure 2 represents the best estimate available for the structure of the Co-W-C phase diagram at temperatures between 750 and 1150 C. To the extent that such an extrapolation is justified, we believe that the final equilibrium state for a W/Co ratio of 4.7 and an a_c of 0.9 at temperatures between 750 and 1150 C lies in the two-phase WC-Co field. The equilibrium pathway to this state from a composition having W/Co = 4.7 is defined by the tie line connecting the W = .825 point on the Co-W line to the carbon vertex.

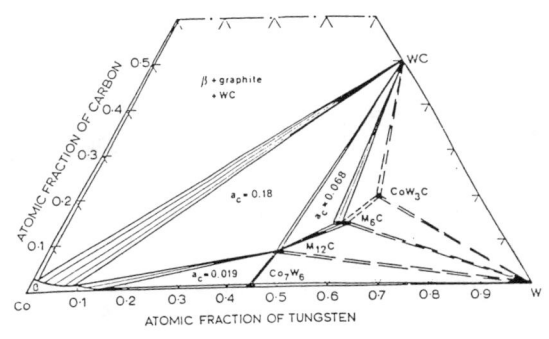

Figure 2. Isothermal section of W-Co-C phase diagram at 1150 C [5].

In these experiments, the reduced intermediate material lies near the intersection of the equilibrium pathway with the $M_{12}C$-W tie line. The carbon activity of the intermediate is probably less than 0.1. When the carbon activity of the gas phase was abruptly jumped to and held at 0.9, the system responded by seeking a new equilibrium state. The TGA experiment recorded the transient kinetic response of the system as it sought this new equilibrium.

The TGA and XRD results clearly indicate that the approach to equilibrium is along a non-equilibrium pathway and passes through two metastable states. Our interpretation of these results is as follows. The first metastable state has a W_2C-type crystal structure, but is carbon deficient and has a substantial amount of W replaced by Co: $W_{1.65}Co_{0.35}C_{1-x}$. At the end of the carburization reaction, the system is trapped in a second metastable state, which has the crystal structure of WC, but is carbon deficient and also has a substantial amount of W replaced by Co: $W_{0.825}Co_{0.175}C_{1-x}$. It appears that the true equilibrium state is not readily attainable at temperatures below 1000 C. It is interesting to note that the final product is quite stable with respect to carbon uptake. We were unable to add anymore carbon to the final product produced at 850 C at an a_c = 0.9 by increasing the temperature to 950 C for 4 hours while maintaining the same carbon activity.

The metastable cobalt-substituted tungsten carbides, described in this paper, may offer new ways to synthesize nanophase composite materials by controlled phase decomposition. Such experiments are ongoing in our laboratory.

ACKNOWLEDGMENT

This research was made possible, through the generous support of the Office of Naval Research under contract number N00014-87-K-0285.

REFERENCES

[1] R.S. Polizzotti and L.E. McCandlish, Research Report, Exxon Research and Engineering Company, 1985 (unpublished).

[2] L.E. McCandlish and R.S. Polizzotti, Proceeding of the 11th International Symposium on the Reactivity of Solids, Princeton University, NJ, to be published in **Solid State Ionics**, 1989 (in press).

[3] F.V. Lenel, Powder Metallurgy, (Metal Powder Industries Federation, Princeton, NJ, 1980), pp. 383-400.

[4] Army-Navy-Air Force Thermochemical Panel, JANAF Thermochemical Data, (The Dow Chemical Co., Midland, MI, 1963).

[5] J. Johansson and B. Uhrenius, **Metal Science**, 83, 1978.

FABRICATION OF FERROMAGNETIC Fe/MULLITE COMPOSITE MATERIALS VIA SOL-GEL PROCESSING

DEANNE P. YAMATO, ABRAHAM L. LANDIS, AND TEH S. KUAN
Lockheed Aeronautical Systems Company, P. O. Box 551, Burbank, CA 91520-7013

ABSTRACT

The sol-gel technique was used to fabricate α-Fe/mullite composite materials with iron concentrations ranging from 20-60 wt.% Fe. Microscopy studies showed the microstructure to consist of particulate α-Fe dispersed uniformly throughout a mullite matrix. Iron particle size, estimated by TEM, ranged from 20-500 nm. The materials exhibited ferromagnetic hysteresis with saturation magnetization M_s values as high as 115 EMU/g.

INTRODUCTION

Recent efforts in sol-gel processing have been directed toward the preparation of di- or triphasic materials such as dispersed metal / ceramic matrix composite materials. Roy and Roy [1] discuss two preparation routes used in making ceramic/metal xerogels. The first method involves mixing all the components in solution simultaneously. The second method involves adding a solution to a pre-made sol before gelation. They studied several ceramic/metal systems using Al_2O_3, ZrO_2, SiO_2, and TiO_2 as oxides and Cu, Ni, Pt, and Sn as metals. Materials with metal concentrations up to 10 wt.% were prepared. Other related work in the area of ceramic/metal nanocomposites includes that of Breval, Dodds, and Macmillan [2] and Breval, Dodds, and Pantano [3]. They studied the physical, mechanical, and microstructural properties of Ni/Al_2O_3 composite materials prepared by the sol-gel method. Materials with nickel concentrations of 3-10 wt.% Ni were prepared. Datta, et al. [4] prepared Ni/SiO_2 materials with a nickel concentration of 0.82 wt.% Ni and studied their magnetic properties.

Our work in the area of metal/ceramic composites prepared by sol-gel techniques is unique in that we incorporate Fe as the metallic species in concentrations much higher than that which has been previously reported in the liturature (up to 60 wt.% Fe). Additionally, the reduction of iron to its elemental state, α-Fe, is thermodynamically more difficult than that of Cu, Pt, Sn, and Ni. The incorporation of iron through chemical processing is an efficient method to uniformly disperse a ferromagnetic species in a ceramic matrix and, thus, to produce a lightweight, high-temperature ferromagnetic material.

EXPERIMENTAL PROCEDURE

The alumino-silicate sols were prepared by mixing Al sec-butoxide and tetraethoxysilane (TEOS) with a suitable solvent in a molar ratio so as to produce stoichiometric mullite ($3Al_2O_3 \cdot 2SiO_2$). Iron was incorporated into the sol in various concentrations through the addition of solutions made by dissolving iron (III) 2,4-pentanedionate in tetrahydrofuran (THF). The sols were gelled in a 100% humidity chamber at room temperature. The gels were initially dried at 100°C in air. Samples containing 30 wt.% Fe or less were then further dried at 300°C, producing homogeneous particles of amorphous iron alumino-silicate. The gels containing 40 wt.% Fe or more decomposed at ~200°C because the alumino-silicate network was not strong enough to retain structural integrity. These high iron concentration samples were pyrolyzed at 300°C, producing diphasic particles consisting of γ-Fe_2O_3 and amorphous alumino-silicate. Both the dried gels and the pyrolyzed gels were used as starting materials for further heat treatments in a reducing atmosphere.

To reduce the iron to its elemental state, α-Fe, and hence, to render the ceramic magnetic, the iron alumino-silicate dried gels and pyrolyzed gels with iron concentrations ranging from 20-60 wt.% Fe were heat treated in flowing 95%N$_2$/5%H$_2$ (forming gas) or 100%H$_2$ at temperatures ranging from 550° to 1200°C. Table I summarizes the heat treatment conditions under which the materials were prepared.

TABLE I
SAMPLE PREPARATION

wt.% Fe	Gas (%H$_2$)	Time (hrs)	Temperature (°C)
20	5	2	550
20	5	2	700
20	5	2	1000
20	5	2	1200
20	100	1	700
20	100	1	1000
20	100	4	1000
20	100	4	1200
30	100	4	1000
40	100	4	1000
50	100	4	1000
60	100	4	1000

Particle size distributions were measured using the laser light scattering method. Surface area and pore size analyses were conducted using a nitrogen BET static volumetric sorption system. X-ray diffraction (XRD) analyses were made on a standard diffractometer. Magnetic measurements were made using a vibrating sample magnetometer (VSM). Optical microscopy and scanning electron microscopy (SEM) were performed on polished particle cross-sections. Energy dispersive x-ray spectroscopy (EDS) was used in the SEM to obtain elemental maps for phase identification. Transmission electron microscopy (TEM) was performed on samples prepared by the ion beam thinning technique.

RESULTS AND DISCUSSION

Physical Properties

The dried gels (those containing 30 wt.% Fe or less) consist of single phase homogeneous particles made up of an ultraporous iron alumino-silicate network. Some physical properties of these dried gels include: extremely high surface areas (up to 647 m^2/g); high porosity (.350-.430 cc/g); and a small uniform pore size (14.33 Å average pore radius). Homogeneity of the dried gels was demonstrated by conducting EDS elemental mapping on polished particle cross-sections. The maps show a homogeneous distribution of Al, Si, O, and Fe throughout the volume of each particle. The pyrolyzed gels (those containing 40 wt.% Fe or more) consist of diphasic particles. The two phases were shown by XRD and EDS to be γ-Fe$_2$O$_3$ and amorphous aluminosilicate. The surface areas of the pyrolyzed gels were less than 6m^2/g. All of the heat treated powders exhibited a unimodal particle size distribution. The overall average particle diameter was 23.3 μm.

X-ray Diffraction Analyses

X-ray diffraction was used to determine the phases present in the materials following reduction heat treatment. A 20 wt.% Fe sample that was heat treated at 550°C for 2 hours in 95%N$_2$/5%H$_2$ (forming gas) was still completely amorphous. As the temperature was increased, formation of FeAl$_2$O$_4$ occured at 700°C, followed by mullite crystallization at 1000°C. It is interesting to note that the mullite phase was completely crystallized at a temperature of 1000°C, which is considerably lower than that needed to crystallize pure mullite prepared by sol-gel techniques (1250°C)[5]. This suggests that the presence of Fe enhances the crystallization kinetics of mullite. The XRD plot of a 20 wt.% Fe sample heat treated at 1200°C for 2 hours in forming gas clearly indicates the presence of both FeAl$_2$O$_4$ and mullite and no elemental Fe (see Figure 1). Selected peaks in this XRD plot, and in subsequent plots, are identified by their corresponding d-spacing number. It is apparent from these analyses that forming gas is not strong enough as a reducing gas to reduce the iron to its elemental state, α-Fe.

The same starting material, 20 wt.% Fe dried gel, was heat treated at 700°C for 1 hour in 100% H_2. The XRD analysis combined with EDS analysis showed that this material consisted of $FeAl_2O_4$ and amorphous aluminosilicate. When the temperature was increased to 1000°C to facilitate the reduction of the iron, the material consisted of α-Fe, $FeAl_2O_4$, and mullite. As the time was increased to 4 hours, more of the $FeAl_2O_4$ was reduced to α-Fe, as can be seen by the change in relative peak heights between the two XRD plots (see Figures 2 and 3). Increasing the heat treatment temperature to 1200°C did not increase the amount of α-Fe, so subsequent samples were heat treated at 1000°C for 4 hours in 100% H_2.

When the Fe concentration was increased to 30 wt.% Fe, the XRD plot indicated the presence of only α-Fe and mullite (see Figure 4). The XRD plots for the 40, 50, and 60 wt.% Fe pyrolyzed gels which had been reduced at high temperature also indicate complete reduction of the iron to α-Fe. In fact, the α-Fe peak intensities were even more exaggerated for the higher Fe concentration samples.

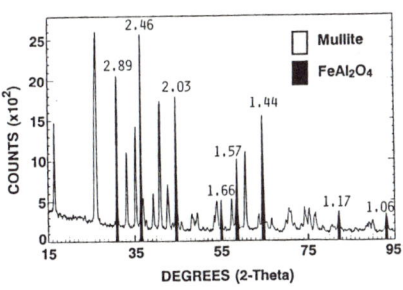

Figure 1. XRD plot of a 20 wt.% Fe sample heat treated at 1200°C for 2 hours in 95%N_2/5%H_2.

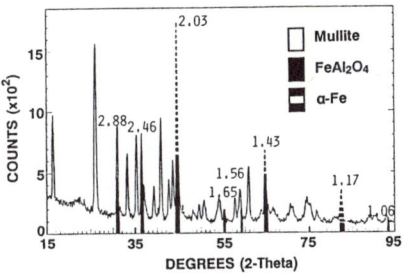

Figure 2. XRD plot of a 20 wt.% Fe sample heat treated at 1000°C for 1 hour in 100% H_2.

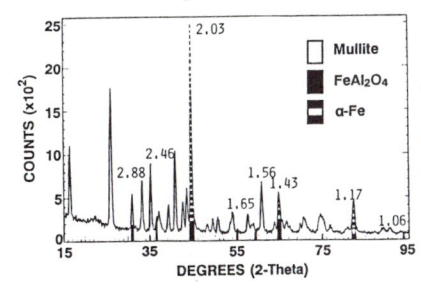

Figure 3. XRD plot of a 20 wt.% Fe sample heat treated at 1000°C for 4 hours in 100% H_2.

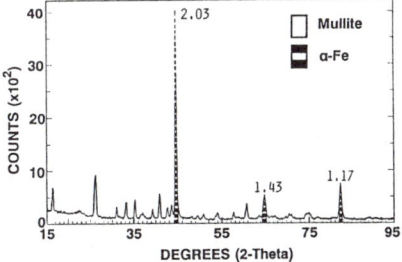

Figure 4. XRD plot of a 30 wt.% Fe sample heat treated at 1000°C for 4 hours in 100% H_2.

Magnetic Measurements

Plots of magnetization (M) vs. magnetic field (H) were generated using a vibrating sample magnetometer (VSM). The magnetization curves for the 20 wt.% Fe samples heat treated for 2 hours in forming gas show that saturation magnetization (M_s) increases as the heat treatment temperature is increased. The maximum M_s is 12.2 EMU/g for the sample heat treated at 1200°C. The low M_s is consistent with our findings that 95%N_2/5%H_2 gas is not strong enough to completely reduce the iron to α-Fe and, instead, causes $FeAl_2O_4$ to form.

The magnetization curves for the 20 wt.% Fe samples heat treated in 100% H_2 show that the M_s values increase as the temperature and time are increased. The maximum M_s is 22.6 EMU/g for the sample heat treated at 1000°C for 4 hours. This trend is consistent with the XRD data which showed an increase in the amount of α-Fe formed as the temperature and time were increased.

Figure 5 shows the magnetization curves for the samples containing 20-60 wt.% Fe which were heat treated at 1000°C for 4 hours in 100% H_2. The M_s values increase as the Fe concentration is increased, as expected. The maximum M_s is 115 EMU/g for the 60 wt.% Fe sample.

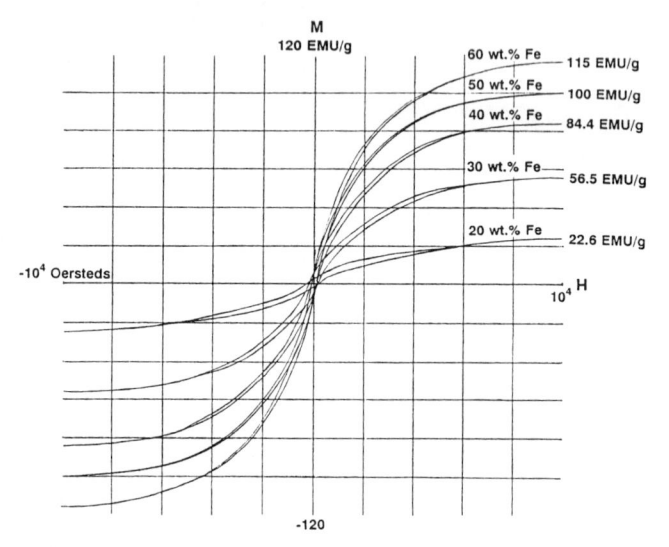

Figure 5. Magnetization curves for samples heat treated at 1000°C for 4 hours in 100% H_2.

Microscopy Studies

Optical microscopy and transmission electron microscopy were conducted on Fe/mullite composite powders heat treated at 1000°C for 4 hours in 100% H_2 to study microstructure. Specimens for optical microscopy were prepared by mounting the heat treated powders in epoxy resin, polishing with 6μm and 1 μm diamond paste, and fine polishing with a 0.3 μm alumina polishing medium. Specimens for TEM were prepared by dispersing the heat treated powders in a high-vacuum resin. Thin sections were prepared by the ion beam thinning technique.

The microstructure of samples containing 20 and 30 wt.% Fe consists of particulate Fe uniformly dispersed throughout a mullite matrix. The Fe particles (bright spots in the optical photographs) are distinct, individual particles of relatively uniform size (see Figures 6a and 7a). The TEM micrographs more accurately depict the microstructure. The 20 wt.% Fe material contains small spherical particles approximately 20-100 nm in diameter (see Figure 6b). The distinction between α-Fe and $FeAl_2O_4$ particles was not made. The 30 wt.% Fe material contains small, uniformly dispersed α-Fe particles; and, it also contains a low percentage of large, relatively spherical α-Fe particles approximately 100-300 nm in diameter (see Figure 7b). The presence of α-Fe was verified by electron diffraction.

The microstructure of the 40 wt.% Fe sample is similar to that of the 30 wt.% Fe material, except that it contains a greater percentage of the large Fe particles (see Figure 8). As the concentration is increased to 50 and 60 wt.% Fe, the Fe particles are still distributed throughout the entire particle volume; but, we begin to see some clustering (see Figures 9a and 10a). The TEM micrographs show the presence of small Fe particles (20-100 nm) as well as large particles (100-500 nm) (see Figures 9b and 10b). The Fe particles in the 50 wt.% Fe material are relatively spherical, whereas some of the large Fe particles in the 60 wt.% material are irregularly shaped which may indicate that agglomeration is taking place.

In general, all the Fe/mullite composite materials exhibit a uniform distribution of Fe throughout the mullite matrix. At low Fe concentrations (20-30 wt.% Fe), only small spherical Fe particles exist. As the Fe concentration is increased (40-60 wt.% Fe), the amount of large Fe particles, which may or may not be spherical, increases.

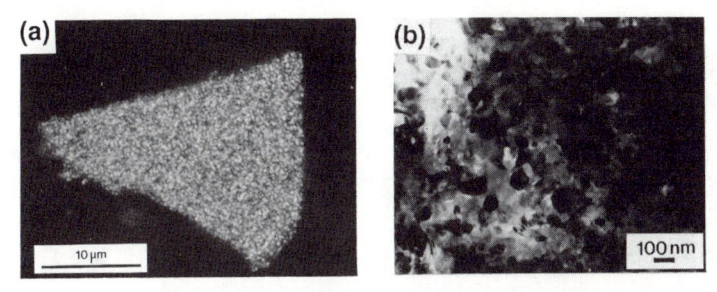

Figure 6. (a) Optical and (b) TEM micrographs of a 20 wt.% Fe sample heat treated at 1000°C for 4 hours in 100% H_2.

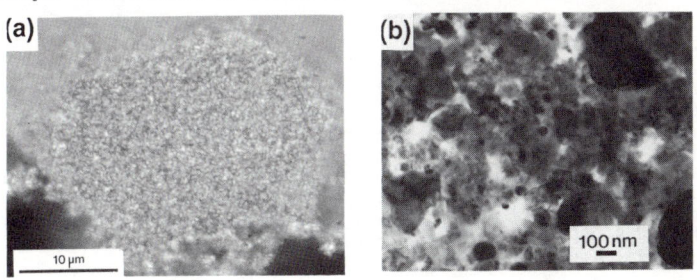

Figure 7. (a) Optical and (b) TEM micrographs of a 30 wt.% Fe sample heat treated at 1000°C for 4 hours in 100% H_2.

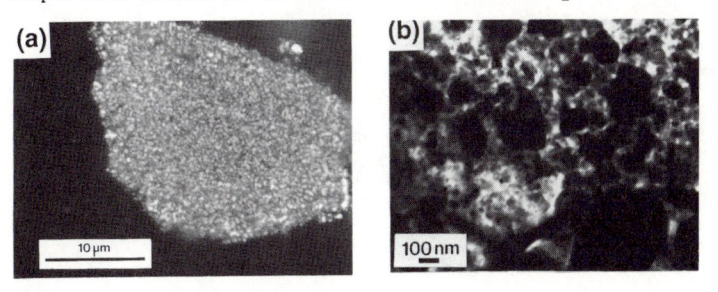

Figure 8. (a) Optical and (b) TEM micrographs of a 40 wt.% Fe sample heat treated at 1000°C for 4 hours in 100% H_2.

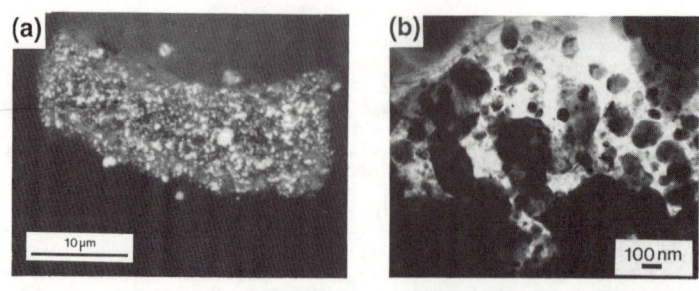

Figure 9. (a) Optical and (b) TEM micrographs of a 50 wt.% Fe sample heat treated at 1000°C for 4 hours in 100% H_2.

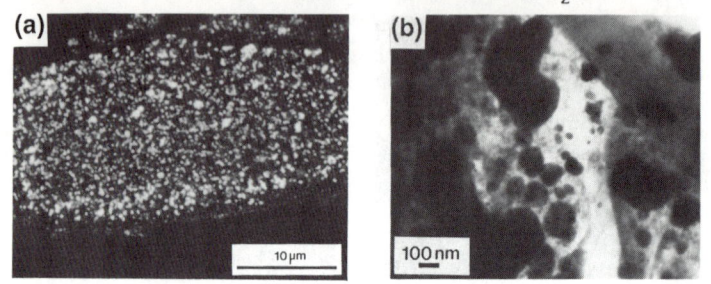

Figure 10. (a) Optical and (b) TEM micrographs of a 60 wt.% Fe sample heat treated at 1000°C for 4 hours in 100% H_2.

SUMMARY

Ferromagnetic particulate α-Fe/mullite matrix nanocomposites with iron concentrations ranging from 20-60 wt.% Fe were fabricated via sol-gel processing. Physical, microstructural, and magnetic properties were studied. Further research needs to be conducted in this area to gain control over the microstructure and, thus, to tailor the magnetic properties. Efforts should be directed toward obtaining a more uniform microstructure in the materials containing a high Fe concentration. Controlling the microstructure could result in improved magnetic properties such as increased coercivity or permeability.

REFERENCES

1. R. A. Roy and R. Roy, Mat. Res. Bull. 19, 169 (1984).
2. E. Breval, G. C. Dodds, and N. H. Macmillan, Mat. Res. Bull. 20, 413 (1985).
3. E. Breval, G. Dodds, and C. G. Pantano, Mat. Res. Bull. 20, 1191 (1985).
4. S. Datta, S. S. Mitra, D. Chakrovorty, S. Ram, and D. Bahadur, J. Mat. Sci. Lett. 5, 89 (1986).
5. N. Shinohara, D. M. Dabbs, and I. A. Aksay, SPIE 683 Infrared and Optical Transmitting Materials, 19 (1986).

CRYOMILLING OF NANO-PHASE DISPERSION STRENGTHENED ALUMINUM

M.J. Luton, C.S. Jayanth, M.M. Disko, S. Matras and J. Vallone

Corporate Science Research Laboratories,
Exxon Research and Engineering Company,
Annandale, NJ 08801

ABSTRACT

In recent years considerable effort has been expended on the development of dispersion strengthened alloys by mechanical alloying. Our research has shown that considerable improvement in microstructure control and properties can be gained by carrying out milling at cryogenic temperatures. We have found that aluminum and dilute aluminum alloys can be dispersion strengthened with aluminum oxy-nitride particles by the use of a slurry milling technique where the fluid medium is liquid nitrogen. The alloyed powders produced by this technique are strengthened by aluminum oxy-nitride particles which are typically 2-10 nm in diameter and with a mean spacing of 50-100 nm. The dispersoids are generated during the milling process by adsorption and reaction with components of the liquid nitrogen bath. On thermal treatment prior to consolidation, the alloyed powders recrystallize to a grain size which is typically in the range 0.05 to 0.3 μm. The alloys exhibit a yield stress in excess of 325 MPa at room temperature and a virtually temperature independent yield stress of about 130 MPa at temperatures greater than 375°C. The paper describes the preparation of dispersion strengthened aluminum by cryomilling, the characteristics of the microstructure and discusses some aspects of the mechanical properties.

INTRODUCTION

Dispersion strengthened alloys can be visualized as microcomposite materials consisting of insoluble non-metallic particles embedded in a metallic matrix. The particles impart strength to the material by interaction with dislocations in much the same way as in conventional precipitation hardened alloys. Unlike these latter materials, the particles are not derived from the matrix so that the strength is essentially decoupled from the matrix chemistry; affording greater flexibility in alloy design. Furthermore, since the particles are insoluble in the metal, the strengthening persists to high fractions of the alloy melting temperature.

It is generally found that as the applied stress in creep is reduced, an apparent threshold is reached below which the creep rate becomes essentially zero [1-5]. The threshold stress is found to be athermal in character; exhibiting the same temperature dependence as the elastic modulus [2,5]. It is observed that the threshold stress typically takes a value which is less than the Orowan stress, so that this stress sets an upper bound for the threshold. Although several models [1,6-10] have been proposed to explain the origin of the threshold stress, no reliable model for high temperature deformation has been developed incorporating these mechanisms. Nevertheless, it is clear that the threshold stress and, in turn, the high temperature strength of dispersion strengthened alloys can be enhanced by a reduction in the size and spacing of the dispersoids. To date, the most successful approach to the manufacture of dispersion strengthened alloys is mechanical alloying, where metallic and non-metallic powders are ball-milled to produce a fine distribution of non-metallic particles in a metallic matrix [11,12]. In the case of aluminum alloys, it is necessary to add significant quantities of organic surfactant to the powder charge to control pressure welding of the particles. This results in alloys which are strengthened by oxides and carbides. Several investigators [13-16] have studied oxide dispersion strengthened aluminum alloys produced by mechanical alloying at room temperature and shown that these alloys have a relatively coarse dispersion of the oxides and carbides which gives rise to a threshold stress of about 0.8×10^{-3} times the elastic modulus. This paper describes an attempt to augment the high temperature strength of dispersion strengthened aluminum by the use of a low temperature slurry milling method known as cryomilling.

MATERIALS PREPARATION

The dispersion strengthened aluminum described in this paper was prepared by ball milling a slurry of aluminum and alumina powders in a liquid nitrogen medium; cryomilling. The operation was performed in a modified Model I-S Szegvari 10 liter attritor mill manufactured by Union Process Inc. The vessel was cooled by a continuous flow of liquid nitrogen through the "water cooling" jacket of the mill. In addition, liquid nitrogen was introduced to the milling chamber itself, throughout the run, to maintain the level of the slurry bath. Thermocouple probes, suspended in the bath, served to monitor the temperature of the slurry and the level of the bath. The powder charge consisted of 44 μm aluminum powder (-325 mesh) of 99.5% purity supplied by Cerac Inc. and 0.05 μm alumina manufactured by Linde. A small amount of surfactant was also added to help inhibit over-consolidation of the powders.

At the completion of the 8 hour milling cycle, the powder was removed from the mill in the form of a slurry through a specially designed valve at the base of the vessel. The slurry was then transferred to a glove-box containing dry argon. Here the nitrogen was allowed to evaporate leaving the alloy powder as a residue. The powder accumulated from a series of milling runs was combined and charged into AA 6063 alloy cans. The cans were evacuated while being slowly heated to 400°C and then sealed. The sealed cans were hot isostatically pressed (HIPped) at 510°C for 5 hours at a pressure of 210 MPa. In early experiments, the HIPped material was then hot swaged to round bars. Later the HIP consolidated alloys were used as billets for extrusion. In this case, the billets were preheated to 350°C or 400°C and extruded using a conforming die designed to induce a near constant true strain rate in the die zone [17]. Extrusion was carried out at a ram speed of 11.8 mm/s, with an extrusion ratio of 16:1, which translates to a constant strain rate of 5 s^{-1} for the particular die geometry that was used. The purpose of using the conforming die was to attempt to exploit the ability of fine-grain dispersion strengthened alloys to deform superplastically at relatively high strain rates and thereby produce essentially texture-free extruded bars [17]. This process is discussed in more detail below. The diameter of the as-extruded bars was 18 mm and these were subsequently cold swaged to 6mm diameter.

RESULTS AND DISCUSSION

1. <u>Influence of Alumina Content</u>

The first set of experiments performed with the cryomilling equipment was to produce a series of alloys, where between 3 and 15% alumina was milled with the aluminum. The alloyed powders were consolidated by HIP and hot swaged to 6 mm round bars. Right cylinders measuring 5.0 mm x 6.6 mm were machined from the bars and were deformed in compression at 3×10^{-3} s^{-1} at temperatures between room temperature and 425°C. The results of these tests are summarized in Figure 1, where the 0.2% proof stress is plotted against temperature. It is notable that, although there is a modest increase in the room temperature strength with increase in alumina content, the same effect is not observed at higher temperatures. Instead, the flow stress for all alloys, at 400°C and above, is approximately 100 MPa. Since the flow stress in this regime shows little temperature dependence, this plateau stress level can be associated with the threshold stress in these alloys. The lack of any significant variation in the threshold with alumina content, however, appears to be in conflict with all models for the threshold stress.

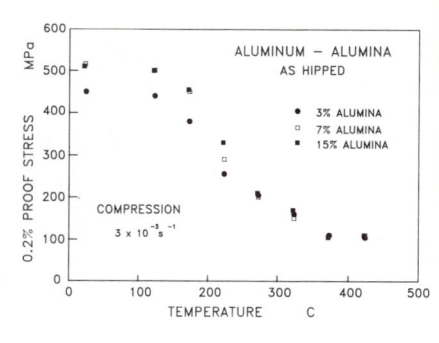

Figure 1.
Plot of 0.2% Proof Stress versus temperature alloys containing between 3% and 15% alumina.

Figure 2.
Electron micrograph of a sample of the aluminum - 3% alumina alloy.

2. Microstructure

To resolve this apparent contradiction TEM studies were performed on the materials. In all cases, TEM revealed that the microstructure consisted of aluminum grains in the size range 0.05 to 0.3 μm, coarse particles with a mean size of 0.05 μm and a reasonably uniform distribution of exceedingly fine dispersoids in the size range 2 to 10 nm. A typical example of the observed microstructures is shown in Figure 2. Examination of the coarse particles by convergent beam electron diffraction showed these particles to be α-alumina. The coarse particles have essentially the same size and composition as the added alumina; suggesting that little attrition of the ceramic occurs during milling. It is feasible, however, that the fine particle fraction originates by some attrition of the added alumina but, in this case, a more continuous distribution of particle sizes between 2 and 50 nm would be expected. The volume fraction of the alumina, as estimated from the electron micrographs was found to be somewhat less than expected if all of the alumina had been dispersed during milling. Some losses evidently occur by carry-over with the vaporized nitrogen.

Since the fine particles were too small to allow their identification by electron diffraction methods, it was decided to examine them using electron energy loss spectroscopy (EELS). A parallel detection data acquisition system was used in this analysis. In this way, the acquisition time is minimized which reduces effects due to beam drift and specimen contamination while still producing acceptable signal-to-noise levels [18]. The parallel EELS profiles were obtained using a Phillips EM400 FEG transmission electron microscope operated in the STEM mode at an accelerating voltage of 100 kV. The specimens were samples of cryomilled aluminum containing 3% alumina which had been electropolished and ion-milled.

An annular dark field STEM image of a region of the specimen containing a 100 nm Al_2O_3 particle, with adjacent fine particles is shown in Figure 3. The large mass thickness variations induced by ion-milling gave rise to the irregular details in the image. Separate low energy loss (10 < E < 140 eV) and core loss (250 < E < 600 eV) EELS profiles were obtained along the lines labelled AB and CD in Figure 3. The feature of interest is the region of relatively low contrast adjacent to C in Figure 3. The low loss signal shows qualitative differences between regions of differing chemistry and provides a relative measure of thickness that can be compared with the intensity in the annular dark field image. The core loss peak intensities for the oxygen and nitrogen K-edges are plotted against position in Figure 4a. Careful analysis of this data is required because of the complex morphology of the sample. The aluminum matrix itself is covered with a scale of amorphous Al_2O_3. Thus at any point in the profile there are unknown fractions of pure Al, oxide scale, particulate

Al$_2$O$_3$ and the fine particles. Details of the procedures used to quantify the data are described elsewhere [19,20]. Figure 4b shows the result of the analysis in terms of the atom ratio of nitrogen to oxygen along the scan path. It is evident that the fine particle contains equal amounts of O and N. This shows that the fine particles differ in chemistry from the added alumina and are, in fact, aluminum oxy-nitrides. Several small regions containing similar nitrogen levels were found in each of the samples. In many cases, however, the foil was too thick to allow quantitave analysis. It is speculated that oxygen and nitrogen are co-adsorbed onto the clean aluminum surfaces during the cryomilling process. These patches are then trapped as the matrix powder particles are re-welded. The Al(ON) is then formed during high temperature consolidation of the alloyed powders.

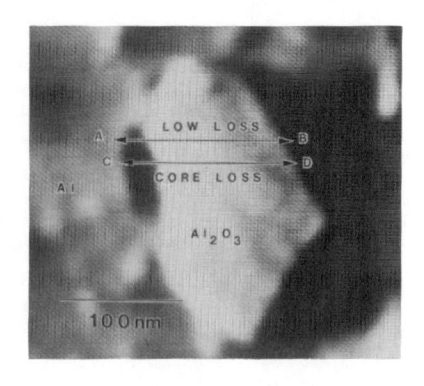

Figure 3.
Annular dark field STEM image of a sample of aluminum - 3% alumina showing the region of the sample subjected to the EELS analysis. A fine particle is evident near point C, in the figure.

In each of the alloys, a volume fraction of 0.15 to 0.2% of the aluminum oxy-nitride dispersoids was present. The spacing of these particles is approximately 80 nm. If only these dispersoids is considered, then the Orowan stress, based on the Ashby-Orowan model [21], is approximately 200 MPa at 400°C. Accordingly, the threshold stress would be expected to be between 100 and 200 MPa; that is the same order as the observed plateau stress. This observation suggests that the high temperature strength is dominated by the aluminum oxy-nitride dispersoids, which are generated in-situ. Also it is consistent with the lack of dependence of the flow stress on the α-alumina content.

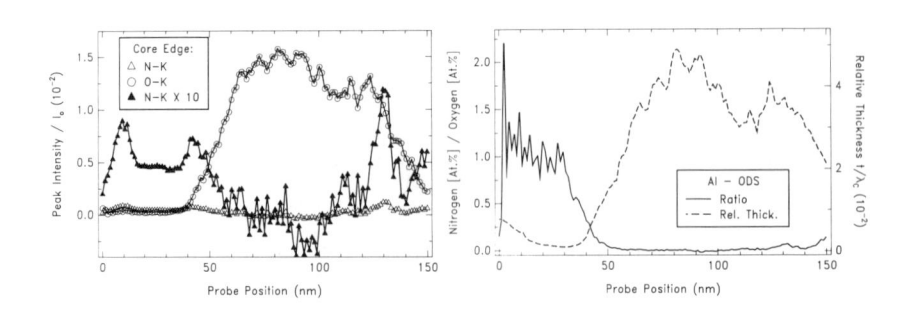

Figure 4(a).
Core loss peak intensities as a function of probe position along the scan CD in Figure 3.

Figure 4(b).
Nitrogen to Oxygen ratio as a function of probe position. Note the value of 1 for the probe scan across a 30 nm particle.

3. Absence of Texture after Extrusion

An aspect of the materials preparation technology explored in this study was the use of a special conforming die in extrusion. This method of consolidation was first demonstrated for iron-base dispersion strengthened alloys as an approach to the reduction of texture generated by extrusion [17]. The concept is that, if the material can be caused to flow superplastically, an essentially random grain orientation distribution can be generated even after the large strains produced by extrusion. A minimal texture in extruded billets is desirable in dispersion strengthened alloys because recrystallization of highly textured polycrystals tends to form highly anisotropic grain shapes [22]. Figure 5 shows <200> pole figure of a sample of the aluminum-3% alumina alloy which had been extruded at 400°C at a near constant true strain rate of 5 s^{-1} and it can be seen that it exhibits random grain orientation. The almost total lack of texture in the extruded bars suggests that deformation during extrusion is dominated by diffusional, superplastic mechanisms rather than dislocation plasticity.

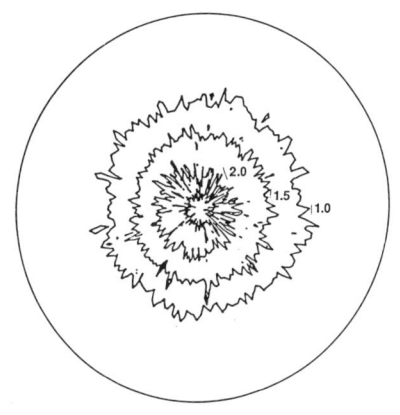

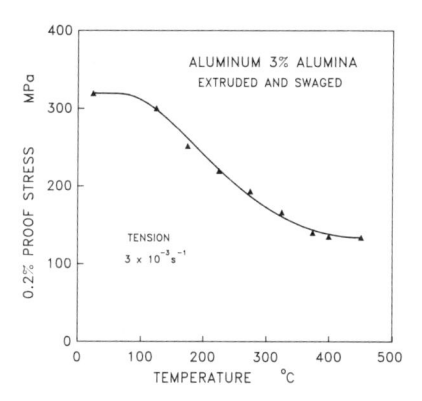

Figure 5.
A <200> axial pole figure obtained from a sample of aluminum - 3% alumina alloy after extrusion.

Figure 6.
The tensile 0.2% proof stress as a function of temperature for extruded and swaged aluminum - 3% alumina.

4. Mechanical Properties

The temperature dependence of the yield stress in tension for the alloy containing 3% alumina is shown in Figure 6. The curves show essentially the same features as the compression results on the as-HIPped material in Figure 1. The material exhibits a room temperature yield stress of 350 MPa. As the temperature is increased, the flow stress decreases continuously to a plateau stress of 130 MPa at about 375°C. In order to assess whether this stress does, in deed, correspond to the threshold stress for creep, constant load creep tests were performed at 400°C and at applied stresses between 70 and 125 MPa. Examples of the creep curves are shown in Figure 7. It is clear that, after an initial transient region the creep rate goes to zero at 70 MPa and decreases to rates of the order of 10^{-10} s^{-1} at stresses above 100 MPa. The total amount of creep strain observed at these stress levels is only slightly greater than the calculated elastic strain. It is concluded that essentially no plastic deformation takes place. This result is consistent with the concept of a threshold stress for dislocation creep, but because of the fine grain size it is expected that some creep by grain boundary sliding - diffusion creep should occur [23]. It should be recalled that no texture was generated in a true strain interval of 2.8 during extrusion at 400°C and a strain rate of 5 s^{-1}. It may be concluded that the strain rate regime in which superplastic processes dominate extends to, at least, 5 s^{-1} at 400°C, i.e. to much higher strain rates than were used in the tension tests. This suggests that the threshold of 125 MPa is associated with a superplastic flow mechanism.

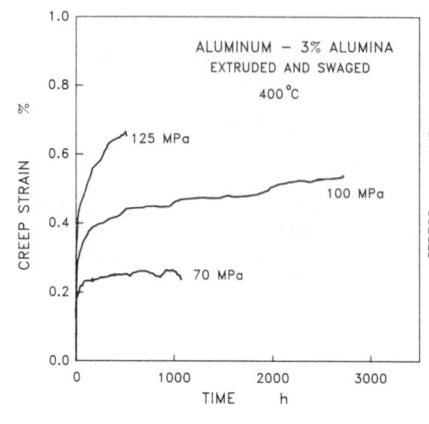

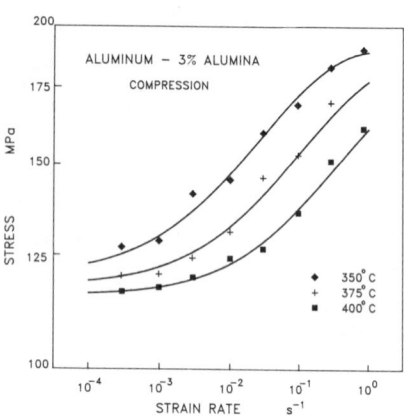

Figure 7.
Samples tensile creep curves for aluminum - 3% alumina in the stress range 70 to 125 MPa at 400°C.

Figure 8.
Strain rate dependence of the 0.2% compressive proof stress for aluminum - 3% alumina at temperatures between 350° and 400°C.

To explore this possibility, compression tests were conducted at strain rates varying from $10^{-5}s^{-1}$ to 1 s^{-1} at temperatures between 350° and 400°C. The results of these experiments are displayed in Figure 8 as a plot of the 0.2% proof stress against strain rate. These curves have the familiar form of the stress-strain rate curves obtained in superplastic alloys. Here the "Region I" is associated with the observed threshold stress. Also the rate sensitivity in "Region II" takes a value of 0.06, which is too low to impart superplastic extension to the specimen. This is consistent with the observed elongation in tension which is typically less than 4% at the high temperatures.

The underlying mechanisms of plasticity in this regime can be examined by considering the temperature and strain rate dependence of the effective stress. Figure 9 shows a plot of the normalized effective stress, $(\sigma - \sigma_{th})/E$, versus the temperature corrected strain rate, $\dot{\varepsilon}\exp(Q/kT)$, where σ is the flow stress, σ_{th} is the threshold stress, E is the Youngs modulus, $\dot{\varepsilon}$ is the strain rate, Q is the apparent activation enthalpy and k and T have their usual meaning. The value of Q was determined from plots of $\dot{\varepsilon}$ versus 1/T at constant $(\sigma - \sigma_{th})/E$. It takes a value of 167 KJ/mol, which is in good agreement with the activation enthalpy for self-diffusion in aluminum [24]. It is notable that curve is reasonably linear at low values of $(\sigma - \sigma_{th})/E$. Furthermore, the strain rate sensitivity of this stress takes a value of 0.5. The inference from this observation is that the underlying mechanism of plasticity has the same characteristics as that which controls superplastic flow in fine grain superplastic alloys and is likely to be associated with grain boundary sliding-diffusional creep. However, the presence of

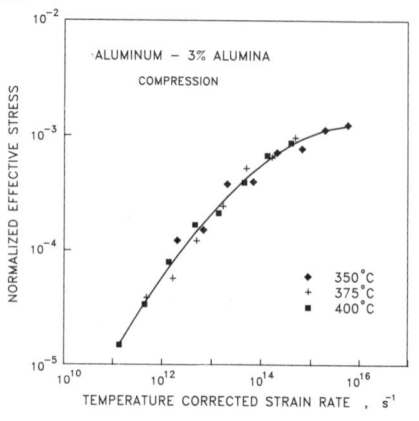

Figure 9.
A plot of the normalized effective stress versus temperature corrected strain rate for aluminum - 3% alumina. The linear portion of the curve at low effective stresses has a slope of 0.5.

the aluminum oxy-nitride dispersoids leads to a threshold stress for these processes resulting in the observed macroscopic behavior. Clearly more work is required to develop an understanding of the plastic flow mechanisms and the basis for the strong influence of the fine dispersoids.

CONCLUSIONS

The cryomilling process, described in this paper, produces alloys which are dispersion strengthened with aluminum oxy-nitride particles. These particles are formed in-situ during the milling process by co-adsorption of nitrogen and oxygen onto clean aluminum surfaces. During subsequent annealing the mechanically trapped nitrogen and oxygen interact with the aluminum to form Al(ON) particles which are typically 2 to 10 nm in diameter. These particles are effective in inhibiting grain growth in the aluminum matrix, so the alloys typically have a grain size in the range 0.05 to 0.3 µm which is stable even on prolonged annealing at high temperatures. In spite of the fine grain size the material exhibits high strength at temperatures up to 450°C. It is speculated that the high strength at elevated temperatures is due to a particle generated threshold stress for superplastic flow in the fine grain alloy.

AKNOWLEDGEMENTS

The authors wish to thank their colleagues, R. Petkovic-Luton, M.P. Anderson, R. Ayer for their many contributions to the study of dispersion strengthened alloys. Also thanks are due to H. Shuman of the Pennsylvania Muscle Institute of the University of Pensylvania for his contributions to the analysis of the aluminum oxy-nitride particles.

REFERENCES

1. R.S.W. Shewfelt and L.M. Brown, *Phil. Mag.* **30**, 1135 (1974).
2. R.W. Lund and W.D. Nix, *Acta metall.* **24**, 469 (1976).
3. G.M. Pharr and W.D. Nix, *Scripta metall.* **10**, 1007 (1976).
4. J.H. Hausselt and W.D. Nix, *Acta metall.* **25**, 1491 (1977).
5. R. Petkovic-Luton, D.J. Srolovitz and M.J. Luton, in *Proc. Frontiers of High Temperature Materials II*, edited by J.S. Benjamin and R.C. Benn (INCO Alloys International, New York,1984), p. 73.
6. D.J. Srolovitz, R. Petkovic-Luton and M.J. Luton, *Phil. Mag. A* **48**, 795 (1983); *Scripta metall.* **18**, 1063 (1984); *Acta metall.* **31**, 2151 (1983).
7. D.J. Srolovitz, M.J. Luton, R. Petkovic-Luton, D.M. Barnett and W.D. Nix, *Acta metall.* **32**, 1079 (1984).
8. E. Artz and D.S. Wilkinson, *Acta metall.* **34**, 1893 (1986).
9. J. Rösler and E. Artz, *Acta metall.* **36**, 1043 (1988).
10. E. Artz and J Rösler, *Acta metall.* **36**, 1053 (1988).
11. R. Sundaresan and F.H. Froes, *J. of Metals*, August, 1987, p 22.
12. P.S. Gilman and J.S. Benjamin, *Ann. Rev. Mater. Sci.* **13**, 279 (1983).
13. J.S. Benjamin and R.D. Schelleng, *Met. Trans.* **12A**, 1827 (1981).
14. W.C. Oliver and W.D. Nix, *Acta. Metall.* **30**, 1335 (1982).
15. Y.M. Kim, in *PM Aerospace Materials*, Vol. I, Berne, Switzerland, November 12 - 14, 1984.
16. S.J. Bane, J. Bradefield and M.R. Edwards, *Matl. Sci. Tech.* **2**, 1025 (1986).
17. M.J. Luton, U.S. Patent No. 4 599 214 (8 July 1986); U.S. Patent No. 4 601 650 (22 July 1986).
18. H. Shuman and P. Kruit, *Rev. Sci. Instr.* **56**, 231 (1985).

19. M.M. Disko and H. Shuman, *Ultramicroscopy* **20**, 43 (1986).
20. R.F. Egerton, *Electron Energy Loss Spectroscopy*, (Plenum Press, New York, 1986).
21. M.F. Ashby, in *Oxide Dispersion Strengthening*, (AIME, New York, 1966), p. 143
22. M.P. Anderson, Unpublished Research, Exxon Research and Engineering, Annandale, NJ 08801.
23. C.M. Sellars and R. Petkovic-Luton, *Mater. Sci. Eng.* **46**, 75 (1980).
24. S.L. Robinson and O.D. Sherby, *Phys. Status Solidi* **A1**, K199 (1970).

MATERIAL WITH NOVEL COMPOSITIONS AND FINE MICROSTRUCTURES PRODUCED VIA THE MIXALLOY PROCESS

ARTHUR K. LEE[*], LUIS E. SANCHEZ-CALDERA[*], JUNG-HOON CHUN[*], AND NAM P. SUH[**]

[*]Sutek Corporation, 14 Brent Drive, Hudson, MA 01749
[**]MIT, Dept. of Mechanical Engineering, 77 Mass. Ave., Cambridge, MA 02139

ABSTRACT

A new processing method, the Mixalloy process, has been developed to process alloys with novel microstructures and compositions. In this process, microstructural control is achieved through the use of turbulent mixing of liquid metals in addition to controlling solidification rate and chemical composition. Boride dispersion strengthened copper alloys were produced using the Mixalloy process. Thermally stable and fine (average less than 100 nm) boride dispersoids were formed by in-situ chemical reaction in the copper alloy matrices during mixing. The uniform mixture of the matrix and dispersoids was then rapidly solidified to maintain the fine microstructure. The consolidated material shows exceptional thermal stability and an excellent combination of strength, ductility, and electrical conductivity. Furthermore, the flexibility of the process allows the matrices of these dispersion strengthened coppers to be easily alloyed to fulfill specific needs. The versatility and simplicity of the Mixalloy process provide an economical alternative to other processing means in the manufacturing of high performance alloys such as dispersion strengthened alloys.

INTRODUCTION

Dispersion strengthened (DS) alloys, particularly those strengthened by a nanoscale refractory particle dispersion, exhibit remarkable stability at high temperatures. Ever since its discovery around 1950, major research efforts have been made to develop means of manufacturing commercial quantities of DS alloys in various metallic systems. Unfortunately, practical and competitively priced DS alloys have not been made available up to now.

A new material processing method, the Mixalloy process [1,2,3,4,5], has been developed to process a variety of innovative alloys and compositions at low cost. This process utilizes the turbulent mixing of impinging metal streams, in-situ chemical reaction, and control of kinetics. In the processing of DS alloys, for example, the refractory particles are generated by in-situ chemical reaction when the liquid metals are mixed, thus avoiding the tedious and time consuming steps necessary to incorporate or generate these paricles in the matrix in methods like mechanical milling and internal oxidation (e.g., in the Cu-Al$_2$O$_3$ system). The Mixalloy process has been very successfully applied to produce boride dispersion strengthened (BDS) copper alloys. DS copper has good strength, excellent thermal stability, and high electrical and thermal conductivity, and is used in electrical applications such as spot welding electrodes, light bulb filament supports, motor commutators, and so on.

This paper will briefly describe the Mixalloy process and present some mechanical and electrical properties obtained from Mixalloy processed BDS copper alloys.

BACKGROUND OF THE MIXALLOY PROCESS

The Mixalloy process is schematically illustrated in Fig. 1. Briefly, two or more turbulent streams of molten metal alloys are directed into a mixing region, known as mixing chamber, at predetermined velocities. There the streams are impinged at high velocities. Because the flow is highly turbulent, small scale eddies in the flow field provide rapid and effective mixing. In the manufacture of BDS copper, for example, molten material 1 in Fig. 1 will be a copper-boron alloy, while molten material 2 should be a copper alloy containing a strong boride-forming alloying element, e.g., titanium. When the molten materials are mixed in the mixing chamber, TiB$_2$ particles are formed in-situ. The resultant mixture then exits the

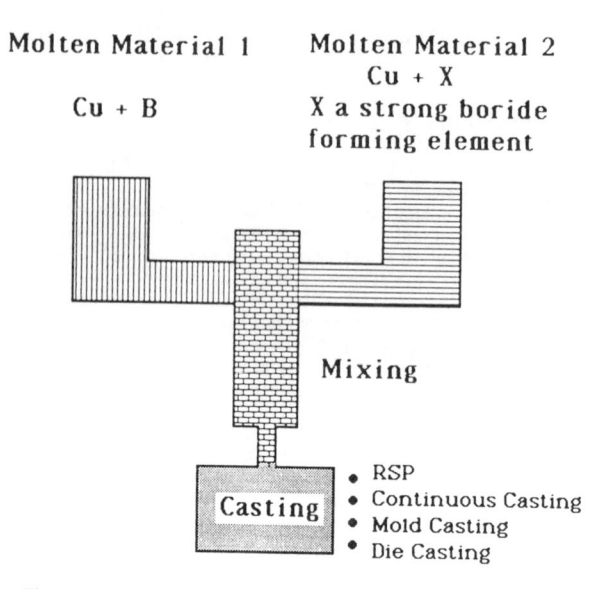

Fig. 1 - Schematic representation of the Mixalloy process.

mixing chamber and is immediately cast using an appropriate casting apparatus. For further improvement in properties, the pure copper matrix of a BDS copper can be alloyed accordingly to fulfill specific needs. The additional alloying element(s) could be introduced as part of molten material 1 or 2.

The degree of turbulent mixing in an impingement mixing chamber is a function of the Reynolds number of the impinging streams [6,7]. Since liquid metals typically have low kinetic viscosities, efficient turbulent mixing can readily be obtained at relatively low velocities. In the traditional metal processing technology, mixing is commonly accomplished by electromagnetic stirring and gas agitation. While these mixing methods are adequate for certain limited applications, they are inappropriate means as mixing methods for microstructural control. The energy density (intensity of agitation) provided by these mixing methods is relatively low, and the residence time is quite long. In impingement mixing, however, residence time can be extremely short and a reasonably high energy density can be obtained. In the manufacturing of BDS copper alloys, for example, the precise control of residence time is extremely important to ensure a fine boride particle size and a uniform distribution in the final product. Also, the hydrodynamic nature of the impingement mixer allows easy scale-up for commercial applications.

MATERIALS PREPARATION AND EXPERIMENTAL PROCEDURE

BDS copper alloys with a pure copper matrix, alloyed dilutely with titanium, and alloyed dilutely with zirconium and chromium were prepared using the Mixalloy process in combination with chill block melt spinning. The melt-spun ribbons were pulverized, freed of reducible surface oxide by annealing in hydrogen, canned and hot extruded into round rods of approximately 1.50 cm diameter.

Room temperature electrical conductivity and Vicker's hardness (200g load) were measured on all the consolidated materials. Room temperature tensile properties and thermal stability were also measured on selected BDS coppers. Thermal stability is measured by the drop of room temperature Vicker's hardness after isochronally annealing (1 hour duration) at different temperatures. Microstructures of as-extruded TiB$_2$ DS copper were observed under the SEM.

RESULTS AND DISCUSSIONS

The thermal stability of cold-rolled (95% reduction) BDS copper strengthened by approximately 5 volume percent TiB$_2$ (MXT5) is shown in Fig. 2. For comparison, data of other high strength, high conductivity precipitation hardened alloys, Cu-0.15Zr and Cu-0.8 Cr (both ingot wrought products and commercially hardened), and dispersion strengthened alloy, Cu-Al$_2$O$_3$ (Al-60, cold rolled 95% reduction), were also included. The remarkable thermal stability of the TiB$_2$ DS copper is evident from Fig. 2, and it compares favorably with the Cu-Al$_2$O$_3$ (Al-60), an oxide DS copper (2.7 volume percent Al$_2$O$_3$) manufactured by the method of internal oxidation and powder metallurgy. The precipitation hardened Cu-0.15Zr and Cu-0.8Cr, on the other hand, overaged rapidly when subjected to temperatures beyond about 500°C.

Fig. 3(a) shows the fine microstructures of the as-extruded MXT5 material, comprising ultra fine (less than 100 nm average) TiB$_2$ particles uniformly distributed in the copper matrix. Fig. 3(b) shows that after annealing at 900°C for 1 hour, no apparent coarsening of the TiB$_2$ particles has been observed, which is in agreement with the thermal stability data shown in Fig. 2.

Table 1 shows that cold rolled MXT5 has higher strength than wrought Cu-0.15Zr, Cu-0.8Cr, and cold rolled AL-60. The higher strength of the MXT5 material over that of the AL-60 can be partially attributed to the higher dispersoid content in this material versus that of AL-60. However, the DS copper by internal oxidation is not capable of providing a much higher dispersoid content than that in AL-60, due to the formation of a continuous Al$_2$O$_3$ film within the powder particles which inhibits further internal oxidation [8]. At the same time, electrical conductivity of the MXT5 compares favorably with the other alloys, and ductility is adequate.

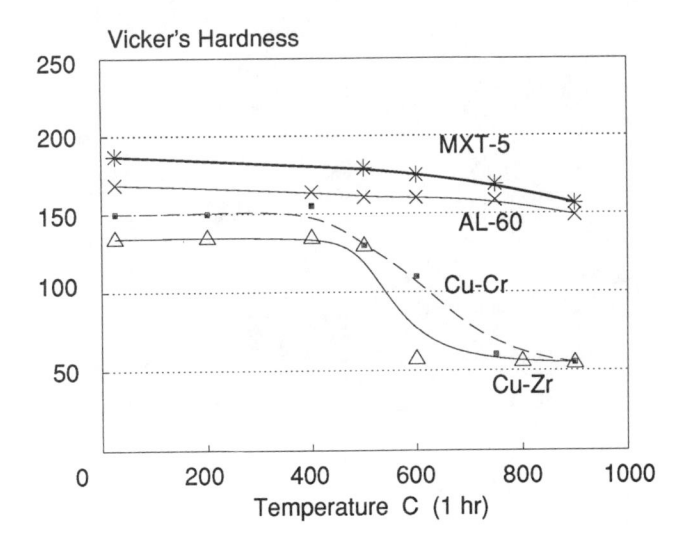

Fig. 2 - Thermal stability of TiB$_2$ DS copper versus selected high performance copper alloys: MXT5, as extruded and cold rolled 95%; AL-60, as-extruded and cold rolled 95%; Cu-0.8Cr and Cu-0.15Zr, commercially hardened, see Table I.

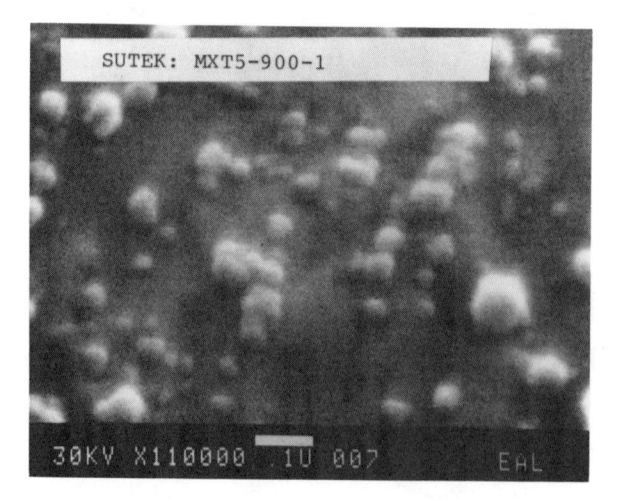

Fig. 3 - Microstructures of MXT5 viewed under the SEM showing the fine TiB$_2$ phase, (a) as-extruded condition and (b) after annealing at 900ºC for 1 hour.

Table I - Tensile and Electrical Properties of Selected High Conductivity Copper Alloys.

Material	Tensile Strength	Yield Strength 0.5% Ext. Under Load	Yield Strength 0.2% Offset	Elongation in 2"	Elongation in 1"	Electrical Conductivity
	MPa(KSI)	MPa(KSI)	MPa(KSI)	%	%	%IACS
Cu-0.15Zr [a]	523(76)	496(72)		1.5		90
Cu-0.8Cr [b]	592(86)	530(77)		14.0		80
AL-60 [c]	620(90)		599(87)	14.0		78
MXT5 [d]	675(98)		620(90)		7.0	76

(a) CDA 150, 1mm (0.040 in) wire, solution heat treated, cold worked 98% and aged.
(b) CDA 182, 4mm (0.150 in) rod, solution heat treated, cold worked 90% and aged.
(c) CDA 15760, 7mm (0.275 in) rod, as-extruded, cold worked 74%.
(d) 0.76mm (0.040 in) thick strip, as-extruded, cold rolled 95%.

The thermal stability of a BDS copper strengthened by about 3 volume percent TiB_2 and alloyed with about 1 percent by weight of titanium (MXT3T), is shown in Fig 4. Although this alloy has a lower TiB_2 content, and thermomechanical treatment was not optimized, there is an increase in hardness of about 15 percent over that of the MXT5. Titanium was added to improve the spring properties of BDS copper. Fig. 4 also shows that the titanium alloyed BDS copper combines the strengthening effect from Ti at low temperatures, and the stabilizing effect by the boride dispersion at high temperatures. This combination of strengthening mechanisms resembles that used in nickel-base superalloys fabricted via the mechanical alloying route. These impressive properties, however, are gained at the expense of electrical conductivity; 25% IACS of the MXT3-T vs. 76% IACS of the unalloyed BDS copper, MXT5.

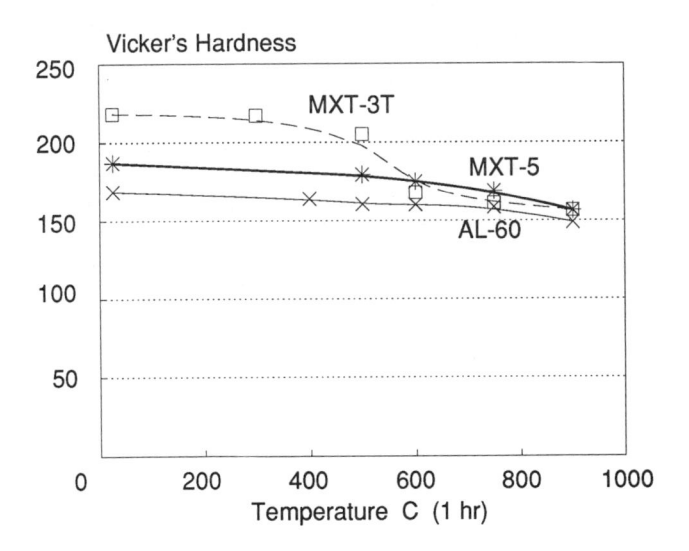

Fig. 4 - Thermal stability of MXT3T, a 3 volume percent TiB_2 DS copper alloyed dilutely with titanium versus selected high performance copper alloys: MXT3T, drop forged 68% reduction with 650°C intermediate anneal, MXT5 and AL-60, as-extruded and cold rolled 95% reduction.

To gain high strength while preserving electrical conductivity, a BDS copper can be alloyed with other elements. A ZrB_2 DS copper containing approximately 2.5 volume percent ZrB_2 and dilute amounts of zirconium and chromium exhibits a Vicker's hardness of 210 after solutionizing, cold rolling and peak aging. For comparison, comparable wrought zirconium copper, chromium copper, and ZrB_2 DS copper all exhibit Vicker's hardness in the 140-160 kg/mm² range. At the same time, a high electrical conductivity value of 77% is maintained. This compares favorably with 90%, 80% and 85% of equivalent zirconium copper, chromium copper, and zirconium diboride DS copper, respectively.

CONCLUSIONS

1) The Mixalloy process, in combination with chill block melt spinning, is capable of producing unique high performance copper base DS materials that have not been available up to now.

2) In the processing of DS materials, the Mixalloy process has a low energy cost since in-situ formation of refractory particles, alloying, and casting can all be accomplished in a substantially continuous fashion.

3) The versatility and simplicity of the Mixalloy process provide an economical alternative to the other processing means in the manufacturing of dispersion strengthened materials.

REFERENCES

[1] N. P. Suh, ASME Journal of Engineering for Industry, 104, 327, (1982).

[2] N. P. Suh, N. Tsuda, M. G. Moon, N. Saka, ASME Journal of Engineering for Industry, 104, 332, (1982).

[3] N. P. Suh, U.S. Patent No. 4 278 622 (14 July 1981).

[4] N. P. Suh, U.S. Patent No. 4 279 843 (21 July 1981).

[5] L. E. Sanchez-Caldera, N. P. Suh, J-H Chun, A. K. Lee, F. S. Blackall IV, U.S. Patent No. 4 706 730 (17 November 1987).

[6] C. L. Tucker and N. P. Suh, Polymer Engineering and Science, 3 (13), 875, (1980).

[7] L. T. Nguyen, "Processing of Interpenetrating Polymer Networks by Reaction Injection Molding", Ph.D. Thesis, Mechanical Engineering Dept., MIT, Cambridge, 1984.

[8] A. V. Nadkarni, E. Klar and W. M. Shefer, Metals Engineering Quarterly, 10, August 1976.

MICROSTRUCTURE OF AN Al_2O_3/METAL COMPOSITE CONTAINING AN Al_2O_3 FILLER MATERIAL

E. Breval* and A. S. Nagelberg**
*The Pennsylvania State University, University Park, PA 16802
**Lanxide Corporation, Tralee Industrial Park, Newark, DE 19711

ABSTRACT

A ceramic matrix composite consisting of 100-200 μm Al_2O_3 grains (90 grit) embedded an Al_2O_3/metal matrix was examined. The matrix was produced by the directed oxidation of a molten Al alloy. The microstructure was studied by optical methods, x-ray diffraction and transmission electron microscopy (TEM) in conjunction with energy dispersive spectrometry (EDS). The matrix Al_2O_3 was interconnected and present as ~50-500 nm grains in ~10-50 μm regions. Within these regions, the Al_2O_3 was separated by low angle Al_2O_3/Al_2O_3 grain boundaries. The boundaries between regions consisted either of high angle Al_2O_3/Al_2O_3 boundaries with no grain boundary phase or of tortuous metal channels from 50 nm up to 1 μm in width. The metallic constituent within the regions was only partly interconnected, consisting of similar channels or isolated spheres of 1-500 nm. The metal consisted of Al and Si. Large fractions of the filler Al_2O_3 surface (~50%) were directly bonded to the matrix Al_2O_3.

INTRODUCTION

A new ceramic matrix composite technology has recently been described [1-4]. The matrix formation process involves the directed oxidation of a molten metal with a gaseous oxidant to form a matrix consisting of interconnected ceramic and at least partially interconnected metal and proceeds in the absence or presence of a substantially inert reinforcement or filler (e.g., Al_2O_3 or SiC). The metal constituent of the matrix has been shown to significantly affect the tensile strength, toughness and thermal shock resistance of the composites [2,5], and it can be anticipated that the distribution of phases is important for the mechanical properties.
In earlier structural investigations of the Al_2O_3/metal matrix produced without a filler phase [2,3], a preferred orientation of Al_2O_3 grains with a tendency of the <001> direction to align parallel to the growth direction was found. Typically, the Al_2O_3 phase was found as millimeter sized columnar regions wherein all Al_2O_3 grains have the same or almost the same orientation. The columnar regions have different orientations from neighboring regions, but still have <001> close to the growth direction. The boundaries between these regions were either high angle Al_2O_3/Al_2O_3 grain boundaries or consisted of 0.05 - 3 μm wide metal channels.
This paper is a similar study for an Al_2O_3 particulate reinforced composite, describing the distribution of phases and their interconnectivity with particular emphasis placed on the boundaries between filler Al_2O_3 and matrix Al_2O_3.

EXPERIMENTAL

The composite investigated consisted of 90 grit fused Al_2O_3 (100-200 µm) embedded in an Al_2O_3/metal matrix. The matrix was formed by the directed oxidation of an Al-alloy (Al-10 wt. % Mg-3 wt. % Si) in air at 1250°C. Three plates (~2 to 3 mm thick) were cut from the composite. One plate was oriented with its large face perpendicular to the growth direction, the other two were oriented parallel to the growth direction but mutually perpendicular. The plates were polished with diamond powder down to 1 µm for metallographic analysis of the microstructure. The possible preferred orientation of the matrix microstructure was examined by x-ray diffraction of the three oriented plates and a crushed powder obtained from one of the plates. For this analysis, it was assumed that the Al_2O_3 filler particles were randomly oriented. Particular x-ray reflections were chosen for these orientation studies: 1) hkl parallel to the main axis, 2) hkl with 1>>h and 1>>k, 3) peaks with reasonable intensities and 4) peaks which are not interfered with by other phases in the specimen. The selected Al_2O_3 peaks were normalized to I = 100 for hkl = 113.

From each of the three plates four to six disks (3mm diameter and ~0.5 mm thick) were cut. The disks were polished on each side with diamond powder to 1 µm and ion beam milled. The specimens were observed in a TEM fitted with an EDS system at an acceleration voltage of 120 kV.

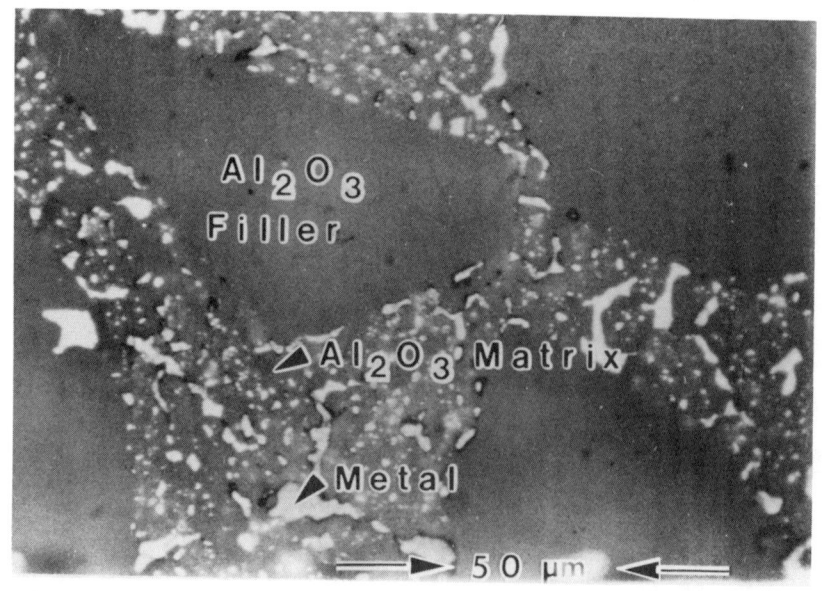

Fig. 1. Optical Micrograph of a polished surface. One filler Al_2O_3 grain is marked. The matrix consists of µm size Al_2O_3 (grey), and metal (white).

RESULTS AND DISCUSSION

The composite growth process leads to the presence of alpha-Al_2O_3 from two different sources in the composite microstructure. The first is the large Al_2O_3 filler grains, clearly visible in Fig. 1. The Al_2O_3 filler is embedded in a matrix of metal and fine grained interconnected Al_2O_3 (and metal) which represents the second type of Al_2O_3. The nature of the boundary between the Al_2O_3 filler and composite matrix was studied in more detail. The matrix Al2O3 was directly bonded to 40+10 % of the filler surface. The typical dimension for the filler-matrix Al_2O_3/Al_2O_3 boundary length was 6+3 μm, but in extreme cases boundaries up to 22 μm and down to 0.5 μm have been observed. The filler/metal boundary length was slightly larger, typically 8+4 μm. There was no significant difference in the extent of bonding between the filler Al_2O_3 and matrix Al_2O_3 with orientation relative to the growth direction.

A slight preferred orientation of the matrix Al_2O_3 was clearly seen by x-ray diffraction (Table I). Crushed composite powder exhibited peak intensities close to those expected for randomly oriented Al_2O_3. Conversely, the intensity of the hk0 diffraction lines for the composite plate cut parallel to the growth direction were significantly enhanced while the lines from planes with l>>h and l>>k were decreased. An opposite trend was observed for the plane cut perpendicular to the growth direction. If the filler Al_2O_3

Table I

Composite Orientation Effect on

X-ray Diffraction Peak Intensity

Selected hkl	Random Orientation (JPCD 10-173)	Al_2O_3 Composite Crushed	Parallel to Growth	Perpendicular to Growth
113	100	100	100	100
110[a]	40	43	106	20
030[a]	50	50	118	55
220[a]	8	5	12	3
208[b]	4	2	<1	1
1.0.10[b]	16	13	<1	23
119[b]	8	9	1	11

[a] planes parallel to main axis
[b] planes with l>>h, l>>k

Table I. X-ray intensities of Al_2O_3 reflections normalized to the most intense Al_2O_3 reflection (113) (I = 100). The reflection listed are either parallel to the main axis or close to perpendicular to the main axis.

can be assumed to be randomly oriented in the filler bed, these results indicate a slight preferred orientation of the matrix Al_2O_3 with <001> parallel to the growth direction.

The extension of the microstructural analysis of the composite structure to the nanometer level was performed by TEM. The dark field technique was utilized to easily distinguish between filler Al_2O_3 grains, with the same crystallographic orientation over hundreds of μm, and the matrix Al_2O_3 which was subdivided into 50 to 500 nm size grains separated by low angle grain boundaries. Regions of similarly oriented grains typically extended for 10-50 μm. Low magnification TEM imaging confirmed the interconnectivity of the matrix Al_2O_3. The boundaries between regions were either high angle Al_2O_3/Al_2O_3 boundaries with no apparent grain boundary phase or metal channels typically ranging from 50 nm up to 1 μm. Figure 2 shows the distribution of isolated metal regions within the matrix Al_2O_3 grains. The metal appeared as circular inclusions (1-500 nm) in a large number of regions regardless of the relative orientation to the growth direction. Since these isolated metal regions were generally of a smaller size than the metal channels, were present in a large number, and always appeared almost circular in cross-section, it was assumed that this enclosed metal was present as spheres.

The grain boundaries between filler Al_2O_3 and matrix Al_2O_3 were studied thoroughly. A majority were high angle Al_2O_3/Al_2O_3 grain boundaries but a few low angle Al_2O_3/Al_2O_3 grain boundaries were also observed. In all cases, the grain

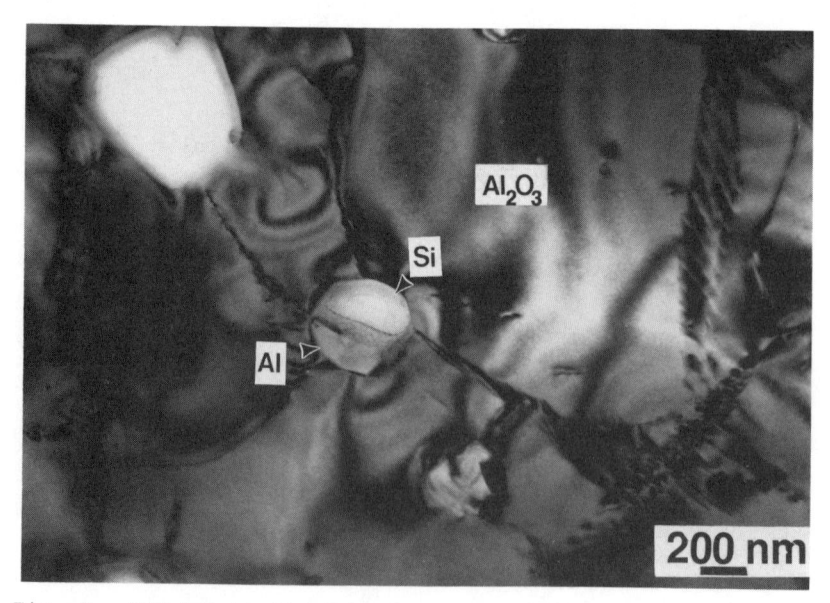

Fig. 2. TEM bright field image of a typical region in the matrix. The Al_2O_3 contains low angle Al_2O_3/Al_2O_3 grain boundaries and many metal inclusions. One marked contains both Al and Si.

boundary between the filler and matrix Al_2O_3 was found to be void of a metal phase. At numerous triple points between metal, filler Al_2O_3, and matrix Al_2O_3, the metal phase was studied thoroughly to see if metal extended into a grain boundary phase between filler and matrix Al_2O_3. The dark field technique was widely used to highlight the metal phase since even nanometer size grain boundary phases could easily be made visible by this technique. The absence of a grain boundary phase or metal penetration between the filler grains and the Al_2O_3 matrix is clearly illustrated in the dark field micrograph shown in Fig. 3.

In comparison to the microstructure in the absence of a filler phase [2], these composites exhibit numerous similarities. For both composites, the matrix Al_2O_3 is clearly interconnected, while the metal is present in both interconnected and isolated regions. Grain boundaries in both cases were found to be devoid of any grain boundary phases. While the composite grown into an Al_2O_3 filler exhibits slight preferred orientation of the matrix Al_2O_3 over small regions (~ 10-50 μm), the matrix Al_2O_3 grains formed in the absence of a filler had similar orientations over regions that extended millimeters. In general, the size of the metal regions was refined in the case of the matrix grown through a filler.

SUMMARY

The microstructure of a composite material consisting of filler Al_2O_3 grains embedded in a Al_2O_3/metal ceramic matrix formed by the directed oxidation of a molten Al alloy was

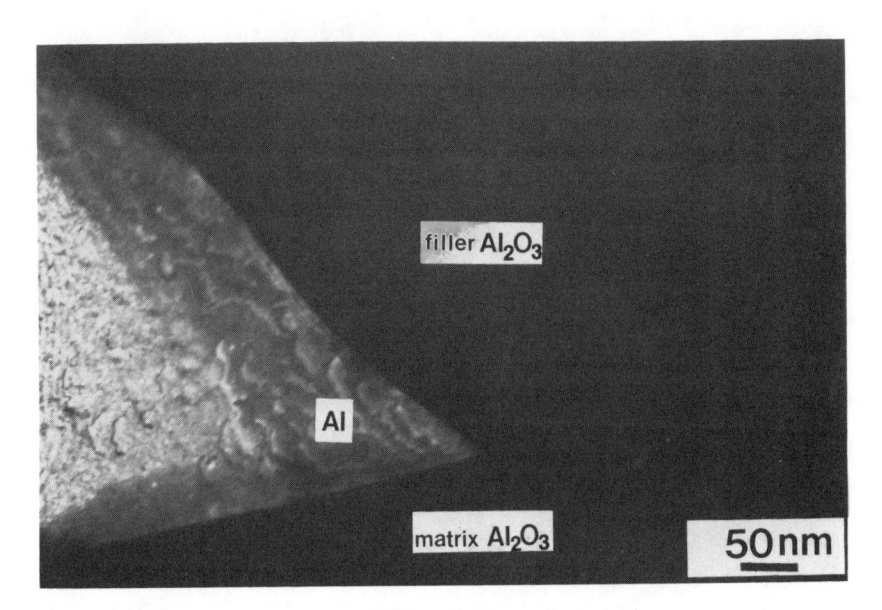

Fig. 3. TEM dark field image of the metal phase separating filler and matrix Al_2O_3. The metal does not extend into the grain boundary between the two types of Al_2O_3.

examined. X-ray diffraction showed that the matrix consisted of α-Al_2O_3, Al, and Si. Analysis of variations in the x-ray diffraction peak intensities of composite plates cut at various orientations to the growth direction showed a tendency for the <001> axis of the matrix Al_2O_3 to lie in the growth direction. TEM investigations showed that the matrix Al_2O_3 is interconnected and is present as ~50 to 500 nm grains in ~10 to 50 μm regions. Within these regions, the Al_2O_3 is separated by low angle Al_2O_3/Al_2O_3 grain boundaries or nanometer to micrometer size metal channels. The boundaries between regions consist either of high angle Al_2O_3/Al_2O_3 boundaries with no grain boundary phase or metal channels 0.05 to 1 μm in width. The metallic constituent, which is present as Al and Si, is only partly interconnected consisting of similar channels or isolated spheres of 1 to 500 nm.

Optical microscopy showed that one third to one half of the surfaces of each Al_2O_3 filler grain is directly bonded to the matrix Al_2O_3. The boundaries between matrix and filler Al_2O_3 typically are 6 μm in diameter but up to 4 times larger boundaries have been observed. TEM investigations have shown that they are void of any grain boundary phases and that these grain boundaries typically are high angle Al_2O_3/Al_2O_3 boundaries.

REFERENCES

1. M. S. Newkirk, A. W. Urquhart, H. R. Zwicker, E. Breval, J. Mater. Res. 1(1), 81-89, Jan./Feb. 1986.

2. M. K. Aghajanian, N. H. Macmillan, C. R. Kennedy, S. J. Luszcz, R. Roy, J. Mater. Sci. 1988 (in press).

3. E. Breval, M. Aghajanian, (unpublished results).

4. M. S. Newkirk, H. D. Lesher, D. R. White, C. R. Kennedy, A. W. Urquhart, T. D. Claar, Ceram. Eng. and Sci. Proc. 8[7-8], 879 (1987).

5. C. A. Andersson and M. K. Aghajanian, to be published in Ceramic Engineering and Science Proceedings.

INFRARED PROPERTIES OF THIN Pt/Al$_2$O$_3$
GRANULAR METAL-INSULATOR COMPOSITE FILMS

M. F. MacMILLAN*, R. P. DEVATY*, and J. V. MANTESE**
*University of Pittsburgh, Department of Physics and Astronomy
100 Allen Hall, Pittsburgh, PA 15260
**General Motors Research Laboratories, Electrical and Electronics
Engineering Department, Warren, MI 48090

ABSTRACT

The reflection of mid- and near-infrared radiation (400-15000 cm^{-1} by thin Pt/Al$_2$O$_3$ cermet films was measured using a Fourier transform spectrometer. The data were compared with predictions of three models for the effective optical constants of heterogeneous materials: Maxwell-Garnett, Bruggeman, and a simplified version of a probabilistic growth model due to Sheng. Sheng's model provides the best description of the data over the complete range of metallic volume fraction. This result is expected based on the most likely topology of the films and is in agreement with other work on similar systems at higher frequencies.

INTRODUCTION

The relationship between the microstructural topology, the optical properties of a heterogeneous system, and their description by an homogeneous effective dielectric function has been made clear in recent years [1]. The prediction of the optical response of a granular metal/insulator composite material over the complete range of composition provides an especially stringent challenge to effective medium theories.

In this paper, we compare the predictions of three simple effective medium models, which describe different microstructural topologies, with the measured relative reflectivity of thin Pt/Al$_2$O$_3$ films in the mid- and near-infrared. These models have been successfully applied to similar cermet systems in the near infrared and visible regions [1,2]. We find that a simplified version of a probability growth model introduced by Sheng [2,3] provides the best agreement with the data over the complete range of volume fraction. This result is not surprising since this model was developed specifically for application to cermet films. The greatest disagreement between theory and experiment occurs for metallic volume fractions near the percolation threshold for the dc conductivity. Effective medium theories are known to fail in this region[4].

EXPERIMENTAL

The Pt/Al$_2$O$_3$ granular films were prepared by coevaporation of Pt and Al$_2$O$_3$ onto single crystal sapphire wafers in a dual e-beam evaporator with a low base pressure (mid 10^{-9} torr). The volume fraction of Pt in the films, f, which ranges from 0.23 to 1.00, was determined by monitoring the relative deposition rates of the two materials using quartz crystal oscillators inside the evaporation chamber [5]. Film thicknesses, measured with a Dektak surface profilometer, ranged from 1100 to 1800 Å. The orientations of the surfaces of the sapphire substrates were determined by back reflection Laue photographs to be about 20° from the ordinary axis. The percolation threshold, f$_c$, as determined by the temperature dependence of the dc resistivity, is in the range of 50-59% Pt by volume [6]. Although no direct determination of microstructure by electron microscopy has been performed on the films, the high value of f$_c$ is a strong indicator of a coated-grain topology. Electron micrographs on cermet films prepared by the same method support this conclusion [7,8].

Room temperature relative reflection data were obtained using a Nicolet 740 Fourier transform infrared spectrometer with a nitrogen purge. The spectral range of 400 to 15000 cm^{-1} was covered using three combinations of sources, beamsplitters, and detectors. The resolution was 4 cm^{-1}. The samples were mounted on a Spectra Tech Model 500 variable angle specular reflectometer. The angle of incidence was 5° from the normal. A 100% Pt film from the set of samples was used as the reference.

THEORY

The predictions of three effective medium theories, the Maxwell-Garnett (MG) [9], Bruggeman (BR) [10], and the Sheng probabilistic growth model (PG) [3], were compared with the data. Each model is based on a specific microstructural topology. In using these models, we assume spherical grains and homogeneity of the cermet on the scale of a wavelength.

The topology of the MG model [9] is a dielectric-coated metal sphere (one may also choose a metal-coated dielectric sphere). The thickness of the coating is determined by the metallic volume fraction of the medium, f. The effective dielectric function is

$$\epsilon_{MG} = \epsilon_o \frac{2\epsilon_o(1-f) + \epsilon_1(2f+1)}{\epsilon_o(f+2) + \epsilon_1(1-f)} \tag{1}$$

where ϵ_0 and ϵ_1 are the complex dielectric functions of the host and metal, respectively. This model is inherently asymmetric in its treatment of the constituent materials. There is no percolation transition since the metal spheres have an insulating coating for any f.

The BR model [10] treats the cermet as a collection of uncoated metal and dielectric spheres. The effective dielectric function is obtained by solving

$$\frac{3f}{2 + \epsilon_1/\epsilon_{BR}} + \frac{3(1-f)}{2 + \epsilon_o/\epsilon_{BR}} = 1. \tag{2}$$

This model treats the constituents in a symmetric fashion. The percolation threshold is $f_c = 1/3$.

The simplified PG model [2,3] can be regarded as a generalization of the MG model that is symmetrical in the constituent materials.
Metal-coated oxide spheres and dielectric-coated metal spheres coexist in the medium. The effective dielectric function, ϵ_{PG}, is determined by

$$\frac{3p_1}{2 + \epsilon_1/\epsilon_{PG}} + \frac{3(1-p_1)}{2 + \epsilon_2/\epsilon_{PG}} = 1 \tag{3}$$

where p_1 is the fraction of oxide-coated metal spheres in the composite. p_1 is related to f by means of a free volume argument by

$$p_1 = \frac{(1 - f^{1/3})^3}{(1 - f^{1/3})^3 + (1 - (1-f)^{1/3})^3}. \tag{4}$$

ϵ_1 and ϵ_2 are the dielectric functions for the oxide-coated metal spheres and the metal-coated oxide spheres [11], respectively. This model predicts percolation at $f_c = 1/2$.

The model treats the reflection of normally incident electromagnetic radiation by a thin film on a substrate according to standard methods [12,13]. Multiple reflections within the substrate crystal were treated as incoherent. Published optical data on bulk Pt [14] and evaporated Al_2O_3 films [15] were used to calculate the optical properties of the cermet film. The substrate crystal was assumed to be aligned with the axis of symmetry normal to the surface. The oscillator parameters from Barker's fit to reflectivity data [16] were used to model the complex dielectric function of the substrate.

RESULTS AND DISCUSSION

Figure 1 compares the measured and calculated relative front reflection for three selected films. Figure 1a shows that both the MG and PG models provide good fits to the measured relative reflectivity of a f=.23 film, which has the lowest metallic volume fraction of any film in our set. The MG and PG models make similar predictions at this relatively low f because most of the coated spheres in the PG model are dielectric-coated metal spheres. The oscillatory nature of the reflectivity arises from an interference effect due to multiple reflections within the cermet film. These oscillations are less apparent in films with greater metallic composition. To obtain the fit, the thickness of the film was assumed to be 2200 Å, whereas the measured value was 1525 Å. The structure below 1500 cm^{-1} is due to optic phonons in the Al_2O_3 host and substrate.

Figure 1b shows the results for a film with f=.58. This volume fraction is near the percolation threshold for this system [6]. The measured film thickness of 1475 Å was used in the calculations. None of the models describes the data very well. This result is in agreement with recent work showing that effective medium theories do not apply near percolation [4]. Details of cluster morphology not included in the effective medium treatment, such as nonspherical shape, are most important near f_c. The rather flat frequency dependence of the relative reflectivity is also a characteristic feature near f_c [4,17]. The MG model best fits the magnitude of relative reflectivity, but the BR and PG models provide a better description of the frequency dependence.

Figure 1c shows that the BR and PG models provide the best fits to a 1300 Å thick film with f=.83. The agreement with the MG model is poor. A reverse MG model using metal-coated dielectric spheres would provide a better description. In fact, this is the dominant topology of the PG model for f near unity.

Berthier and Lafait [18] have studied the infrared properties of Pt/Al_2O_3 cermet films prepared by rf cosputtering. They were able to fit the measured optical constants using the BR model over a broad range of composition. However, they treated the volume fraction as an adjustable parameter and introduced an effective depolarization factor as well. The BR model, with fewer adjustable parameters, does not apply to our films, which were prepared by a different method.

In conclusion, we find that the PG model provides the best description, among three simple effective medium theories, of the relative reflectivity of thin Pt/Al_2O_3 granular cermet films over the complete range of composition, although the agreement is not quantitative, particularly near f_c. We intend to extend this work by measuring the absolute reflectivity using the W-V method and to extract the frequency dependent complex index of refraction of films for which the transmission is also measurable.

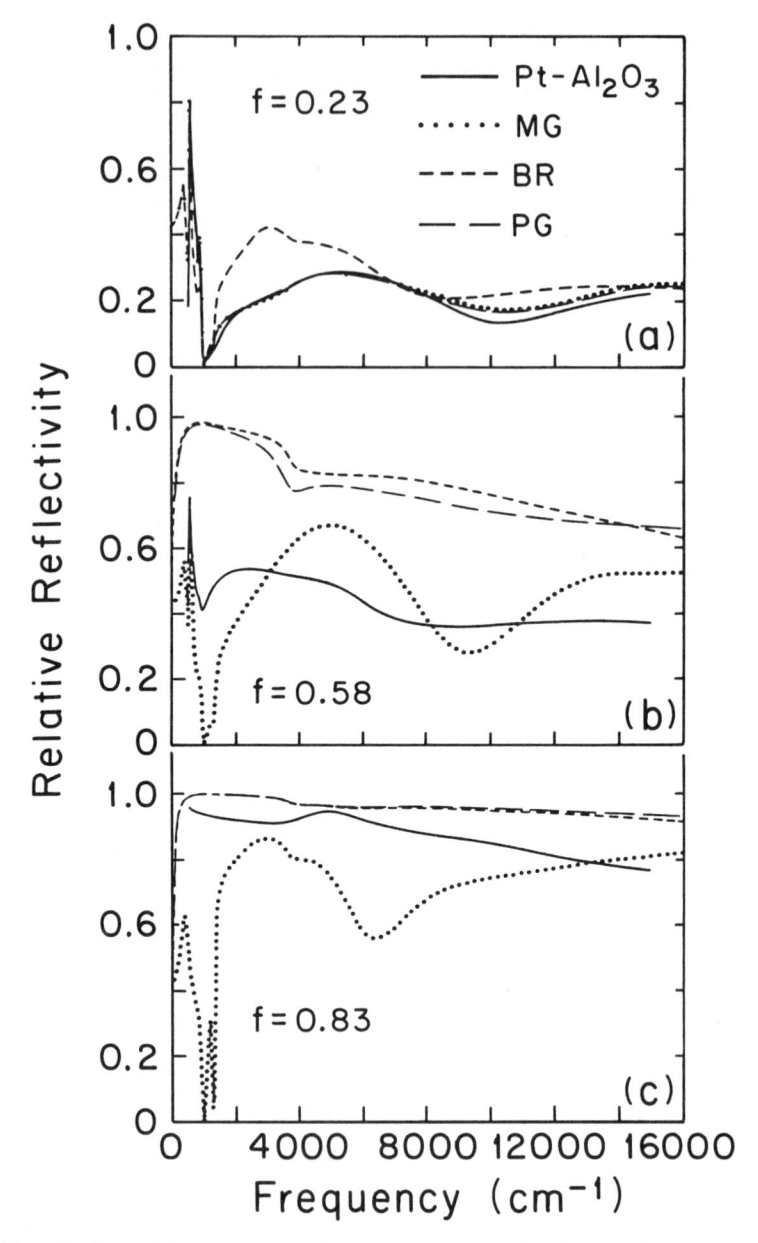

Figure 1: Composition dependence of relative front reflectivity of Pt/Al$_2$O$_3$ granular composite films. A Pt film is the reference. The predictions of the three models discussed in the text are also shown.

ACKNOWLEDGEMENT

This work was supported by ONR Contract #N00014-85-K-0808.

REFERENCES

[1] U. J. Gibson, H. G. Craighead, and R. A. Buhrman, Phys. Rev. B25, 1449 (1982).

[2] U. J. Gibson and R. A. Buhrman, Phys. Rev. B27, 5046 (1983).

[3] Ping Sheng, Phys. Rev. Lett. 45, 60 (1980).

[4] Y. Yagil and G. Deutscher, Appl. Phys. Lett. 52, 374 (1988), and references therein.

[5] J. V. Mantese, Ph.D. Thesis, Cornell University, 1985.

[6] J. V. Mantese, W. A. Curtin, and W. W. Webb, Phys. Rev. B33, 7897 (1986).

[7] H. G. Craighead, Ph.D. Thesis, Cornell University, 1980.

[8] U. J. Gibson, Ph.D. Thesis, Cornell University, 1982.

[9] J. C. M. Garnett, Phil. Trans. R. Soc. London, Ser A203, 385 (1904).

[10] D. A. G. Bruggeman, Ann. Phys. (Leipzig) 24, 636 (1936).

[11] H. C. van der Hulst, Light Scattering by Small Particles (Wiley, New York, 1957).

[12] Z. Knittl, Optics of Thin Films (Wiley, New York, 1976).

[13] O. S. Heavens, Optical Properties of Thin Solid Films (Butterworths Publications, Ltd., London, 1955).

[14] D. W. Lynch and W. R. Hunter, in Handbook of Optical Constants of Solids, edited by E. Palik (Academic Press, New York, 1985).

[15] T. S. Eriksson, A. Hjortsberg, G. A. Niklasson, and C. G. Granqvist, Appl. Opt. 20, 2742 (1981).

[16] A. S. Barker, Jr., Phys. Rev. 132, 1474 (1963).

[17] P. Gadenne, A. Beghdadi, and J. Lafait, Optics Commun. 65, 17 (1988).

[18] S. Berthier and J. Lafait, J. Physique 47, 249 (1986).

PROPERTIES OF TRANSPARENT SILICA GEL _ PMMA COMPOSITES

EDWARD J. A. POPE, MINUO ASAMI, AND JOHN D. MACKENZIE
Department of Materials Science and Engineering,
University of California, Los Angeles, CA 90024

ABSTRACT

Transparent silica gel - polymer composites have been prepared by the impregnation of porous silica gels with fluid organic monomer followed by in situ polymerization. These materials constitute an entirely new class of transparent composites.

INTRODUCTION

Recently, the sol-gel method has received widespread interest for the preparation of bulk glassses, porous solids, and thin films. Unlike traditional glass and ceramic processing techniques, in which powders are reacted at high temperatures, the sol-gel process relies upon hydrolysis and polymerization reactions in liquid solution at temperatures near ambient [1]. After gelation, the gel can be dried and heat treated rendering a porous solid or dense glass or ceramic. Theoretically, almost any oxide can be prepared in a porous form by the sol-gel method. Secondly, the pore size distribution and porosity can also be varied over wide ranges. Thus, by the use of porous oxides derived from gels, an entirely new family of composites can be prepared through the impregnation of different organic polymers.

With the exception of composites composed of alternate layers of glass and polymer, such as those laminated composites used in automobile windows, and some glass-ceramics, such as those commercially marketed by Corning, almost all presently known composite materials are opaque . The main reason most glass-polymer composites are opaque is due to light scattering. In our composites, the phase dimensions are on the order of 100 angstroms, much smaller than the wavelengthes of visible light (3700 to 7000 angstroms). Hence, this new class of glass--polymer composites is quite transparent, which opens up a wide variety of applications for which composite materials have been previously excluded.

It has been previously demonstrated that a wide variety of organic polymers can be successfully polymerized inside silica gels, including polymethylmethacrylate(PMMA), silicone, polystyrene, butyl acrylate, and dimethyl butadiene(DMB)[2-4]. In addition, six different copolymers of these polymers have also been impregnated into silica gels. In figures 1 and 2, photographs of transparent silica gel - PMMA and silica gel - silicone composites are presented.

Applications

Through the impregnation of porous silica gels with organic monomers, such as methyl methacrylate, followed by in situ polymerization, transparent silica gel - PMMA composites can be prepared. Such transparent composites, prepared at relatively low temperatures, have potential applications as

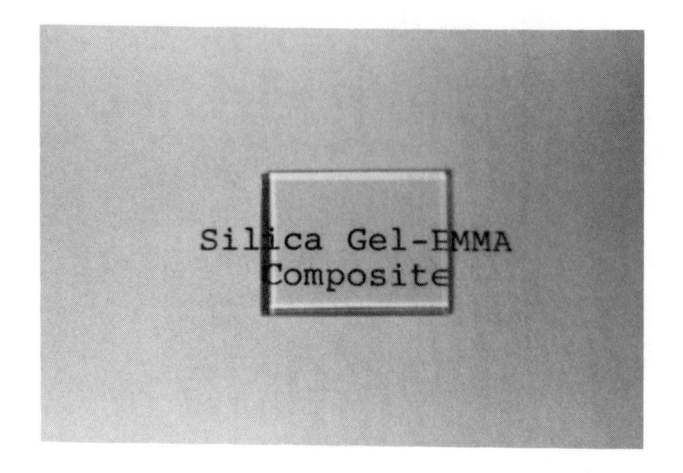

Figure 1: Transparent Silica Gel - PMMA Composite.

Figure 2: Transparent Silica Gel - Silicone Composite rod. Sample length is 3 cm and sample diameter is 6 mm.

optical hosts for organic dye molecules[5]. Many organic dyes possess unusual optical properties, such as fluorescence and high non-linear index of refraction, which can be employed in passive and active optical components. Several researchers have successfully incorporated organic dye molecules into porous gels[5]. The incorporation of organic dyes into gels and gel-derived composites offers several device potentials:

        *** Optical Filters,
        *** Solid-state Dye Lasers,
        *** Nonlinear Optical Filters,
        *** Optical Data Storage, and
        *** Photoconductive Devices and Films.

While many organic dyes have been successfully incorporated into porous silica gels, the open porosity seriously limits potential applications and deletereously affects physical, mechanical, and optical properties. Through the impregnation of porous gels with organic monomer followed by polymerization, open porosity can be eliminated rendering a transparent, non-porous composite. In table I, transparent composites incorporating organic dyes are listed which have been prepared by our group. In figure 3, a photograph of an mNA-Silica-PMMA is shown.

Table I: Organic dyes successfully incorporated into Silica Gel - PMMA Composites.

Organic Dye	Potential Application	Color
Rhodamine b	Solid-State Dye Laser	Red
2-methyl-4-nitroanaline	Nonlinear Optics	Yellow
4-methylembelliferone	Dye Lasers	Colorless
1',3',3'-trimethyl-6-hydroxyspiro-(2H-1-benzo-pyran2,2'-indoline)	Photochromic Lens	orange

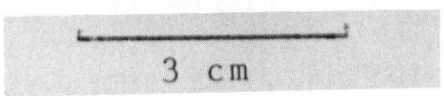

Figure 3: A composite lens of methyl nitroanaline doped silica gel - PMMA composite.

SELECTED PROPERTIES OF SILICA GEL - PMMA COMPOSITES

Through the impregnation of porous silica gels with methyl methacrylate monomer followed by polymerization, silica gel - PMMA composites can be prepared. The relative ratio of silica to PMMA can be varied across the entire compositional range by varying the porosity of the silica phase prior to impregnation. In an upcoming publication, properties, such as density, refractive index, modulus of rupture, compressive strength, elastic modulus, abrasion rate, and Vicker's hardness is examined as a function of polymer to oxide ratio for silica gel - PMMA composites[6]. In this paper, selected properties, such as index of refraction and transparency are presented.

Transparency

Despite being a mixture of two phases, the composites prepared in this study are highly transparent. The reason these composites are transparent is primarily due to the ultra-fine phase dimensions of approximately 150 angstroms, much smaller than the wavelength of visible light. Also important is the close match in indices of refraction. The morphology of this composite is essentially that of two continuous, interpenetrating phases of very fine diameter. There is no known detailed theoretical analysis of the scattering produced by such an unusual morphology. Therefore, simple calculations can be made using some approximation, such as that of discrete spherical particles uniformly distributed in a matrix. The equation is given by:

$$I/I_o = \exp[-3V_p x r^3 (n_p/n_m - 1)/4\lambda^4] \qquad (1)$$

where V_p = vol. fraction second phase, x = sample thickness, r = "phase dimension", n_p = index of refraction of second phase, n_m = index of refraction of matrix phase, and λ = wavelength of light. For the purpose of this approximation, the second phase is taken to be the minor constituent. For a 1 mm thick plate of 70 percent porous silica gel with 156 angstrom pore diameter, the transmission of red light (5000A) is 70 percent when porous and 99.2 percent when impregnated with PMMA. These figure agree well with transmission measurements made for such samples in our laboratory.

Index of Refraction

The index of refraction vs. PMMA content is presented in figure 4. Also presented is a theoretical plot using the Clausius-Mossotti relationship. Within a few percent scatter, the theoretically calculated curve agrees well with experimentally obtained data. The Clausius-Mossotti relation is based upon the net contributions of the electronic polarizabilities of the two phases present as a function of volume content.

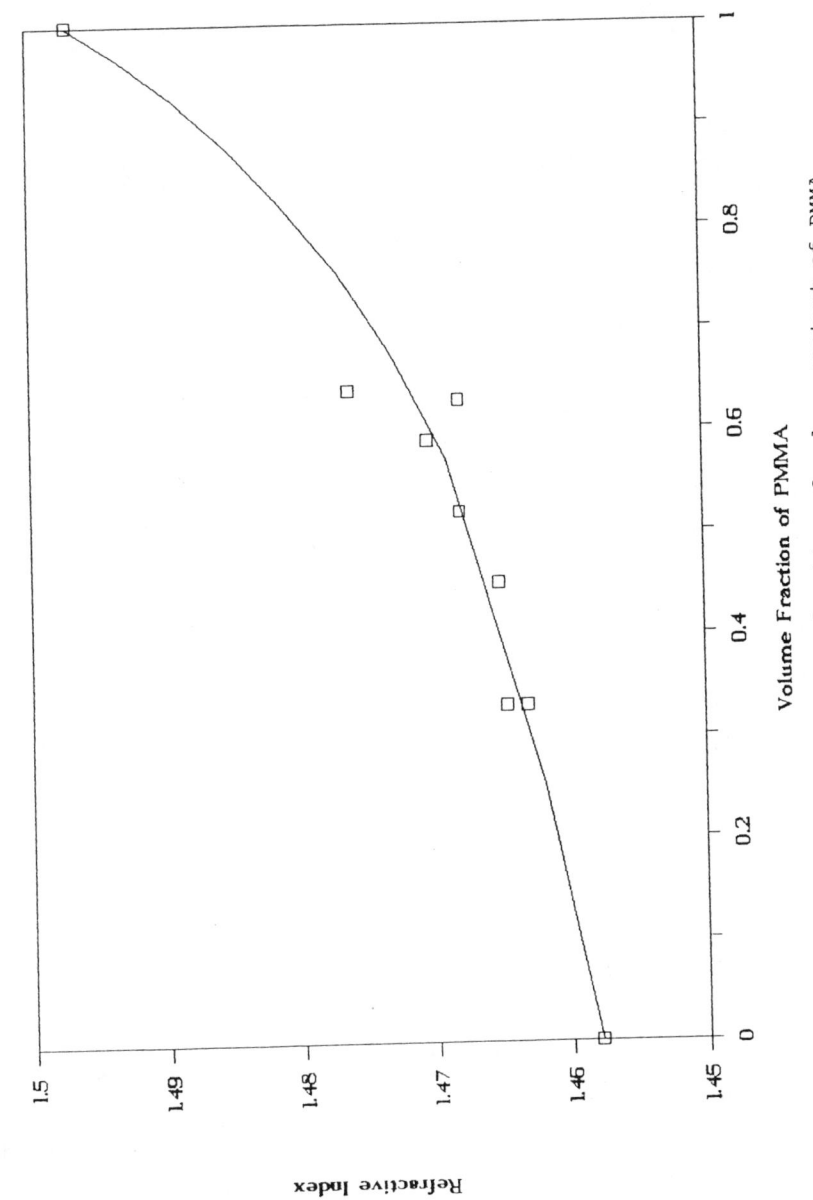

Figure 4 : Index of Refraction plotted as a function of volume content of PMMA.

CONCLUSIONS

In the fabrication of transparent silica gel - PMMA composites, the relative amount of each phase can be adjusted over the entire compositional range. Optical transparency and index of refraction of this new composite material have been presented.

ACKNOWLEDGEMENTS

We wish to acknowledge the support of the Air Force Office of Scientific Research, Directorate of Chemical and Atmospheric Sciences.

REFERENCES

[1] E. J. A. Pope and J. D. Mackenzie, *J. Non-Cryst. Sol.* 87(1986)185-198.

[2] E. J. A. Pope and J.D. Mackenzie, in *Better Ceramics Through Chemistry II*, edited by C. J. Brinker, D. E. Clark, and D. R. Ulrich(Mater. Res. Soc. Proc. 73, Pittsburgh PA 1986)pp. 809-814.

[3] E. J. A. Pope and J. D. Mackenzie, publ in *Tailoring Multiphase and Composite Ceramics*, ed. by R. E. Tressler, et al. (Plenum Press, New York, 1986) pp. 187-192.

[4] E. J. A. Pope and J. D. Mackenzie, in *Proc. of the 32nd Int'l SAMPE Symposium*, ed. by R. Carson, et al,(SAMPE, Anaheim, 1987)pp. 760-771.

[5] E. J. A. Pope and J. D. Mackenzie, *Mat. Res. Soc. Bull.*, 12[3[(1987) pp. 29-31.

[6] E. J. A. Pope, M. Asami, and J. D. Mackenzie, *J. Mater. Res.*, submitted.

PREPARATION OF ULTRAFINE COPPER PARTICLES IN POLY(2-VINYLPYRIDINE)

ALAN M. LYONS[*], S. NAKAHARA[*] AND E. M. PEARCE[**]
[*]AT&T Bell Laboratories, 600 Mountain Avenue, Murray Hill, NJ 07974
[**]Polytechnic University, Brooklyn, NY

ABSTRACT

Ultrafine copper particles were prepared by the thermal decomposition of a copper formate-poly(2-vinylpyridine) complex. At temperatures above 125°C, a redox reaction occurs where Cu^{+2} is reduced to copper metal and formate is oxidized to CO_2 and H_2. The decomposition reaction was studied by thermogravimetric analysis, differential scanning calorimetry and mass spectrometry. Copper concentrations up to 23 wt% have been incorporated into the polymer by this technique. The presence of the polymeric ligand induces the redox reaction to occur at a temperature 80°C lower than in uncomplexed copper formate.

Incorporation of the reducing agent (formate anion) into the polymer precursor enables the redox reaction to occur in the solid state. Films of the polymer precursor were prepared and the formation of metallic copper particles were studied by visible and infrared spectroscopy, x-ray diffraction techniques, and transmission electron microscopy. Results from these measurements indicate that spherical copper particles with an average diameter of 35angstrom are isolated within the polymer matrix. The particles are thermodynamically stable at temperatures up to the decomposition of the polymer matrix ($\approx 350°C$), but oxidize rapidly upon exposure air.

INTRODUCTION

Nanocomposite materials can be prepared where ultrafine metal particles are dispersed in a polymer matrix. One of the first reports of metal dispersions in polymers was the preparation of cobalt colloids in solution by the thermal decomposition of cobalt carbonyl.[1] Cobalt particles in the size range from 5-100 nm were formed. Smith and co-workers [2] [3] studied a similar system, i.e. the thermolysis of iron pentacarbonyl solutions, to form colloidal iron dispersions. The polymer was found to serve as a loci for the catalytic decomposition of iron carbonyl and subsequent particle nucleation. The formation of ultrafine transition metal particles in polymer solutions has also been investigated for several other metals, including copper,[4] by Hirai and coworkers.

This paper describes a unique method for the preparation of copper-polymer nanocomposites in the solid state. A soluble precursor was synthesized by complexing poly(2-vinylpyridine) with copper formate in methanol. Thermal decomposition of the complex results in a redox reaction whereby Cu(II) is reduced to copper metal and the formate anion is oxidized to CO_2 and H_2. By incorporating a reducing agent into the complex, the thermal decomposition reaction is not diffusion limited, and nanocrystalline copper particles can be prepared in the solid state. Thermal analytical techniques, such as thermogravimetric analysis (TGA), differential scanning calorimetry (DSC), TGA-Mass Spectrometry (TGA-MS) as well as temperature programmed visible and infrared spectroscopy (IR), x-ray diffraction and transmission electron microscopy (TEM) were used to characterize the formation of the copper-polymer nanocomposite.

EXPERIMENTAL

Complexes were prepared by addition of $Cu(HCO_2)_2 \cdot 2H_2O$ crystals directly to stirred methanol solutions of poly(2-vinylpyridine)(P2VPy). A small amount of $Cu(HCO_2)_2$ precipitate formed during the first hour after addition of the copper salt. These solutions were filtered and remained stable to further precipitation for 1-2 days after which additional precipitation occurred. After 24 hours of stirring, the complexes were isolated by freezing the solutions in liquid nitrogen and removing the solvent under a vacuum of $\approx 100\mu m$ of Hg as the solution thawed. Films were prepared on either quartz or CaF_2 substrates by pipeting a few drops of solution on the substrate and evaporating the solvent under vacuum.

Electronic spectra were recorded using a Hewlett Packard model 8452A diode array spectrophotometer. Infrared spectra were recorded on a Niclolet Model 5DX Fourier transform spectrophotometer. Temperature programmed spectroscopy was performed both in the visible as well as the infrared regions. Films prepared on the appropriate substrates were mounted into a heated cell (Janis Research) equipped with either quartz or potassium chloride windows. Experiments were conducted under both static and dynamic vacuum at 1.5×10^{-2} torr at a heating rate of $10^{\circ}C/min$.

The TGA and DSC experiments were performed on a Perkin Elmer TGS-2 thermobalance and DSC-4 calorimeter, respectively, in purified N_2 as previously described. [5] In certain experiments, an Inficon model IQ200 quadrupole mass spectrometer was connected to the thermobalance by a 10 foot, differentially pumped, 0.020 inch i.d. stainless steel capillary tube. In this manner, the volatile species generated during decomposition of the sample in an inert gas ambient were monitored.

TEM micrographs were taken of the 1:2 copper formate complex using a Phillips 420 electron microscope operated at the accelerating voltage of 120kV. Samples were prepared by placing one drop of the 1:2 solution onto a 3mm copper grid coated with a 200angstrom carbon film. The sample was decomposed directly in the TEM under a vacuum of $\approx 1 \times 10^{-7}$ torr by heating through a series of steps at 46, 76, 111, 150, 193, 237 and $306^{\circ}C$ over several hours. Particle size distributions were determined using a Joyce-Loeble imaging system.

RESULTS AND DISCUSSION

Products were isolated with nominal copper to pyridine ratios of 1:4 and 1:2. Results from the chemical analysis of the 1:2 complex were in good agreement with theoretical values. A weak complex is formed as indicated by small shifts in the visible and infrared spectra as well as the enhanced solubility of copper formate in P2VPy-methanol solutions. In the case of poly(vinylpyridine), where the heterocyclic amine is attached to a polymer backbone, complexation of two amines about the

113

Figure 1   Proposed structure of the 1:2 Cu(HCOO)$_2$:P2VPy complex.

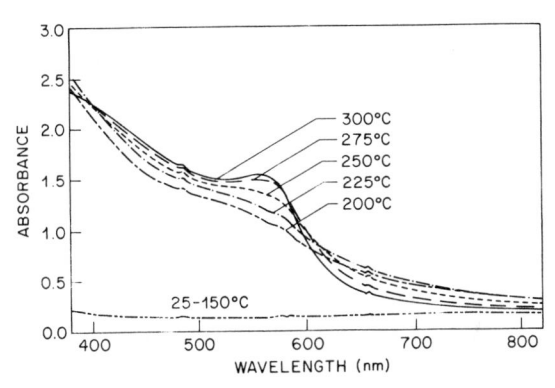

Figure 2   Visible spectra of the 1:2 complex recorded from 25-300°C in a static vacuum.

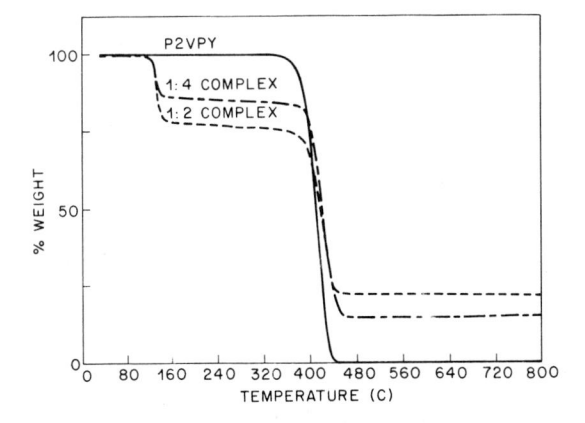

Figure 3   Thermograms of P2VPy and complexes between copper formate and P2VPy (1:4 and 1:2 ratios).

same copper atom would lead to crosslinking. This is indeed observed for the sterically unhindered P4VPy. In the case of P2VPy, however, soluble complexes are formed. As a result, only one pyridine moiety is complexed to each copper since the sterically bulky polymer chain blocks further coordination. A structure of the complex is postulated where copper is chelated by two formate anions, with the pyridyl group occupying the apex of a distorted square pyramid, as shown in Figure 1.

At 125°C, the complex undergoes a redox reaction where $Cu^{(II)}$ is reduced to copper metal and formate is oxidized to CO_2 and H_2. This reaction was observed by changes in the optical spectra (uv-visible and infrared), and by transitions in the calorimeter, thermobalance and mass spectrometer. The formation of CO_2 and H_2 during the thermal decomposition has been observed for all forms of copper formate investigated. [6]

Spectra of a 1:2 film heated from 25 to 300°C in static vacuum are shown in Figure 2. No change in the spectra is observed below 150°C. Above this temperature, the films turn a deep red color and the spectra exhibit an intense ultraviolet absorption, with a shoulder at ≈550 nm. The intensity of the shoulder increases with increasing temperature until a well defined maxima at 554 nm appears above 250°C. Upon cooling the film, the intensity of the maxima increases and the position shifts to 568 nm. Spectra of uncomplexed P2VPy show no absorption in the visible for films heated to temperatures from 30-350°C. The nature of the visible band at ≈565 nm has been postulated to arise from plasma resonance [7] [8] or interband transitions. [9]

Thermograms for P2VPy and the 1:4 and 1:2 complexes are shown in Figure 3. A new weight loss event is observed for the complexes with an extrapolated onset temperature of 125°C. A second, much smaller, event occurs at ≈ 250°C. The weight lost in the 125-250°C interval corresponds closely with calculated value of formate in the complex. The onset temperature for formate decomposition in the complexes is 80°C lower than for either $Cu(HCO_2)_2 \cdot 2H_2O$ or a 1:2 physical mixture of $Cu(HCO_2)_2 \cdot 2H_2O$ with P2VPy. The decomposition of formate to CO_2 and H_2 was confirmed by mass spectroscopy. Peaks at 2 and 44 a.m.u., corresponding to H_2 and CO_2, were observed at 130°C in the TGA-MS system. No evidence of CO was observed. The second principle region of weight loss at approximately 400°C is due to P2VPy decomposition.

A micrograph of the 1:2 copper formate complex, heated to 306°C within the TEM, is shown in Figure 4. The particles are uniformly distributed throughout the film and isolated from one another. Selected area diffraction results demonstrate that these particles are crystalline. The particle size distribution of the copper particles formed in the TEM reveals a mean particle diameter of 35angstrom (Figure 5). Although the results for such a thin film (<500angstrom) are not necessarily representative of a bulk specimen, the optical properties of thick (≈5μm) films are consistent with particles of this dimension, based on correlations to previous studies of the optical properties of copper colloids. [7] [10] [11] X-ray diffraction results, and the rapidity by which the particles are oxidized, also supports this contention.

The lack of agglomeration of these particles at elevated temperatures probably results from interactions between the copper particles and the polymer. Polymers containing ligands which are able to form complexes have been found to stabilize colloidal metal particles prepared by thermolysis [1] [2] [3] and chemical reduction [4] of metal salts in solution. The agglomeration of silver particles prepared by metal vapor techniques [12] was also found to be retarded most effectively by polymers. The

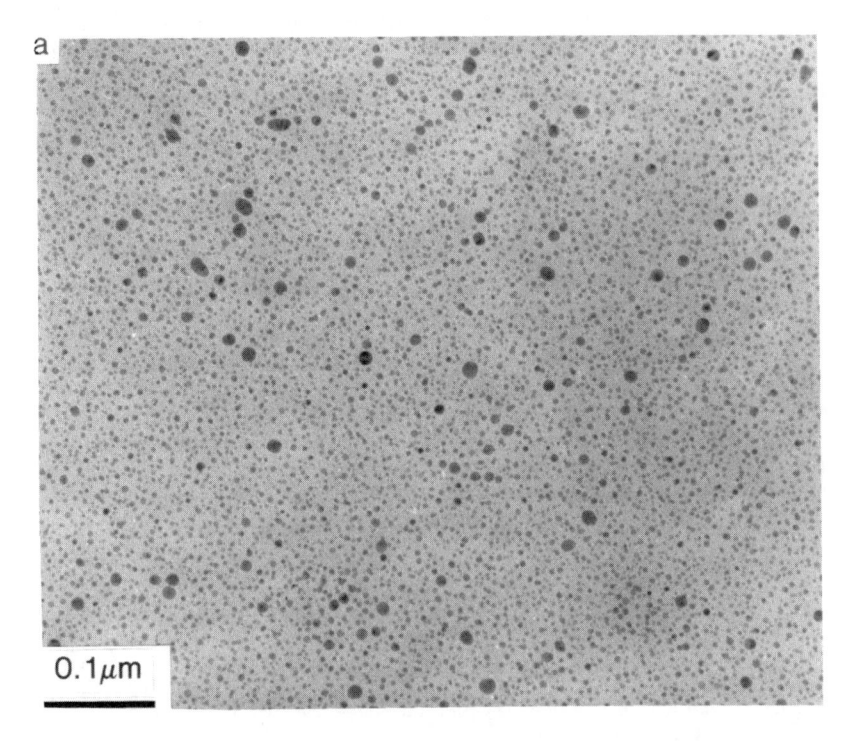

Figure 4   TEM micrograph of a copper-P2VPy nanocomposite prepared by heating a 1:2 copper formate:P2VPy complex to $\approx 310°C$.

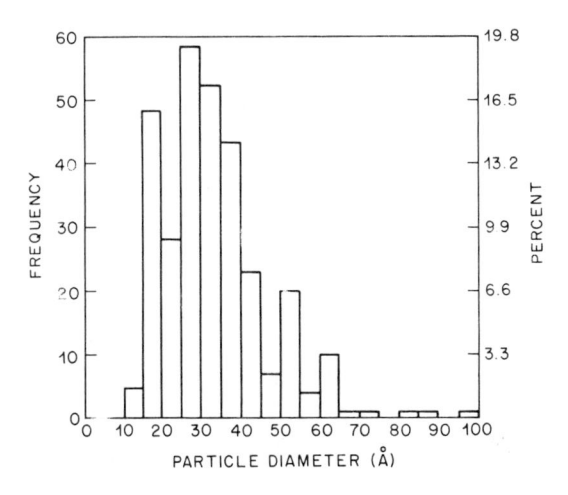

Figure 5   Distribution of particle sizes determined from Figure 4.

formation of a protective barrier about the growing particle was postulated as the mechanism which prevented further growth, once the particle reached a critical dimension. As the presence of small copper particles in P2VPy matrices does not affect the T_g or the IR spectrum of P2VPy, the interactions between the polymer and copper are either weak or infrequent.

CONCLUSION

Small copper particles, 35angstrom in diameter, were formed via the thermal decomposition of complexes between copper formate and poly(2-vinylpyridine). Soluble copper-polymer complexes, containing high weight % copper, were prepared by preventing crosslink formation through the use of a sterically hindered polymeric ligand. At 125°C, copper formate decomposes in a rigid, glassy matrix to evolve CO_2 and H_2 and create copper nucleation sites. As the decomposition proceeds, the glass transition temperature in the uncomplexed regions is exceeded, and the mobility of the copper particles is increased. [2] [3] Particles grow until a critical size is achieved, further growth being limited by a protective polymer coating. [12]

ACKNOWLEDGEMENTS

We would like to thank Mark Andrews for inspiration and many helpful discussions. The kindness of Joe Abys for the long-term loan of the Inficon mass spectrometer is greatly appreciated.

REFERENCES

1. P. H. Hess and P. H. Parker, Jr., J. Appl. Polym. Sci. **10,** 1915 (1966).

2. T. W. Smith and D. Wychick, J. Chem. Phys. **84,** 1621 (1980).

3. C. H. Griffiths, M. P. O'Horo, and T. W. Smith, J. Appl. Phys. **50,** 7108 (1979).

4. H. Hirai, H, Wakabayashi, and M. Komiyama, Bull. Chem. Soc. Jpn. **59,** 367 (1986).

5. A. M. Lyons, M. J. Vasile, E. M. Pearce and J. V. Waszczak, Macromolecules, **21,** (1988).

6. A. K. Galwey, D. M. Jamieson, and M. E. Brown, J. Phys. Chem. **78,** 2664 (1974).

7. M. Moskovits and J. E. Hulse, J. Chem. Phys. **67,** 4271 (1977). M. Moskovits, Acc. Chem. Res. **12,** 229 (1979).

8. M. Moskovits and J.E. Hulse, J. Chem. Phys. **66,** 3988 (1977).

9. H. Abe, K. P. Charle, B. Tesche, and W. Schulze, Chem. Phys. **68,** 137 (1982).
10. S. Garoff and C. D. Hanson, Appl. Optics **20,** 758 (1981).
11. K. Kimura and S. Bandow, Bull. Chem. Soc. Jpn., **56,** 3578 (1983).
12. M. P. Andrews and G. A. Ozin, J.Phys. Chem. **90,** 2929 (1986).

OPTICAL PROPERTIES OF FINELY-STRUCTURED PARTICULATES

PING SHENG*, MIN-YAO ZHOU*[a], ZHE CHEN*[b], AND S.T. CHUI**
*Exxon Research and Engineering Co., Route 22 East, Clinton Township, Annandale, NJ 08801
**Bartol Research Foundation, University of Delaware, Newark, Delaware 19716

ABSTRACT

We show that rod-like particulates structured from interleaving layers of metal and insulator (or semiconductor) can exhibit novel infrared and far-infrared characteristics that are tunable through the variation of their geometric parameters. In particular, a dispersion of these particles can be made to absorb strongly at any prescribed far-infrared frequency by controlling their transverse dimensions and insulator fractions.

INTRODUCTION

It is well known that the structural properties of ceramics are maneuverable through the control of their microstructures, i.e. the size, shape, relative composition, and topological arrangements of the constituents. Similarly, the optical properties of composites are sensitive to their microstructures, and this fact has been utilized traditionally to design multi-layered reflectors and composite absorbers, for examples. In this work, we wish to point out that a composite consisting of a dispersion of rod-like particulates made from interleaving layers of metal and insulator (or semiconductor) can exhibit novel infrared and far-infrared characteristics [1,2]. Compared to the particles in cermets or granular metals, these particulates are "structured" in the sense that the metal-insulator layerings have to be on a scale much smaller than the wavelength of light, and their transverse dimensions also have to be precisely controlled. While these structural requirements may not be achievable by the traditional co-sputtering or co-evaporation methods commonly used for the preparation of cermet films, they are nevertheless entirely within the feasibility of modern microfabrication and lithographic techniques.

STRUCTURED PARTICULATES AND THEIR EFFECTIVE-MEDIUM DIELECTRIC CONSTANT

Figure 1 gives a schematic picture of a structured metal-insulator (or semiconductor) rod-like particle. It has transverse dimension d, length L, and a metal-insulator periodicity ℓ_0 with the metal occupying a fraction p of the unit cell. Here p and d are the only two parameters requiring precise control since the optical properties are sensitive to their variations. The periodicity ℓ_0 is generally in the range of tens to

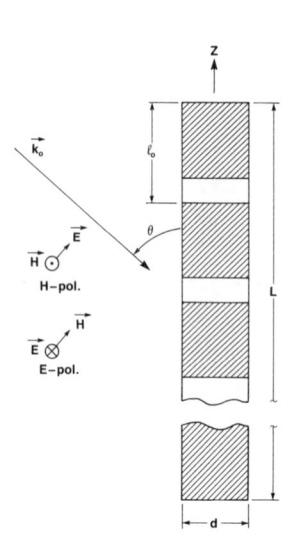

Fig. 1 Schematic illustration of a structured rod-like particulate and the two distinct polarizations for the incident wave.

hundreds of angstroms, and L is generally not controlled as long as it is much larger than d. These particles can be thought of as resulting from a metal-insulator superlattice that has been segmented so as to increase the "edge" area for light exposure.

The key to the special infrared properties of the structured particulates lies in their exposed layered facets. Since $\ell_0 \ll$ wavelength, the electromagnetic properties of the layered composite may be accurately modeled by using its effective medium dielectric constant. That is, since the wave can not resolve the individual layerings, it essentially perceives a homogeneous, anisotropic dielectric. The effective dielectic constant of this "homogeneous" material can be determined from that of the constituents by using the following arguments. For an electric field perpendicular to the layerings, the displacement field D is a constant and the electric field $E_k = D/\epsilon_k$, where the subscript k denotes the kth layer. Since the effective dielectric constant is by definition $\bar{\epsilon} = D/\langle E_k \rangle$, with $\langle \rangle$ denoting averaging, we have in this case

$$\frac{1}{\bar{\epsilon}_\perp} = \frac{p}{\epsilon_m} + \frac{1-p}{\epsilon_i} \quad , \tag{1}$$

where ϵ_m and ϵ_i are the dielectric constants of the metal and insulator, respectively. In the other case where the electric field vector is parallel to the layers, E is a constant and $D_k = \epsilon_k E$. Therefore we have

$$\bar{\epsilon}_\| = \frac{\langle D_k \rangle}{E} = p\,\epsilon_m + (1-p)\epsilon_i \ . \tag{2}$$

The effective dielectric constant is thus anisotropic and characterized by the dielectric tensor

$$\underset{\sim}{\bar{\epsilon}} = \begin{pmatrix} \bar{\epsilon}_\| & 0 & 0 \\ 0 & \bar{\epsilon}_\| & 0 \\ 0 & 0 & \bar{\epsilon}_\perp \end{pmatrix} \tag{3}$$

in a coordinate system where the z-axis coincides with that of the rod axis.

Given the effective dielectric constant $\underset{\sim}{\bar{\epsilon}}$, the problem of calculating the absorption and scattering cross sections becomes that of solving Maxwell equations. This can indeed be done analytically and the details are given in Ref. [1]. By comparing the results with that of the rigorous first-principle calculation [3], which that does not use the effective-medium approximation, it is found that the effective-medium approximation for the dielectric constant is accurate as long as $\ell_0 \ll \lambda$ and d > 500Å. For smaller values of d, the growing importance of the fringing field causes the effective-medium approximation to be less accurate but still qualitatively correct. With this caveat in mind, in the following we will use the effective medium approximation to assess the infrared characteristics of the structured particulates.

INFRARED CHARACTERISTICS

In Figs. 2a and 2b we contrast the absorption spectra of normally-incident E-polarized light with that of H-polarized light for a

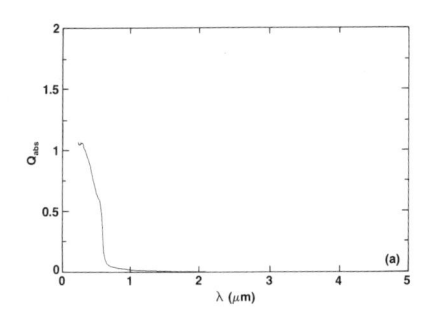

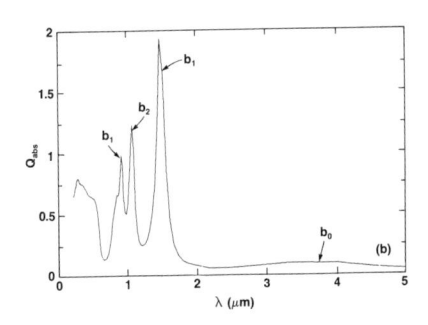

Fig. 2 Absorption spectra of a Cu-SiO$_2$ particulate with d=2000Å and p=0.91. (a) Normally incident E-polarized light. (b) Normally incident H-polarized light. .

Cu-SiO$_2$ particle with d = 2000Å. Here the two polarizations are defined as shown in Fig. 1, and the Q$_{abs}$, the absorption efficiency, is defined as the ratio between the absorption cross section and the geometric cross section. For the E-polarized case, the light perceives a cylinder with $\bar{\epsilon} = \bar{\epsilon}_\parallel$. Since in the infrared the metal dielectric constant is large and negative, $\bar{\epsilon}_\parallel$ is metal-like. Therefore the absorption is practically zero in the infrared. For the H-polarization, on the other hand, the light perceives a $\bar{\epsilon} = \bar{\epsilon}_\perp$, which is now dominated by the insulator dielectric in the infrared with the metal contributing to an imaginary part for absorption. The wave can now penetrate freely into the interior of the particle and set up electromagnetic modes inside. For cylindrical geometry, the radial part of these eigenmodes must behave as Bessel functions. The absorption peaks seen in Fig. 2b correspond to situations where the Bessel functions reach a maximum at the cylindrical surface and thereby allow a large intensity to enter the particle. Once inside, the presence of metal insures that most of the radiation is absorbed.

The understanding of the origin of the infrared absorption peaks enables us to assess the means for controlling their frequency positions. In general, the absorption peak position depends on three parameters: (1) the transverse dimension d, (2) the insulator fraction 1-p, and (3) the insulator dielectric constant ϵ_i. The last two factors may in fact be lumped together as $(1-p)/\epsilon_i$, which is the parameter that controls the insulator capacitance. To see first how the position of a peak varies with d, we observe that each peak has a given value of $\sqrt{\epsilon_\perp}k_0 d$. As d increases, k$_0$ must decrease proportionally in order to keep the product constant. Therefore, the peak frequency position should decrease as d^{-1} when the transverse dimension increases. In Fig. 3 we show a polarization- and

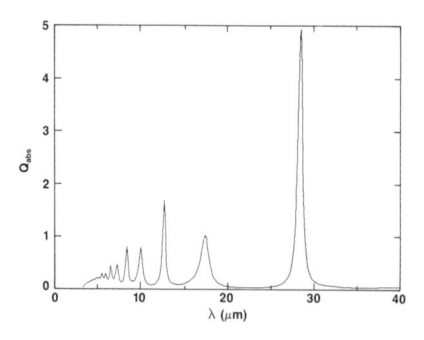

Fig. 3 Angle- and polarization-averaged absorption spectrum for a Cu-Ge particle with d=2μm and p=0.9667.

orientation-averaged absorption efficiency calculation for a Cu-Ge particle with d=2μm. For this value of d, it is seen that there is a large and narrow absorption peak at $\lambda \simeq 28\mu m$. The magnitude of any given peak is also noted to increase as the peak position moves towards lower frequencies. Besides changing the transverse dimension d of the particle, the peak position can also be varied by varying the insulator capacitance $\epsilon_i/(1-p)$. In this case the shift to lower frequencies is also accompanied by increased peak height.

Another effect that should be noted is the appearance of the broad-band absorption as d increases. When the peaks move to the far-infrared range, many more appear in the infrared range and they merge into a continuous band. In the limit of d→∞, the layered surface becomes a broad-band absorber. In Fig. 4 we show the reflectivity spectrum for a layered Cu-Ge surface under H-polarized incident wave. What is not reflected is totally absorbed by the layered material.

COMPARISON WITH CONVENTIONAL MATERIALS

The infrared characteristics of the structured particles should be contrasted with that of metals, insulators, and semiconductors. Infrared light is almost totally reflecting from metals and transmitting in insulators except for some phonon Reststrahlen frequencies. For semiconductors, one relies on the narrow band-gap and impurities for absorption. However, even for narrow band-gap semiconductors the absorption spectrum behaves as a low-pass filter, if there is no free carrier absorption. The infrared properties of the layered metal-insulator particles are therefore unique in the sense that they offer tunable absorption peaks as well as broad band absorption.

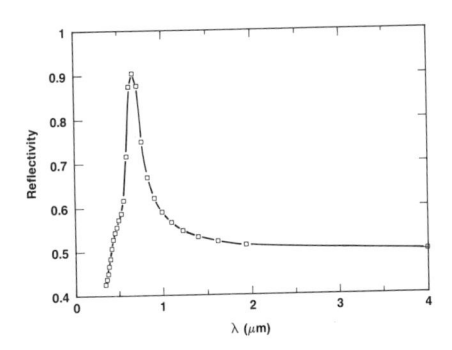

Fig. 4 Reflectivity as a function of wavelength for a layered Cu-Ge surface with p=0.53. The light is normally incident and H-polarized.

In addition, they may be made polarization-sensitive by aligning the cylindrical axes of the particles. However, whereas a semiconductor can directly convert light into electrical signals, here the absorbed light can only be converted into heat.

This work is supported by the Office of Naval Research Contract #N00014-88-K-0003. We wish to thank K. Unruh, J. Beamish, and C. L. Chien for helpful discussions.

REFERENCES AND NOTES

(a) Permanent address: Dept. of Physics, Shanghai Teacher's University, Shanghai, China
(b) Also at Dept. of Physics, City College of New York, NY, NY 10031.
1. M.Y. Zhou, P. Sheng, Z. Chen, and S.T. Chui, to be published.
2. S.T. Chui, M.Y. Zhou, P. Sheng, and Z. Chen, to be published.
3. Z. Chen and P. Sheng, to appear in Phys. Rev. B$\underline{15}$.

PART III

Alloys, Metals, and Magnetic Materials

A NEW LOOK AT AMORPHOUS VS MICROCRYSTALLINE STRUCTURE

FRANS SPAEPEN
Division of Applied Sciences, Harvard University, Cambridge MA 02138

Abstract

Historically, non-periodic models have usually found preference over microcrystalline ones for describing the structure of amorphous materials. Nevertheless, truly microcrystalline materials do exist, and the characteristic, monotonically decreasing isothermal calorimetric signal associated with the growth of the grains, can sometimes be used to identify them unambiguously. An example of the identification of the micro-quasicrystalline structure of some sputtered aluminum-transition metal alloys is discussed.

Introduction

For the diffraction pattern of a material to consist of sharp diffraction peaks, its atomic structure must have translational periodicity (taken to include quasi-periodicity as well in further discussion) that extends over large distances, on the order of the inverse of its peak widths in reciprocal space. When the diffraction peaks become very broad, however, the structural interpretation of the peak widths becomes ambiguous: does the broadening arise from the presence of very small, randomly oriented, fragments of a bulk crystal, or does it result from the translational correlations in a phase-specific, non-periodic structure, unrelated to the periodicity of the crystalline phases of the material? It is therefore not surprising that every time a new "amorphous" material (i.e., one with broad diffraction peaks) is found, a debate ensues on whether it is "only" microcrystalline or "truly" amorphous (in the narrow sense of non-periodic, as we will use the term from here on). Often, intermediate models arise as well, such assemblies of microcrystals disordered by strain, or embedded in an amorphous matrix.

The fundamental reason for the recurrence of debate is that, by itself, diffraction cannot settle the issue fully and unambiguously. Since only the intensity, and not the phase, of the scattered radiation is measured, Fourier inversion yields only the autocorrelation function (or radial distribution function) of the structure, which cannot uniquely be related the actual atomic positions. Nevertheless, the diffraction data certainly allow many structural models to be ruled out. Furthermore, if a model can account for all the details of the diffraction data, it should become the model of choice, and evidence from other sources should be very persuasive if it is to change this. Usually, however, the combination of diffraction data with that of other experiments (e.g., density, spectroscopic, and, as discussed below, calorimetric), leads to reasonable consensus on a model.

In this paper some of the main controversies, all of which led to a consensus around an amorphous structural model, are briefly revisited. Real microcrystalline materials certainly exist as well, and some of them are discussed in the next section. Finally, it is pointed out how isothermal calorimetry can be a powerful additional tool in settling this question, and an example is given of its application to some sputtered Al-transition metals.

Amorphous Structures

The first of many non-periodic structural models that proved successful was the continuous random network model proposed in 1932 by Zachariasen [1] for covalent materials. For example, in the case of SiO_2 glass, the X-ray [2] and neutron scattering [3] data agree well with the predictions from specific random network models for this material [4], and rule out various microcrystalline proposals (β-cristobalite [5,6], distorted crystallites [7]).

When amorphous Si and Ge were discovered, a similar debate on their structure arose in the early 1970's. Figure 1 shows clearly that all attempts at modeling their structure with various kinds of microcrystals were unsuccessful [8]. Figure 2, on the other hand, shows that the continuous random network model proposed by Polk [9] accounts quite accurately for diffraction data [10]. This conclusion was reinforced by the evidence from the optical properties [11] and the transformation behavior [12]. The debate over this system was kept alive for a while by high-resolution microscopy observation of what appeared to be crystalline-like lattice fringes extending over 14Å regions [13]. It was demonstrated later, though, that the fringes were artifacts of the electron optical transfer function [14].

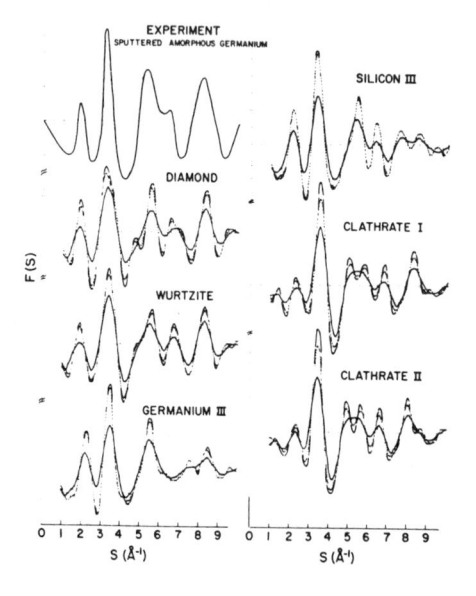

Fig. 1. Comparison of the structure factor of amorphous germanium with that of the microcrystalline models for various crystal structures and crystallite sizes. (From Weinstein and Davis [8].)

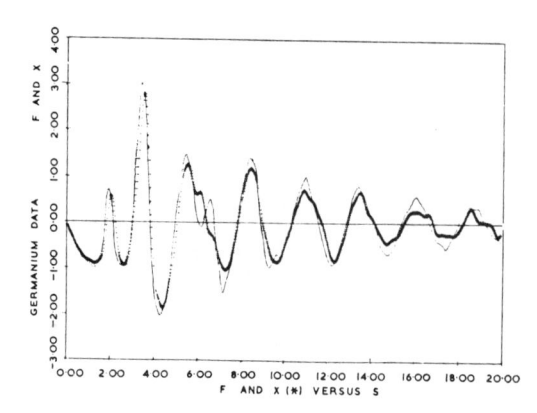

Fig. 2. Comparison of the structure factor of amorphous germanium (dots) with that of Polk's continuous random network model [9] (crosses). (From Chaudhari and Graczyk [10].)

For polymeric melts and glasses, the debate over Flory's random coil model [15] versus various chain-folded crystallite models [16,17] was settled in favor of the former by a combination of wide-angle and small-angle diffraction experiments [18].

The earliest attempts at modeling metallic liquids also used microcrystalline structures [19,20]. The alternative, favored by later diffraction work on glasses [21], by the thermodynamics of the liquids [22], and by the large undercoolings necessary for homogeneous crystal nucleation [23], is the dense random packing of hard spheres (DRP). It was developed first as a physical model by Bernal [24] and Finney [25], and later refined by computer techniques to apply to a variety of alloy glasses [26]. The polytetrahedral character of the DRP is the basis of analytic descriptions of the structure, such as Nelson's [27], that treat it as a "defective" projection of perfect polytetrahedral packing in four-dimensional curved space. The strong short-range order in some amorphous metallic alloys is explicitly taken into account in the "stereochemically defined" models that consist of a non-periodic assembly of nearest-neighbor clusters also found in intermetallic crystalline compounds of a similar composition [28].

Microcrystalline Models

Although the microcrystalline models have usually lost favor in the many debates on the structure of "amorphous" materials, truly microcrystalline materials do nevertheless exist. An simple model example is provided in the initial (t=0) state of the two-dimensional dynamic hard sphere system [29] obtained by first bringing the spheres in the "vapor" state (i.e., agitating them sufficiently so they can freely pass one another in the third dimension), quenching (turning off the agitation), and slightly "annealing" to allow the spheres to form a two-dimensional structure again. The resulting crystals are indeed very small. Even smaller crystals (less than 10 atoms across) have been found in Bragg and Nye's bubble raft immediately after having been stirred up [30]. It is worth noting that in both these models, with very differenct interatomic potentials, the grain boundaries are well-defined and their structure is the same as that found in boundaries between large crystals. This casts doubt on Gleiter's assertion that the interfacial atoms in an assembly of "nanocrystals" relax to a very open, gaslike intercrystalline "phase" [31]. The strong positional correlations imposed by the crystals seem to assert themselves right up to a boundary plane: there is very little ambiguity in assigning atoms to a particular crystallite.

An experimental example of a microcrystalline material is the Ag-Cu alloy vapor-deposited by Wagner et al. [32] in a clean vacuum. Figure 3 shows that its diffraction pattern, which is qualitatively different from that of a metallic glass, can be described quite satisfactorily as a broadened powder pattern from 125-atom f.c.c. crystals.

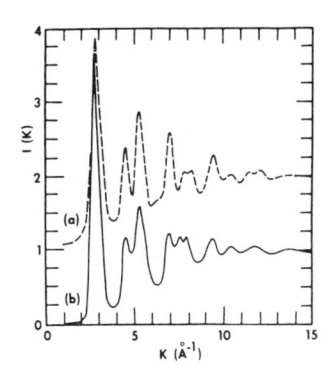

Fig. 3. Structure factors of (a) an assembly of randomly oriented fcc crystallites containing 125 atoms and (b) a vapor-deposited $Cu_{48}Ag_{52}$ film. (From Wagner et al. [32].)

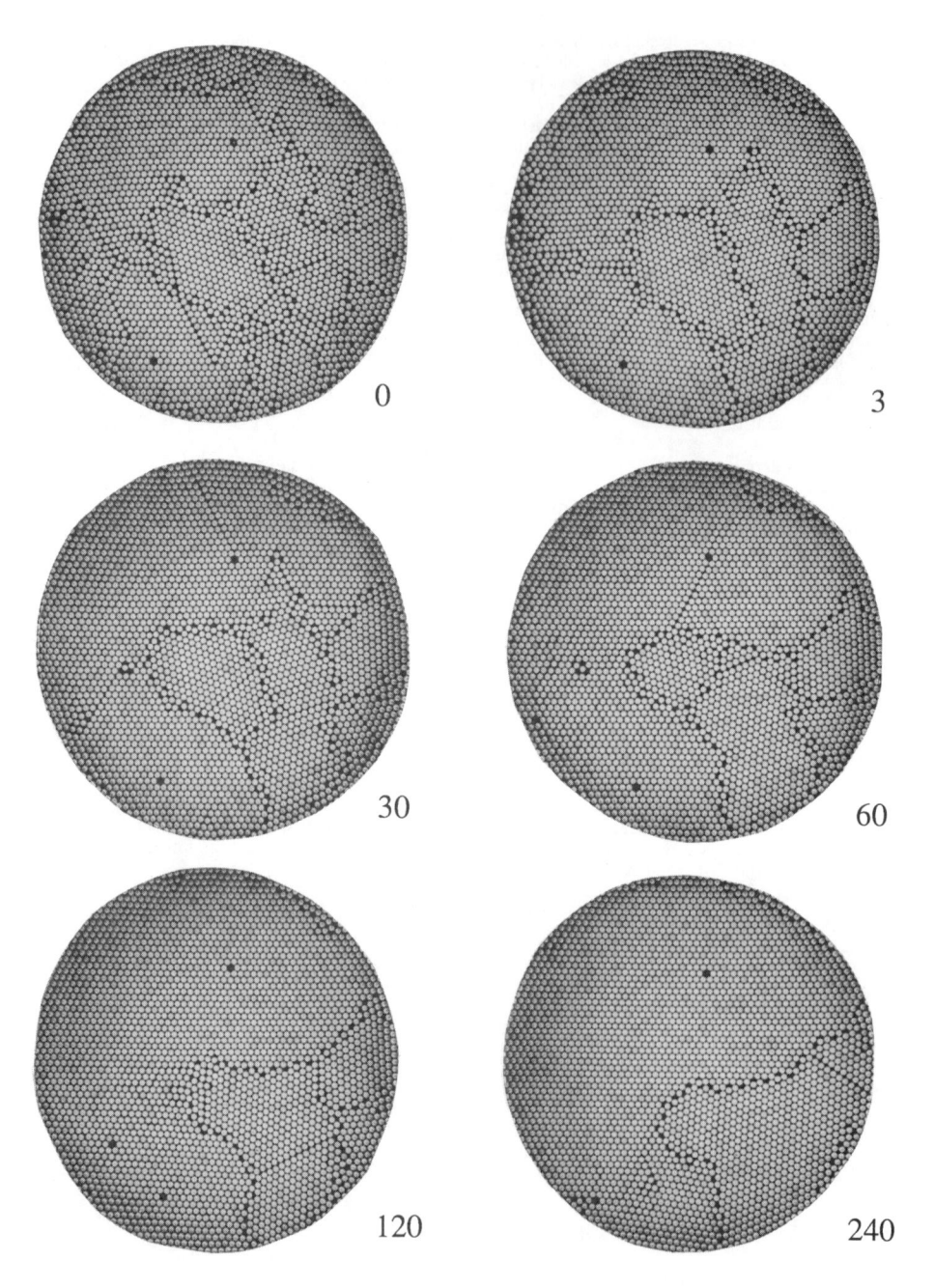

Fig. 4. Evolution of a two-dimensional dynamic hard sphere system, initially "quenched" form the "vapor" (see text). The numbers are the annealing time in seconds.

A more recent example is provided by the Al$_{86}$Mn$_{14}$ alloys quenched from the melt by Bendersky and Ridder [33]. They suggested, based on electron microscopy and diffraction observations, that the droplets are micro-quasicrystalline, with a quasicrystal size on the order of 1 nm for the highest quench rates (10^{6}K/s). Robertson et al. [34] also showed that the X-ray diffraction pattern of "amorphous" sputtered Al$_{72}$Mn$_{22}$Si$_6$ could be fit very well by broadening the quasicrystalline powder pattern from the same material sputtered at a higher temperature with a combination of size (2.5 nm diameter) and phason strain functions. For the purposes of this paper, quasicrystals are no different from regular crystals, in that the boundary between two quasicrystals of different orientation is geometrically as well defined as a regular grain boundary.

Calorimetry as a Method for Identifying Microcrystalline Materials

Calorimetry has traditionally been extremely valuable for providing additional information to distinguish between structural models. For example, the thermal manifestation of a glass transition is strong evidence for a liquid-like, and hence non-periodic structure. If a glass transition is not observed, however, this does not necessarily mean that the structure is microcrystalline, since the transition can be obscured by crystallization. In that case it is still possible to use calorimetry to identify a microcrystalline structure unambiguously by observing the isothermal crystallization kinetics.

The "crystallization", i.e. the exothermic transformation to a structure with sharp diffraction rings, of truly amorphous materials is qualitatively different from that of microcrystalline ones. In the former case, new crystals nucleate and grow, and the heat given off is the difference in enthalpy between the amorphous and crystalline phases. A microcrystalline assembly, however, does not undergo a phase transformation, but simply coarsens by a process of *grain growth*, as illustrated in the evolution of the dynamic hard sphere model in Figure 4. The heat given off during grain growth corresponds to the reduction in interfacial enthalpy. The kinetics of these two processes are fundamentally different.

The fraction of the material transformed by an isothermal nucleation-and-growth process is described by the Johnson-Mehl-Avrami expression $x = 1 - \exp(-bt^n)$, where b is a constant that depends on the nucleation frequency and the growth velocity, t is the time, and n is an exponent, that, depending on the nucleation mechanism, lies between 2 and 4 [35]. Figure 5(a) shows the characteristic sigmoidal shape. The heat given off per unit time in such a

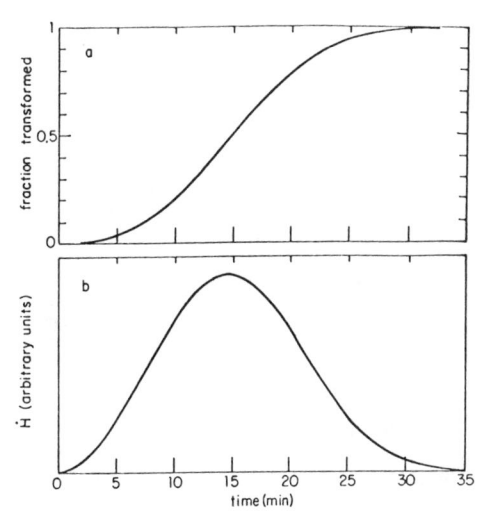

Fig. 5. Crystallization by isothermal nucleation and growth. Top: Fraction crystallized. Bottom: Corresponding enthalpy release. The parameters used in the Johnson-Mehl-Avrami expression for the transformation are those determined by Greer [40] for the crystallization of Fe$_{80}$B$_{20}$ glass. (From Chen and Spaepen [38].)

process is proportional to dx/dt, and hence results in a peak, as illustrated in Figure 5(b). Qualitatively, in a nucleation-and-growth process the transformation starts at a finite number of points, so that the amount of material transformed at early times is always small.

In a grain growth process, the rate of increase of the average grain radius, r, is generally written as $dr/dt = M\gamma/r^m$, where M is a mobility, γ is the interfacial surface tension and m is an exponent, empirically between 0.5 and 3 [36]. Most theoretical grain growth models give m=1 [37]. The average grain radius as a function of time can be written as $r^{m+1}(t) = r^{m+1}(0) + (m+1)M\gamma t$, and is shown in Figure 6(a). Since the corresponding interfacial enthalpy is $H(t)=H(0) [r(0)/r(t)]$, the isothermal calorimetric signal for the grain growth process is $- dH/dt = H(0)r(0)M\gamma/r^{m+2}$. As illustrated by Figure 6(b), this is a *monotonically decreasing* signal, which is qualitatively different from the peak observed in nucleation-and-growth.

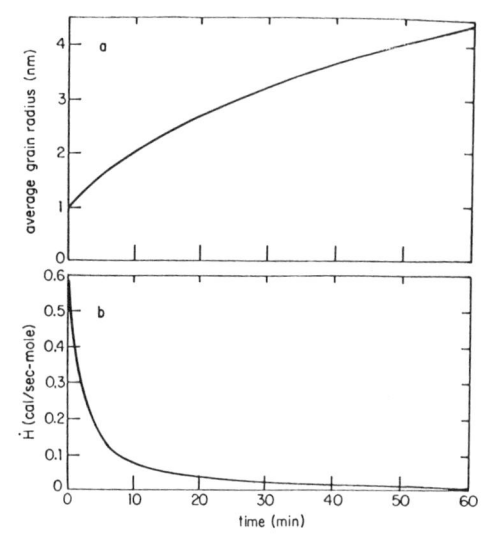

Fig. 6. Simulation of the transformation of a microcrystalline material by isothermal grain growth. Top: Evolution of the average grain size. Bottom: Corresponding enthalpy release. (From Chen and Spaepen [38].)

This technique was recently tested on sputtered $Al_{82.6}Mn_{17.4}$ and $Al_{82-83}Fe_{17-18}$ films [38]. Upon heating in a scanning calorimeter, two exothermic transformations are observed. In the as-prepared state the materials are "amorphous" (i.e., they showed broad diffraction halos and featureless TEM images); they become quasicrystalline after the first transformation, and crystalline after the second one [39]. When the sample is heated to a temperature at the beginning of the first transformation peak, and then held, a monotonically decreasing signal (Figure 7) characteristic of grain growth is observed. Comparison with the simulation of Figure 6(b) shows the isothermal signal can be modeled quite satisfactorily by the grain growth formalism. That the initial part of the data is more difficult to fit is not surprising, since deviations from the simple macroscopic grain growth formalism are most likely to occur in this regime.

Sputtered $Al_{82.6}Mn_{17.4}$ and $Al_{82-83}Fe_{17-18}$ therefore are indeed micro-quasicrystalline, as Bendersky and Ridder had anticipated. This is not entirely surprising since the Al-rich alloys have been notoriously resistant to glass formation by melt quenching. It should be kept in mind, however, that the observation of a peak in an isothermal calorimeter experiment does *not*, by itself, rule out a microcrystalline structure, since abnormal grain growth, in which a few grains grow at the expense of the others, would give a similar signal.

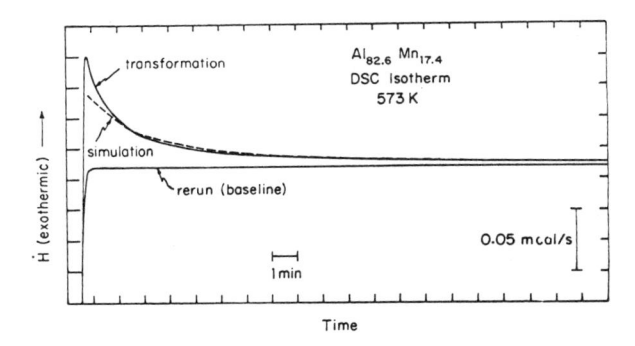

Fig. 7. Differential scanning calorimeter signal (solid line) from a sputtered Al$_{82.6}$Mn$_{17.4}$ film held at 573K. The dashed line is the simulation of Fig. 6. (From Chen and Spaepen [38].)

Conclusion

As long as new materials are being discovered, especially very metastable ones in rapid solidification, mechanical alloying or ion implantation, the debate about microcrystalline versus amorphous structure will continue. Past experience tells us that diffraction experiments are indispensible in providing the main criterion for identifying successful models. However, additional information, especially from such simple techniques as isothermal calorimetry, will continue to be needed to complete the picture.

At the same time, the availability of very fine-grained materials, with a large total interfacial enthalpy, also opens up the possibility of studying grain growth with commercial differential scanning calorimetry (DSC) equipment, which would be a simple and efficient way to determine the average grain boundary enthalpy, the growth exponent, n, and other kinetic parameters [39]. The calorimetry technique is also complementary to direct observations of grain growth, usually with electron microscopy. The smallest grains are difficult to resolve in a TEM foil, but their aggregate has a large interfacial enthalpy that can be detected by DSC; larger grained-samples, on the other hand have a small interfacial enthalpy, but can easily be analyzed in the TEM.

Acknowledgements

It is a pleasure to thank L.C. Chen and S.C. Moss for many useful discussions. Our work in this area has been supported by the Office of Naval Research under contract number 00014-85-K-0023, and by the National Science Foundation through the Harvard Materials Research Laboratory under contract number DMR-86-14003.

References

1. W.H. Zachariasen, J. Amer. Chem. Soc. **54**, 3841 (1932).

2. B.E. Warren, J. Appl. Phys. **8**, 645 (1937); B. E. Warren, H. Krutter, and O. Morningstar, J. Amer. Ceram. Soc. **19**, 202 (1936); R.L. Mozzi and B.E. Warren, J. Appl. Cryst. **2**, 164 (1969).

3. A.C. Wright, J. Non-Cryst. Solids **75**, 15 (1985).

4. R. J. Bell and P. Dean, Phil. Mag **25**, 1381 (1972).

5. J.T. Randall, H.P. Rooksby, and B.S. Cooper, Z. Krist. **75**, 196 (1930).

6. J.C. Phillips, Solid State Physics **37**, 93 (1982).

7. N. Valenkov and E.A. Porai-Koshits, Z. Krist. **95**, 195 (1936).

8. F.C. Weinstein and E.A. Davis, J. Non-cryst. Sol. **13**, 153 (1973).

9. D.E. Polk, J. Non-cryst. Sol. **5**, 365 (1971).

10. P. Chaudhari and J.F. Graczyk, *Proc. 5th Int. Conf. on Liquid and Amorphous Semiconductors*, ed. by J. Stuke and W. Brenig, Taylor and Francis, London (1974), p. 59.

11. R. Zallen, *The Physics of Amorphous Solids*, Wiley, New York (1983).

12. S. C. Moss and J.F. Graczyk, Phys. Rev. Lett. **23**, 1167 (1969).

13. M.L. Rudee, Phys. Sta. Sol. **B46**, K1 (1971); M.L. Rudee and A. Howie, Phil. Mag. **25**, 1001 (1972).

14. D.G. Ast, W. Krakow and W. Goldfarb, Phil Mag. **33**, 985 (1976).

15. P.J. Flory, *Principles of Polymer Chemistry*, Cornell University Press, Ithaca, NY (1953).

16. G.S.Y. Yeh, J. Macromol. Sci. **B6**, 451, 465 (1972).

17. W. Pechhold and S. Blasenbrey, Kolloid Z. u. Z. Polymere, **214**, 955 (1970).

18. E.W. Fischer and M. Dettenmaier, J. Non-cryst. Sol. **31**, 181 (1987).

19. J.A. Prins and H. Petersen, Physica **3**, 147 (1936).

20. N.F. Mott and R.W. Gurney, Trans. Farad. Soc **35**, 364 (1939).

21. G.S. Cargill III, Sol. St. Phys. **30**, edited by H. Ehrenreich, F. Seitz and D. Turnbull, 227-320 (Academic Press, NY, 1975).

22. T. E. Faber, *Introduction to the Theory of Liquid Metals*, Cambridge University Press, (1972), chapter 2.

23. D. Turnbull, J. Chem. Phys. **20**, 411 (1952).

24. J.D. Bernal, Proc. Roy. Soc. **A280**, 299 (1964).

25. J.L. Finney, Proc. Roy. Soc. **A319**, 497 (1970).

26. D.S. Bourdreaux and J.M. Gregor, J.Appl. Phys. **48**, 152, 5057 (1977).

27. D.R. Nelson and F. Spaepen, Solid State Physics **42** (1989), to appear.

28. P.H. Gaskell, in *Glassy Metals II*, Topics in Applied Physics **53**, 5-29 (Springer, Berlin, 1983).

29. D. Turnbull and R.L. Cormia, J. Appl. Phys. **31**, 674 (1960).

30. W.L. Bragg and J.F. Nye, Proc. Roy. Soc. **190**, 474 (1947) Figure 12a.

31. R. Birringer, U. Herr and H. Gleiter, Suppl. Trans. Jap. Inst. Met. **27**, 43 (1986).

32. C.N.J. Wagner, T.B. Light, N.C. Halder, and W.E. Lukens, J. Appl. Phys. **39**, 3690-3693 (1968).

33. L.A. Bendersky and D. Ridder, J. Mater. Res. **1**, 405-414 (1986).

34. J.L. Robertson, S.C. Moss, and K.G. Kreider, Phys. Rev. Lett. **60**, 2062-2065 (1988).

35. J.W. Christian, *The Theory of Transformations in Metals and Alloys*, 2nd edition, 525-548 (Pergamon, Oxford, 1975).

36. H.V. Atkinson, Acta Metall. **36**, 469-491 (1988).

37. C.V.Thompson, H.J. Frost, and F. Spaepen, Acta Metall. **35**, 887-890 (1987).

38. L.C. Chen and F. Spaepen, Nature **336**, 366-368 (1988).

39. L.C. Chen, F. Spaepen, J.L.Robertson, S.C. Moss, and K. Hiraga, to be published.

40. A.L. Greer, Acta Metall. **30**, 171-192 (1982).

41. L.C. Chen and F. Spaepen, to be published.

MECHANISM OF ACHIEVING NANOCRYSTALLINE AlRu BY BALL MILLING

E. HELLSTERN, H. J. FECHT, C. GARLAND AND W. L. JOHNSON
W. M. Keck Laboratory of Engineering Materials, California Institute of Technology, Pasadena, CA 91125, USA

ABSTRACT

We investigated through X- ray diffraction and transmission electron microscopy the crystal refinement of the intermetallic compound AlRu by high- energy ball milling. The deformation process causes a decrease of crystal size to 5– 7 nm and an increase of atomic level strain. This deformation is localized in shear bands with a thickness of 0.5 to 1 micron. Within these bands the crystal lattice breaks into small grains with a typical size of 8– 14 nm. Further deformation leads to a final nanocrystalline structure with randomly oriented crystallite grains separated by high- angle grain boundaries.

INTRODUCTION

Within the last few years nanocrystalline materials have become of primary interest in materials research. It has been suggested by Gleiter and co- workers that nanocrystals form a new category of solids structurally different from crystalline or amorphous materials [1]. Decreasing the grain size to less than 10 nm results in a drastic increase in the number of grain boundaries. The lack of long and short range order at these interfaces leads to new physical properties which are characteristic of a "gas- like" component. Up to now, nanocrystals have been prepared mainly through an inert gas condensation technique by evaporating and condensing material in a noble gas atmosphere [1]. Recently, Hellstern *et al.* have shown that nanocrystals can also be produced by ball milling [2]. More than ten years ago ball milling was developed by Benjamin as a way to circumvent the limitations of conventional alloying [3]. In this metallurgical process, powder particles are trapped by colliding steel balls and are repeatedly deformed, cold welded and fractured. This so- called "mechanical alloying" is mainly used to combine materials that are impossible to combine by other means, i. e. dispersion- hardened superalloys for the aircraft industry or tungsten carbide – cobalt composites. A few years ago it was shown that it is even possible to produce amorphous metals by mechanical alloying of elemental powders [4] or by ball milling various intermetallic compounds [5].
In addition, ball milling can also be used to produce nanocrystalline materials, as was recently shown for a number of pure elements and intermetallic phases [2,6]. In this case one starts with polycrystalline powder with a grain size of several microns. Heavy deformation by high- energy ball milling causes a grain size reduction to the nanometer range. However, the mechanism of refinement has yet to be investigated. The purpose of this paper is to show intermediate stages of crystal refinement in AlRu and to suggest a mechanism for achieving nanocrystalline material by ball milling.

EXPERIMENTAL DETAILS

The intermetallic compound AlRu, which has a CsCl- type structure, was prepared by carefully mixing elemental Ru (99.95 % purity) and Al (99.995 % purity) powders and by annealing the pressed powder mixture at 1000° C for three days in a sealed quartz tube. Ball milling of the AlRu powder was done in a Spex shaker mill with hardened steel balls and vials. After each ball milling time a small amount of powder was removed in an argon filled glove bag and used for further analysis. X- ray diffraction spectra were taken on a Norelco diffractometer in step- scanning mode using Cu Kα- radiation (λ=0.1542 nm). For transmission electron microscopy (TEM) the powder was mixed with epoxy, and thin slices of 40– 90 nm thickness were cut with a diamond microtome blade.

RESULTS

We determined the crystallite size and the atomic level strain of the material by analyzing the broadening of X-ray diffraction peaks and its dependence on diffraction angle (for details see ref. 2). The calculated values for crystallite size and strain are average values over a large sample volume. Fig. 1 shows the average microcrystallite size d as a function of processing time. In the very beginning it decreases quickly to less than 20 nm and finally reaches a size of about 7 nm after 64 hr ball milling. The strain is shown in Fig. 2 as a function of milling time. It increases to a maximum value of approximately 3 % after 8–16 hr and decreases with further milling.

For a better understanding of the deformation process we investigated the powder by transmission electron microscopy. Fig. 3 shows part of an AlRu particle after 10 minutes of ball milling. The total diameter of the particle is approximately 20 microns. It consists of a heavily strained crystal, as can be seen in the corresponding diffraction pattern (see insert in Fig. 3). However, the deformation itself is rather inhomogeneous. The arrows in Fig. 3 point to highly deformed regions running through the crystal. These deformation zones have widths of up to one micron. Most of the particle, however, consists of a less deformed crystal lattice. Fig. 4 shows a high-resolution bright field image of a highly deformed region of the same particle. Individual grains with a diameter of 8–12 nm are slightly rotated against each other with a rotation angle of less than 30 degrees. The insert shows a microdiffraction pattern which was taken of a larger area of the deformed zone. The rings of the diffraction pattern indicate that the investigated part consists of a large number of small crystals. Fig. 5 shows a particle after 2 hr milling. One can see that small crystals with a diameter of typically 8–10 nm break off at the edges. Finally, after 64 hr the material consists only of small grains with a size of 5–7 nm, which are separated by high-angle grain boundaries (Fig. 6). The orientation of neighboring crystals is completely random, as can be seen from the lattice fringes and in the corresponding diffraction pattern (see inset Fig. 6).

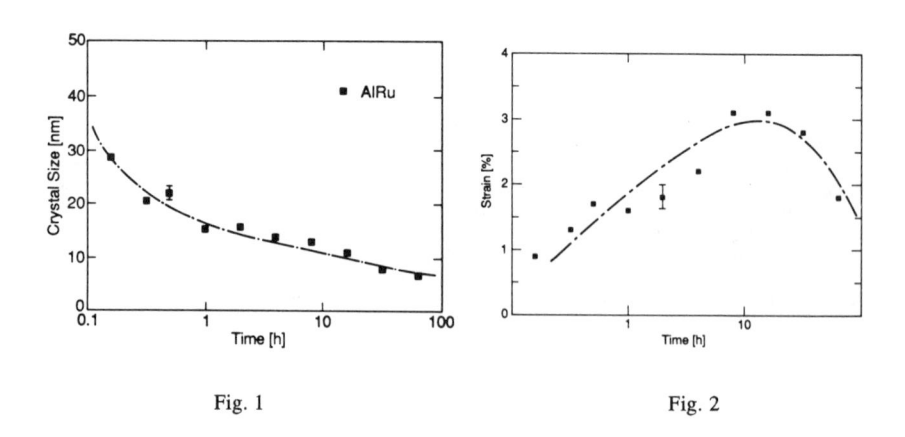

Fig. 1

Fig. 2

Fig. 1. Average crystal size of AlRu as a function of ball milling time.

Fig. 2. Average strain in AlRu as a function of ball milling time.

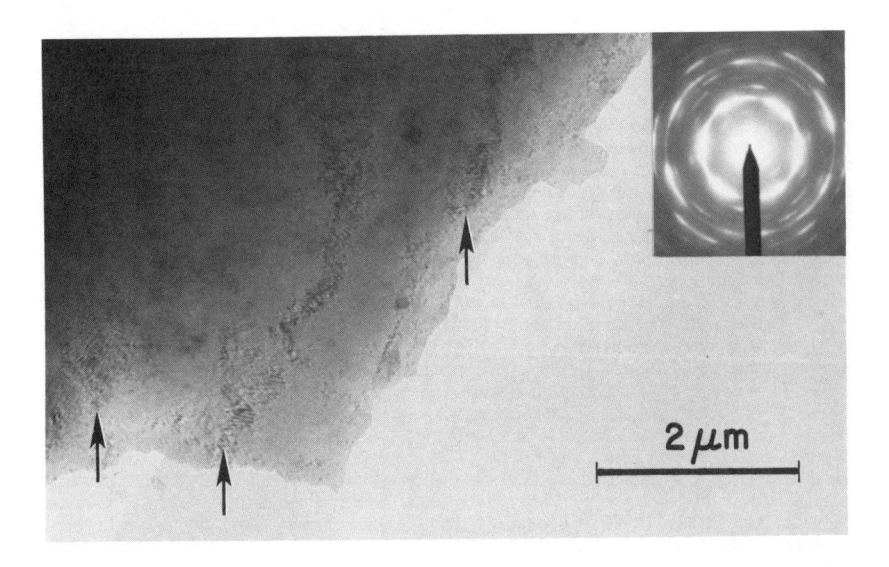

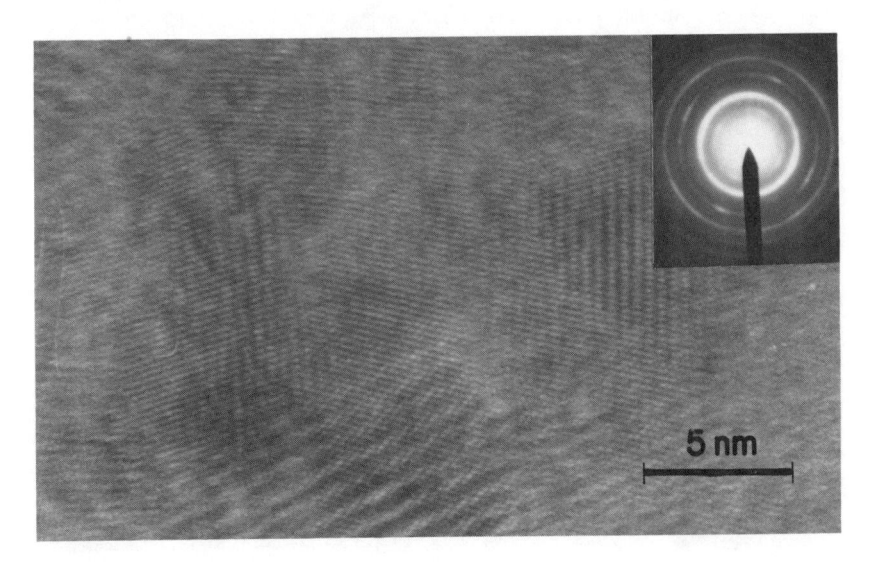

Fig. 3. TEM– bright field image of an AlRu powder particle after 10 minutes milling. The arrows point to highly deformed regions. The inset shows the corresponding diffraction pattern.

Fig. 4. TEM– high resolution bright field image of AlRu after 10 minutes milling and its corresponding diffraction pattern.

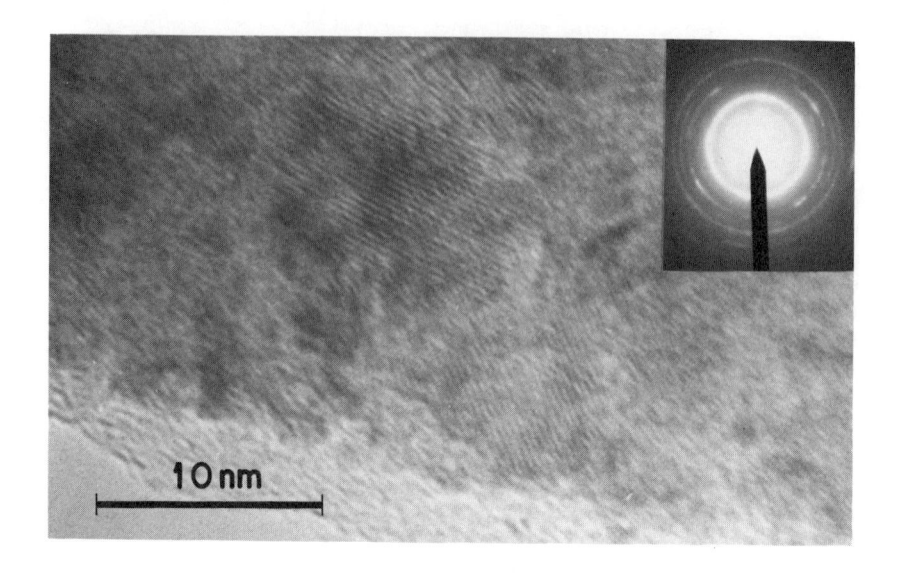

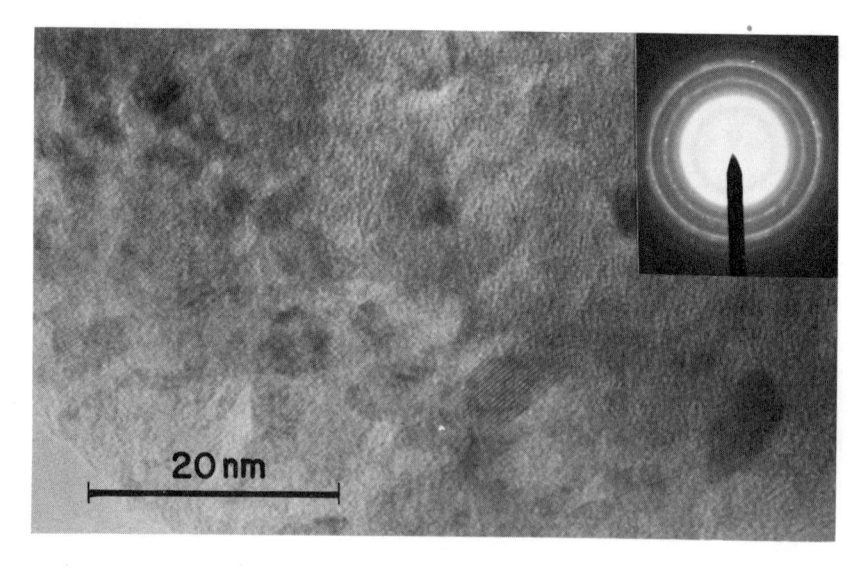

Fig. 5. TEM– high resolution bright field image of AlRu after 2 hr milling. The inset shows the corresponding diffraction pattern.

Fig. 6. TEM– high resolution bright field image of AlRu after 64 hr milling and its corresponding diffraction pattern.

DISCUSSION

Recent studies of material structures and properties at large plastic strains are helpful in understanding the observed deformation mechanism. Whereas plastic deformation is accomplished by slip and twinning at low and moderate strains, the formation of shear bands is the dominant mechanism at higher strain rates [7]. Hatherly and Malin investigated the formation of shear bands in rolled metals [8]. They found that the shear bands have a typical width of 0.1– 1 micron and usually consist of very small subgrains with sizes ranging from 0.02– 0.1 μm, which are rotated against each other. The shear instability of the crystal lattice may be caused by material inhomogeneities and is probably enhanced by thermoplastic instabilities due to non– uniform heating during the cold– rolling process in certain regions [9]. Similar results are also found by Rigney and co– workers, who investigated wear and material transfer in sliding systems [10]. They show that deformation at large plastic strains drastically changes the near– surface microstructure and that the material is unstable to local shear. The wear debris has an ultrafine grain structure with a grain size of less than 10 nm.

Heavy deformation by ball milling and mechanical alloying leads to a similar microstructure. As is shown in the TEM– images, the crystal refinement occurs in shear bands and, initially, in zones close to the surface of the powder particles, i. e. in heavily strained regions. Within these zones the crystal disintegrates into subgrains which are separated by low angle grain boundaries with an angle of less than 30º. Additionally, the overall strain in the material increases in the early deformation stages. It arises predominantly from a high dislocation density, i. e. a large number of antiphase domain boundaries. From a comparison of many TEM– images it appears that the small grains in the shear bands are less strained than the surrounding matrix. The large decrease in strain after long ball milling times is probably due to a reduction in dislocation density in grain structures on a nanometer scale. It is known that dislocations are unstable and annihilate when the spacings between them get very small, such as 50– 500 nm for screw dislocations and 1.6 nm for edge dislocations in copper [11]. Several authors suggest that grains in the nanometer range are dislocation free [12, 13].

Once the nanocrystalline structure is achieved, further refinement seems to be impossible. Probably, the very high stress required for dislocation movement hinders plastic deformation of very small crystallites. Similar results are found by Donovan and Stobbs, who investigated the deformation mechanism in a nanocrystalline PdSi alloy with a typical grain size of 5 nm [14]. They suggest that shear deformation is accomplished in grain boundaries, probably due to a lower shear modulus at the interfaces. They observed void formation during the deformation and defined the shear process as a "movement of crystallites in a sea of dilated grain boundaries." Moreover, recent experiments on nanocrystalline ceramics indicate that plastic deformation at low temperatures occurs by a diffusional flow of atoms along the grain boundaries [15]. This leads to considerably improved ductility of the material.

An interesting phenomenon is that the refinement does not occur gradually; instead, the small grains break off in a rather well– defined size. The crystal lattice can release a part of the strain by attaining nanocrystalline dimensions. Rigney et al. suggest that a balance of strain energy and surface energy might be responsible for the minimum possible grain size, and they estimate a crystal size of 6 nm for copper [10]. Another approach is taken by Darken [12], who proposed that ultrafine grains should have the strength of a perfect dislocation– free crystal. The predicted limited grain size is then in the range of 4– 18 nm for metals [10, 13].

SUMMARY

From TEM– and X– ray diffraction experiments, crystal refinement of AlRu by ball milling can be characterized as follows. The deformation process is localized in zones where the crystal is heavily deformed. These shear bands have a thickness of 0.5– 1 micron. Within these bands the crystal lattice breaks into small grains with a typical diameter of 8– 14 nm, which are initially separated by small angle grain boundaries (10º– 30º). In addition, small crystals break off from the edges of larger particles. Longer ball milling leads to a homogeneously deformed material, in which the particles disintegrate completely into smaller grains. Several successive deformation steps result in rotation of the grains with respect to neighboring grains and lead to the formation of high- angle grain boundaries. The deformation enhances the lattice strain up to 3 %, probably due to a very high dislocation density and the creation of antiphase domain boundaries. The large reduction of strain after long ball milling times may be explained by an instability of dislocations in nanometer-sized grains.

ACKNOWLEDGEMENTS

This work was supported by the U. S. Department of Energy (DOE Contract No. DE- FG03- 86ER45242). One of the authors (E. H.) gratefully acknowledges partial support by an Ernst von Siemens grant. Helpful discussions with Prof. H. Gleiter are gratefully acknowledged.

REFERENCES

1. H. Gleiter and P. Marquardt, Z. Metall. 75, 263 (1984). R. Birringer, H. Gleiter, H.- P. Klein, and P. Marquardt, Phys. Lett. A 102, 365 (1984). H. E. Schaefer, R. Würschum, R. Birringer, and H. Gleiter, J. Less- Comm. Met. 140, 161 (1988).
2. E. Hellstern, H. J. Fecht, Z. Fu, and W. L. Johnson, J. Appl. Phys. (Jan. 89), in print.
3. J. S. Benjamin, Sci. Amer. 234, 40 (1976).
4. C. C. Koch, O. B. Cavin, C. G. McKamey, and J. O. Scarbrough, Appl. Phys. Lett. 43, 1017 (1983).
5. A. E. Ermakov, E. E. Yurchikov, and V. A. Barinov, Fiz. Metal. Metalloved. 52, 1184 (1981).
6. E. Hellstern, H. J. Fecht, and W. L. Johnson, unpublished.
7. A. H. Cottrell, in "Dislocations and plastic flow in crystals", (Caledron Press, Oxford, 1972), p. 162.
8. M. Hatherly and A. S. Malin, Scr. Metall. 18, 449 (1984).
9. R. J. Clifton, J. Duffy, K. A. Hartley, and T. G. Shawki, Scr. Metall. 18, 443 (1984).
10. D. A. Rigney, L. H. Chen, M. G. S. Naylor, and A. R. Rosenfield, Wear 100, 195 (1984).
11. U. Essmann and H. Mughrabi, Phil. Mag. A 40, 731 (1979).
12. L. S. Darken, Trans. ASM 54, 599 (1961).
13. D. A. Rigney, Ann. Rev. Mater. Sci. 18, 141 (1988).
14. P. E. Donovan and W. M. Stobbs, Acta metall 31, 1 (1983).
15. J. Karch, R. Birringer, and H. Gleiter, Nature 330, 556 (1987).

FABRICATION AND PROPERTIES OF GRANULAR Fe-Ni ALLOYS

A. Gavrin and C.L. Chien
Department of Physics and Astronomy,
The Johns Hopkins University, Baltimore, Maryland 21218

ABSTRACT

Granular alloys of the form $(Fe_{50}Ni_{50})_x(Al_2O_3)_{1-x}$, where x is the volume percent, have been produced by sputtering from a single composite target. The microstructure has been verified by X-ray diffraction and by transmission electron microscopy (TEM). The grains are single phase FeNi alloys with the fcc structure and range in size from 15 Å to 50 Å. Superparamagnetism has been observed, and the coercivity has been found to decrease with increasing grain size. The results are compared to previous work on granular $Fe_x(SiO_2)_{1-x}$.

INTRODUCTION

Granular metals are composite materials consisting of metal particles, usually a few nanometers in size, embedded in an amorphous insulating matrix. Granular systems of a single metallic element have been investigated in the past, and shown to exhibit a number of interesting and potentially useful properties [1-3]. In particular, the magnetic behaviors of $Fe_x(SiO_2)_{1-x}$ granular materials have been extensively investigated because of their unique magnetic properties and potential as recording media [4].

In light of the scientific and technological potential of these systems, it is of interest to determine whether the range of granular materials can be expanded to include the richness of alloy systems. Furthermore, work on the $Fe_x(SiO_2)_{1-x}$ system has revealed several surprising results, including anomalously high coercivities. The elucidation of these results may be aided by the study of Fe based granular alloys. We have successfully fabricated a number of granular alloy systems by sputtering techniques. In this work, we report on the magnetic properties of granular Fe-Ni alloys. Important differences and similarities with previous work on granular pure Fe will be discussed. The composition $Fe_{50}Ni_{50}$ (in atomic percent) was chosen for several reasons. Pure Fe and Ni have distinctly different crystal structures, bcc and fcc respectively. If pure Fe and Ni particles were formed rather than alloyed particles, one could readily detect them by x-ray and electron diffraction. The magnetocrystalline anisotropy of $Fe_{50}Ni_{50}$ is much smaller than that of pure Fe ($10^4 ergs/cm^3$ as opposed to $5 \times 10^5 ergs/cm^3$) allowing investigation of the role of magnetocrystalline anisotropy in the coercivity of granular magnetic systems.

SAMPLE PREPARATION AND CHARACTERIZATION

Granular metal films may be produced by a variety of means, including evaporation, co-sputtering of a metal and an insulator, and sequential sputtering of a metal and an insulator [5]. All of the films discussed in this work were sputtered from homogeneous targets made by mixing appropriate amounts of one or more metals together with Al_2O_3 and pressing the material into a disk 1.75 inches in diameter. The base pressure of the vacuum chamber was about 5×10^{-8} Torr. Sputtering was carried out in the RF mode from planar magnetron sources in an atmosphere of 4.5 mT argon. RF power was maintained at 75 W resulting in a power density of approximately 150W/in^2, and a deposition rate of roughly 1 μm/hr. The substrate temperature was controlled in the range $100 \leq T_s \leq 450$ °C. All of the magnetic measurements discussed in this work were made on films approximately 2 to 4 μm in thickness deposited on Corning glass 7059 substrates.

After sputtering, the films were characterized by X-ray diffraction in a Philips APD 3720 diffractometer. Fig. 1 shows the diffraction pattern of a sample of composition $(Fe_{50}Ni_{50})_{40}(Al_2O_3)_{60}$ deposited at T_s=450 °C. Throughout this work, the metal content of the sample (e.g., 40%) denotes volume fraction. Also shown are the patterns for bulk fcc FeNi and bcc Fe metal. The presence of a single fcc phase is clearly indicated.

Transmission electron microscopy (TEM) was carried out on a Philips model EM420 electron microscope. The samples were thin (150 Å) films deposited on holey carbon grids under conditions identical to those used for the deposition of the thick films. Fig. 2 shows a micrograph from which an average grain size of 46 Å has been determined.

MAGNETIZATION MEASUREMENTS

The magnetic characteristics of the samples were determined by using a SHE SQUID magnetometer with a field range of 0 to 50 kG and a temperature range of 2 to 400 K. All measurements to be discussed here were made with the plane of the sample parallel to the

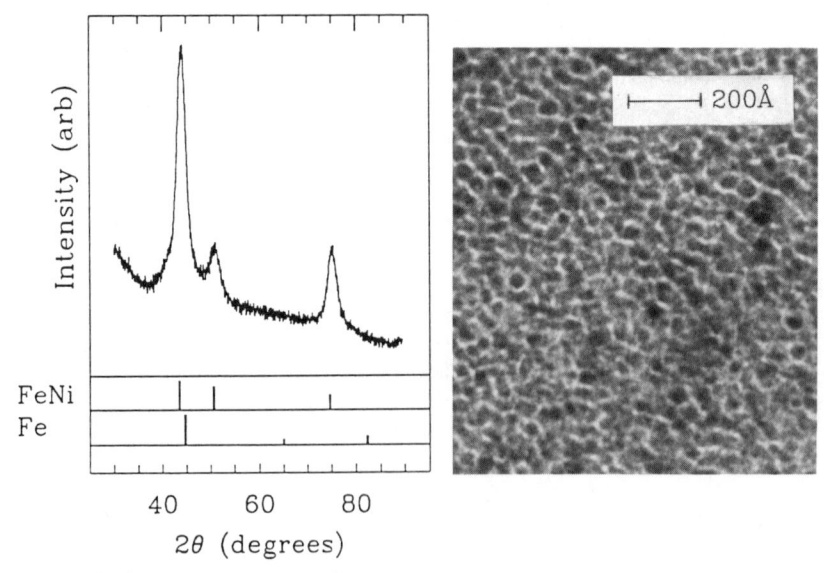

Fig. 1 θ–2θ diffraction pattern of granular $(Fe_{50}Ni_{50})_{40}(Al_2O_3)_{60}$ deposited at 450 °C. Also shown are standard patterns for bulk FeNi and Fe.

Fig. 2 Bright field electron micrograph of granular $(Fe_{50}Ni_{50})_{45}(Al_2O_3)_{55}$ deposited at 450 °C.

applied field, thus minimizing demagnetization effects. Magnetization was measured versus temperature (both field cooled and zero field cooled) under an external field of 5 G in order to determine T_B, the superparamagnetic blocking temperature. Hysteresis loops were obtained at 4.2 K to determine the ground state coercivity and remanence of the materials.

DISCUSSION

The X-ray diffraction pattern shown in Fig. 1 is typical of the patterns for samples deposited at high (≥300 °C) substrate temperature. Samples grown at lower temperatures show peaks which are qualitatively the same, but are considerably more broadened due to the smaller grain sizes. The most important aspect of these results is the lack of any feature at or near the expected positions of the Fe (200) and (211) peaks at 2θ=65° and 82.3° respectively. This is a clear indication that the Fe and Ni atoms have become alloyed, rather than forming separate grains of Fe and Ni. Furthermore, the X-ray pattern shows that the alloy formed is entirely in the fcc phase typical of sputtered $Fe_{50}Ni_{50}$ [6,7]. These facts demonstrate that the granular alloy system has been formed. Grain sizes estimated from the broadening of the X-ray lines are consistent with those derived from electron microscopy.

In the electron micrograph shown in Fig. 2, the particles are seen to be of relatively tight size distribution and to be roughly spherical in shape. Tilting of the sample in the electron microscope showed no indication of preferential orientation of the particles, or of a tendency for the particles to be aspherical in the direction out of the sample plane. Electron diffraction ring patterns were seen to be consistent with the fcc pattern observed by X-ray diffraction.

For a ferromagnetic material, there is a critical size below which a magnetic domain structure is not favorable. Below this size, single domain particles are formed. In such particles, all of the atomic moments remain aligned, even at zero applied field. The critical size for most materials has been estimated to be of the order of 150 Å [8]. Since the granules in the present samples are all well below this critical size, we expect the magnetic behavior of these samples to be that of a collection of single domain particles.

The temperature response of a sample in low magnetic field is illustrated in Fig. 3. The sample is initially cooled to liquid He temperature in zero applied field. At this point the moment of each particle is randomly oriented and frozen in position, similar to the atomic moments in a spin glass below the freezing temperature. As the sample is warmed, the grains acquire enough thermal energy for their moments to move freely. Above the blocking temperature, T_B, the particle acts as a superparamagnet, all of the atomic moments within the grains fluctuating in unison. This superparamagnetic relaxation may be described by the Arrhenius law [9],

$$\tau = \tau_0 e^{CV/k_B T}, \qquad (1)$$

where C is the total magnetic anisotropy energy per unit volume, V is the particle volume, k_B is the Boltzmann constant, and T is the temperature. The time constant, τ_0, has been estimated to lie in the range from 10^{-9} to 10^{-13} sec. The blocking temperature is the temperature at which the rate of fluctuation of the particle moment is comparable to the time scale of the measurement process. Explicitly,

$$T_B = \frac{CV}{k_B \ln(\frac{\tau_{SQUID}}{\tau_0})}. \qquad (2)$$

According to this equation, the blocking temperature should scale with the cube of the particle diameter (D). This variation is observed experimentally (Fig. 4). For the $Fe_x(SiO_2)_{1-x}$ system, τ_0 has been found to be approximately 10^{-13} sec [4]. Taking this value, we may use $\ln(\tau_{SQUID}/\tau_0) \approx 30$, and the line shown in Fig. 4 yields $C \approx 3 \times 10^7 ergs/cm^3$.

This value of C must be compared with the magnetocrystalline anisotropy for bulk $Fe_{50}Ni_{50}$, and with the total anisotropy derived from similar data concerning granular Fe particles. One of the surprising results concerning granular $Fe_x(SiO_2)_{1-x}$ to which we have alluded is the value of approximately $10^7 ergs/cm^3$ obtained for C. The present result is

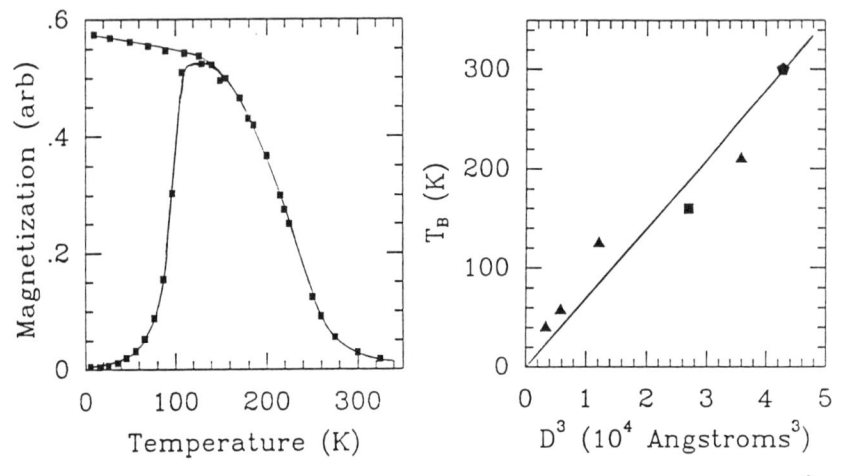

Fig. 3 Field cooled and zero field cooled magnetization of granular $(Fe_{50}Ni_{50})_{40}(Al_2O_3)_{60}$ deposited at 100 °C vs temperature with an applied field of 5 G.

Fig. 4 Blocking temperature versus D^3 for various samples of the general composition $(Fe_{50}Ni_{50})_x(Al_2O_3)_{1-x}$. Triangles, squares and pentagons indicate substrate temperatures of 100 °C 300°C and 450°C respectively.

seen to be of comparable magnitude. Since the magnetocrystalline anisotropy energies of Fe and $Fe_{50}Ni_{50}$ are approximately 5×10^5 and 1×10^4 $ergs/cm^3$ respectively [8], we must conclude that C is much larger than the contribution due to magnetocrystalline anisotropy, and that the blocking of the granular moments is dominated by some other factor (e.g., stress or surface anisotropy) which is of similar magnitude in these systems.

The hysteretic behavior of the $Fe_x(SiO_2)_{1-x}$ system has received a great amount of attention recently [4] because the coercivity is anomalously large, and increases dramatically with particle size. This phenomenon has not yet been satisfactorily explained. Most theoretical predictions of the behavior of a collection of small particles call for the coercivity to be independent of particle size for non-interacting particles, or to fall off linearly with the packing fraction for interacting particles [8]. Further, as the percolation threshold is approached, the grains will begin to touch and magnetic closure loops will be formed. Thus, the single domain particle nature of the sample will be destroyed, and the coercivity drastically reduced. While this percolation behavior has indeed been observed in the $Fe_x(SiO_2)_{1-x}$ system (as seen in Fig. 5), the sharp peak in coercivity at approximately 42% metal fraction remains anomalous.

The coercivity of the alloy system behaves much more as expected as shown in Fig. 5. The value of the coercivity changes smoothly as the metal content is increased, and drops off sharply at the percolation threshold. Another important result in the Fe-Ni alloy is the excellent correlation between coercivity and particle size. This is demonstrated in Fig. 6 where the different symbols indicate different compositions and substrate temperatures. This demonstrates that the decrease in coercivity depends solely on the particle size, rather than directly on composition or substrate temperature. This decrease may be attributable to two factors. The anisotropy pinning the moments may be proportional to the surface area of the particle. Since the energy to be gained by aligning with the field is proportional to the particle's moment and hence to its volume, the decrease in the surface area to volume ratio as particle size increases must yield a decrease in coercivity. Another possibility may be an increase in the strength of the interparticle interaction as the particles increase in size.

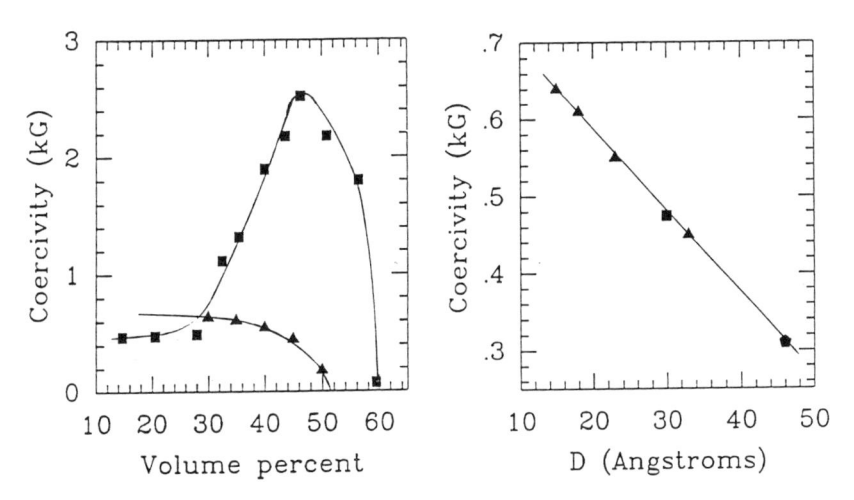

Fig. 5 Coercivity vs volume fraction for granular $Fe_x(SiO_2)_{1-x}$ (squares) and granular $(Fe_{50}Ni_{50})_{40}(Al_2O_3)_{60}$ (triangles).

Fig. 6 Coercivity vs D for granular $(Fe_{50}Ni_{50})_x(Al_2O_3)_{1-x}$. Triangles, squares and pentagons indicate substrate temperatures of 100 °C 300 °C and 450 °C respectively.

CONCLUSIONS

Granular alloy samples have been fabricated, and shown to exhibit single-domain particle ferromagnetism and superparamagnetism. The total anisotropy, as determined from the variation in blocking temperatures, is found to be large (3×10^7 ergs/cm^3) and similar to that of $Fe_x(SiO_2)_{1-x}$. However, the coercive behavior differs dramatically from that of $Fe_x(SiO_2)_{1-x}$, decreasing gradually as the particle size is increased.

ACKNOWLEDGEMENT

This work was supported by the Office of Naval Research under contract No. N00014-85-K-0175

REFERENCES

1. B. Abeles, P. Sheng, M.D. Couts, and Y. Arie, Adv. Phys. **24**,407 (1975).
2. Ping Sheng, Phys. Rev **B31**, 4906 (1985).
3. D. Schoenberg, *Superconductivity* (Cambridge University Press, London, 1962).
4. Gang Xiao, S.H. Liou, A. Levy, J.N. Taylor, and C.L. Chien, Phys. Rev. **B34**, 7573 (1986); Gang Xiao and C.L. Chien, Appl. Phys. Lett **51**, 1280 (1987); S.H. Liou and C.L. Chien, Appl. Phys. Lett **52**, 512 (1988).
5. R.W. Cohen and B. Abeles, Phys. Rev. **168**, 444 (1967); L.G. Feinstein and R.D. Huttemann, Thin Solid Films **20**, 103 (1974); E.M. Logothetis, W.J. Kaiser, H.K. Plummer and S.S. Shinozaki, J. Appl. Phys. **60**, 2548 (1986).
6. K. Sumiyama and Y. Nakamura, J. Magn. Magn. Mater. **35**, 219 (1983).
7. M. Hansen and K. Anderko: *Constitution of Binary Alloys* (Mcgraw-Hill, New York, 1958) p.678
8. A.H. Morrish, *The Physical Principles of Magnetism* (Wiley, New York, 1965), pp. 317-318.
9. I.S. Jacobs and C.P. Bean, in *Magnetism III*, edited by G.T. Rado and H. Suhl (Academic, New York, 1963), p. 275.

SMALL ANGLE SCATTERING FROM NANOCRYSTALLINE Pd

G.Wallner*,E.Jorra*,H.Franz*,J.Peisl*,R.Birringer**,H.Gleiter**,T.Haubold**,W.Petry***
* Sektion Physik, Ludwig-Maximilians-Universität München, D-8000 MÜNCHEN, FRG
** Universität des Saarlandes, D-6600 SAARBRÜCKEN,FRG
*** Institut Laue-Langevin, F-38042 GRENOBLE, France

ABSTRACT

The microstructure of nanocrystalline Pd was investigated by small angle scattering of neutrons and X-rays. The samples were prepared by compacting small crystallites produced by evaporation and condensation in an inert gas atmosphere. The strong scattering signal is interpreted to arise from crystallites embedded in a matrix of incoherent interfaces. Size distributions were deduced from the scattering curves. They consist of two parts: the crystallite size distribution dictated by the production process, and a structureless contribution due to the correlation in the spatial arrangement of the crystallites. The crystallite size distribution may be described by a log-normal distribution centred at R=2nm. The characteristic form of the correlation contribution arises from the dense packing of non-spherical crystallites. From the scattering cross-section in absolute units the volume fraction v_c of crystallites was obtained as $v_c \approx 0.3$, and the mean atomic density ρ_i in the interfaces as $\rho_i \approx 0.52$. The change of structural parameters during thermal annealing of the samples was studied. Up to high temperatures an appreciable volume fraction of crystallites with nearly unchanged size remains along with large particles.

INTRODUCTION

Nanocrystalline solids exhibit a number of novel and interesting physical properties [1], which are related to their unusual structure: nanometer-sized crystallites are embedded in a matrix of interfacial structure of about equal volume fraction. Since small angle scattering is an appropriate method for the study of structures on this length scale, it can help us to interpret and predict physical and technological properties of a nanocrystalline material. In this communication we concentrate on experiments on nanocrystalline Pd. Structural parameters were determined by small angle neutron scattering - these experiments are discussed in more detail in [2] -, their change upon thermal recovery was observed by small angle X-ray scattering.

THEORETICAL ASPECTS

Small angle scattering is diffuse scattering at small scattering angles Θ, or small values of the scattering vector **q** ($|q|=q=4\pi\sin\Theta/\lambda$). It is caused by correlations in the arrangement of scattering centres extending over distances larger than the interatomic spacing. If we consider scattering from a particle of average scattering length density ρ_p embedded in a matrix of average scattering length density ρ_m, the scattering cross-section (SCS) - which is proportional to the observable scattering intensity - is given by

$$\frac{d\sigma}{d\Omega}(q) = \left| \Delta\rho \int_{-\infty}^{+\infty} F(r) e^{iqr} dr \right|^2 \qquad (1)$$

where $\Delta\rho$ is ($\rho_m-\rho_p$), and F(**r**) is the form function of the particle: F(**r**)=1 inside, F(**r**)=0 outside the particle.

If N particles are distributed in the sample in an uncorrelated way, the SCS is enhanced by the factor N. If correlations in the arrangement of different particles are important, the SCS is changed in a way characteristic of the specific correlation. Following Porod [3] one may qualitatively distinguish two types of correlation: "liquid"-type correlation causes a decrease of the SCS at small q and arises e.g. from dense packing of spherical particles, "gas"-type correlation causes an increase of the SCS at small q and may arise from dense packing of non-spherical particles, that touch one another in many places.

EXPERIMENTAL DETAILS

We performed experiments on several "pellets" of about 8 mm diameter and 0.1 to 0.3 mm thickness. They were produced by compacting nanocrystalline Pd powder - generated by evaporation and condensation onto a cold substrate - at room temperature. Measurements of small angle neutron scattering were conducted at the instrument D11[4] at the ILL Grenoble with neutrons of wavelength $\lambda=0.7$nm. The X-ray scattering experiments were conducted at a 12 kW rotating anode source at $\lambda=0.071$nm. The samples were annealed in situ in a furnace implemented in the instrument.

EXPERIMENTAL RESULTS AND DISCUSSION

Figure 1 shows the SCS of neutrons for a representative nanocrystalline Pd sample. The solid line was computed from the size distribution S(R) of scattering units obtained from experimental data by the method given by Glatter[5]. The size distribution is shown in figure 2. It may be divided into two regions: the region at small radii (R≤15nm) - containing the asymmetric peak at 2nm - reflects the size distribution of Pd crystallites dictated by the production process. The contribution at larger R is due to correlations in the arrangement of different particles.

The crystallite size distribution centred at R≈2nm is similar to the log-norm distribution found for isolated crystallites by Granqvist and Buhrman [6]. It dominates the behaviour of the SCS at large q. In this range the SCS calculated from the size distribution of figure 2, truncated at 15nm, coincides with the experimental curve (compare the dash-dotted line in figure 1).

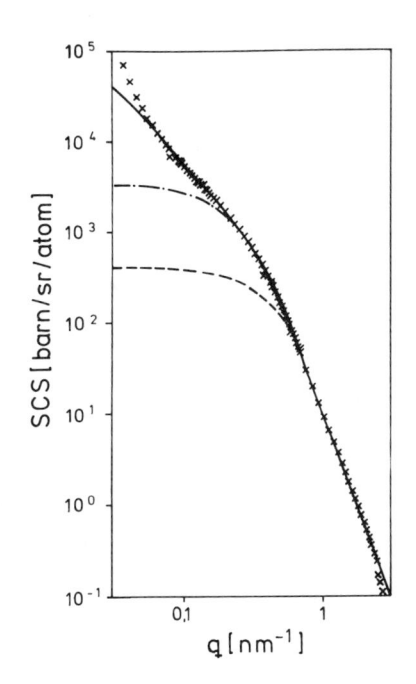

At small q-values the SCS is enhanced above that expected for the crystallites due to correlations. Evidently this correlation is of "gas"-type, i.e. crystallites are not spherical and may touch in many places. At the smallest q-values contributions from very large scattering units, possibly voids of some μm size are observed.

Figure 1:

Small angle SCS of neutrons for nanocrystalline Pd. The solid line corresponds to the SCS calculated from the size-distribution S(R) of figure 2. The dash-dotted line corresponds to S(R) truncated at R=15nm. The dashed line corresponds to the log-normal distribution approximated to the low-R part of S(R)

Figure 2:

Normalized size distribution obtained from the experimental SCS of figure 1. Left: linear-representation, right: louble-logarithmic plot. The dashed line gives the log-normal distribution approximated to the low-R part of S(R).

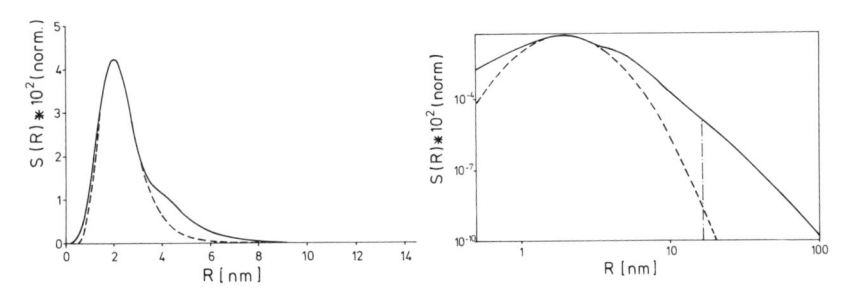

Valuable information on the constitution of the sample may be obtained from the evaluation of the SCS in absolute units. Under the assumption of an "ideal" nanocrystalline solid - sharply bounded crystallites of average scattering length density ρ_c (equal to that of single crystalline Pd) and volume fraction v_c are embedded in a uniform matrix of scattering length density ρ_i and volume fraction $(1-v_c)$ - we obtain $\rho_i \approx 0.53 \rho_c$, $v_c \approx 0.3$. Appreciable variation of these parameters from sample to sample was observed.

The behaviour of a nanocrystalline solid upon thermal recovery is of considerable physical and technological interest. On one hand there is the problem of the thermodynamic properties of a system with large entropical contributions, on the other hand any technological application of nanocrystalline materials will depend on their thermal performance.

We performed small angle X-ray scattering experiments on a nanocrystalline Pd sample, which was annealed for 1800 s at several temperatures increasing consecutively. The results of measurements after annealing at 520 K and at 1023 K are compared with the measurement on the unannealed sample in figure 3. Somewhat surprisingly no dramatic change in the scattering curves is observed.

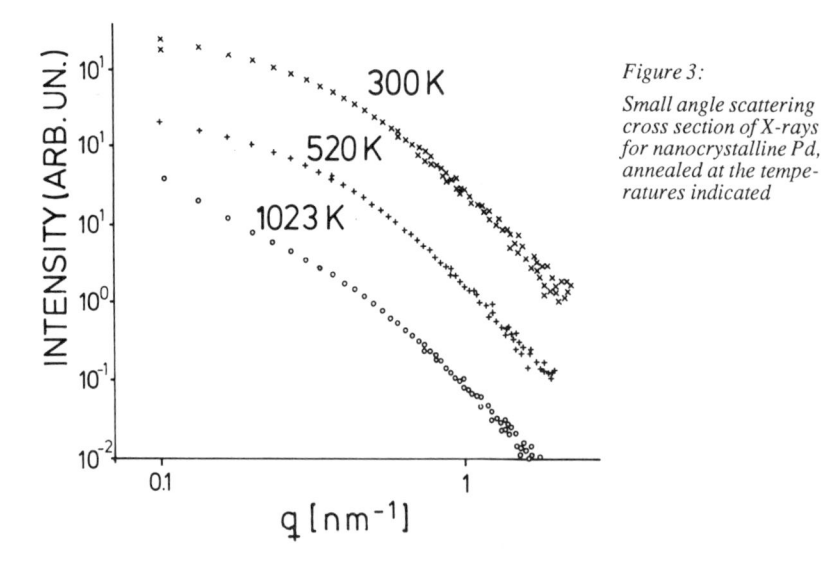

Figure 3:

Small angle scattering cross section of X-rays for nanocrystalline Pd, annealed at the temperatures indicated

On detailed inspection again two regions may be discerned, which behave differently on recovery: at large q-values the functional dependence of the SCS on q remains nearly unchanged, but the scattered intensity decreases by about a factor of 2 at 1023 K. At small values of q the decrease of the SCS with q gets steeper after annealing, so that the intensity remains nearly unchanged at the smallest q-values accessed.

These observations are compatible with the transformation of a part of the crystallites either into some large crystallites, or into a system of crystallites with high "gas"-type correlation. A considerable part of the crystallites remains rather unaffected by recovery up to 1000K, and causes the unchanged scattering signal at large q.

ACKNOWLEDGEMENT

We are indebted to Ch. Landesberger, who conducted experiments of X-ray small angle scattering. This work was in part funded by the German Ministry for Research and Technology under contract number 03PE1LMU3.

REFERENCES

[1] R.Birringer, U.Herr, H.Gleiter, Trans.Jap.Inst.Met.Suppl. **27**,43 (1986).

[2] E.Jorra,H.Franz,J.Peisl,G.Wallner,W.Petry,R.Birringer,H.Gleiter,T.Haubold, Phil.Mag.A (1988), in press.

[3] G.Porod, Small Angle X-ray Scattering, eds. O.Glatter, O.Kratky, Academic Press London-New York (1982) pp 17-51.

[4] K.Ibel, J.appl.Cryst. **9,** 296 (1976).

[5] O.Glatter, J.appl.Cryst. **13,** 7 (1980).

[6] C.G.Granqvist and R.A.Buhrman, J.appl.Phys. **47** ,2200 (1976).

CHARACTERISATION OF ULTRAFINE MICROSTRUCTURES USING A
POSITION-SENSITIVE ATOM PROBE (POSAP)

ALFRED CEREZO, CHRIS R.M. GROVENOR, MARK G. HETHERINGTON,
BARBARA A. SHOLLOCK AND GEORGE D.W. SMITH
*Department of Metallurgy and Science of Materials, Oxford University,
Parks Road, Oxford OX1 3PH, U.K.

ABSTRACT

A new development in the experimental techniques of atom probe microanalysis is described, which involves the use of a position sensitive detector system. This detector subtends a large solid angle (~20°) at the specimen, and therefore permits the collection of ions from a substantial fraction of the whole surface area of the emitter. Progressive pulsed field evaporation leads to the construction of a three-dimensional map of the atomic chemistry of the specimen. The new instrument is ideally suited to the investigation of complex, ultrafine microstructures. Applications to the study of age-hardened aluminium alloys and Alnico permanent magnet materials are described.

INTRODUCTION

The characterisation of nanometre-scale microstructures presents new challenges to materials science. Conventional electron microscopy and microanalysis are difficult, because of image overlap problems, and beam spreading within the specimens. In principle field ion microscopy and atom probe microanalysis provide suitable alternative methods, because they allow the structures to be examined one atom layer at a time. However, in a conventional probe-hole aperture atom probe, the analytical results are only one-dimensional, since ion collection is restricted to a narrow cylinder of material, approximately 1-2nm across [1]. The imaging atom probe permits spatially resolved study of a wider area of the specimen, but the time gating technique used for chemical mapping is restricted to one species at a time, and thus the results are qualitative, or at best only semi-quantitative [2]. In an effort to overcome some of these difficulties, we have incorporated a wide-angle position-sensitive ion detector into our atom probe, as shown schematically in Fig. 1(a) [3]. The detector consists of a double microchannel plate assembly, behind which is located a wedge and strip anode. The principle of operation of this kind of anode was first described by Anger [4]; the geometry is shown in Fig. 1(b). Ions are pulse field evaporated from the specimen by the application of a succession of high voltage or laser pulses. The ions are chemically identified from their time-of-flight to the channel plate. The position of origin of each ion is found from the output charge pulse of the channel plates, which is distributed between the x-, y- and z- electrodes of the wedge and strip anode. Only one signal is processed at a time, and hence operation is sequential, with an average field evaporation rate of less than one ion per pulse. As evaporation proceeds, the distribution of all the elements present on the specimen surface can be mapped out. Since atom probe analysis is a depth-profiling technique, each atomic layer removed from the surface can be analysed in turn. This allows the composition of the sample to be reconstructed in three dimensions at the atomic level. Subsequent computer processing can then be used to obtain analyses from selected regions within the overall volume of material analysed, to rotate the image about any desired axis, and to derive composition profiles along any chosen direction in the specimen.

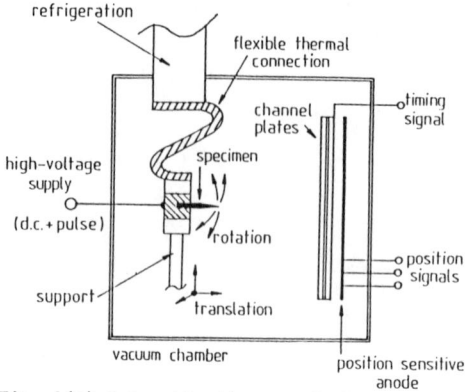

Fig. 1(a) Schematic diagram of the position sensitive atom probe (POSAP).

Fig. 1(b) Geometry of wedge and strip anode assembly for measuring arrival position of field evaporated ions.

In our present instrument, the detector is 35mm in diameter, and is located at a distance of 110mm from the specimen. Under typical operating conditions, a specimen region 10-15nm in diameter is sampled, with about 1500-3000 ions being collected for each atom layer field evaporated. Colour computer graphics are used to represent the complex mass of information which is obtained. For purposes of reproduction in this article the overall composition data have been divided into separate maps for the distribution of each individual element and are displayed either as grey-scale images, or dot maps, in which each dot represents the detection and identification of a single atom. It should be stressed that each set of element-separated maps corresponds to just one analysis area, and not to a succession of different analysis areas, as in the earlier imaging atom probe. Thus fully quantitative analysis is obtainable from any given area.

RESULTS

The prototype instrument has already been used for a range of phase chemical studies of metallurgical systems, for investigation of segregation on Pt-Rh catalysts and Y-Ba-Cu-O superconductor surfaces, and for the investigation of chemical intermixing in multilayer quantum well structures [5-8]. Two examples of current work on the analysis of multi-component ultra-fine microstructures are given below.

(a) Age-hardening of Al Alloys

G.P. zones and more stable precipitates in a variety of aluminium based alloys are being studied, with the long-term objective of improving the understanding of the role of alloying additions and trace elements in these very complex materials. As an example, the Al-Cu-Mg-Ag system has been examined to establish the composition of the orthorhombic Ω-phase precipitates, and to look for the possibility of segregation of Ag at the interphase interface. Fig. 2(a) shows a neon field-ion micrograph of an Al-4wt.%Cu - 0.3wt.%Mg - 0.4wt.%Ag alloy, solution treated and aged for 24h at 170°C, to produce large (~10nm diameter) Ω-phase precipitate plates on (111) Al matrix habit planes [9]. Fig. 2(b) shows elemental maps for the distribution of Cu, Mg and Ag in one of these precipitates. The overall composition of the precipitates is approximately 85at.%Al - 13at.%Cu - 1 at.%Mg - 1 at.%Ag. The distribution of Cu, Mg and Ag appears to be essentially uniform within the precipitates. The pronounced effect

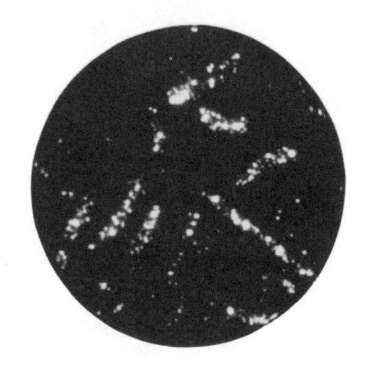

Fig. 2(a) Neon field-ion micrograph of an Al-Cu-Mg-Ag alloy, (5.5kV), showing brightly imaging Ω-phase precipitates on Al(111).

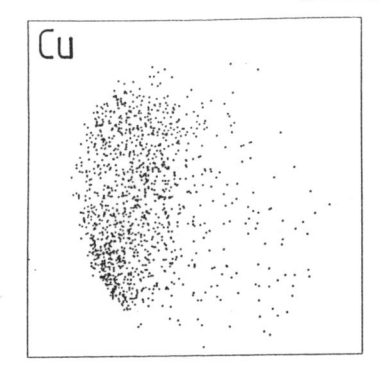

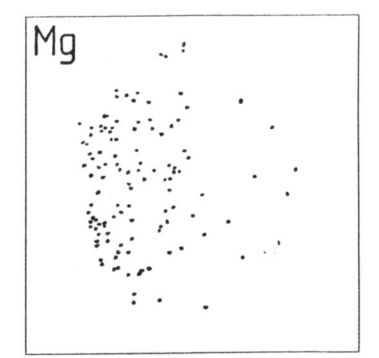

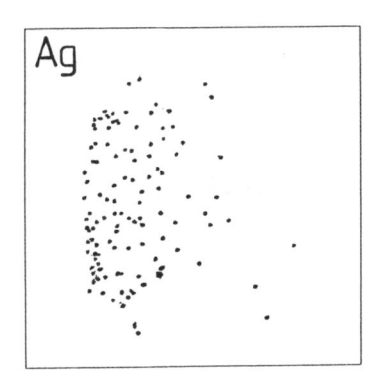

Fig. 2(b) Element-separated POSAP analysis of a single Ω phase precipitate of the kind illustrated in Fig. 2(a). Each dot represents the detection and identification of a single atom of Cu, Mg, or Ag. The distribution of Ag within the precipitate is seen to be essentially uniform.

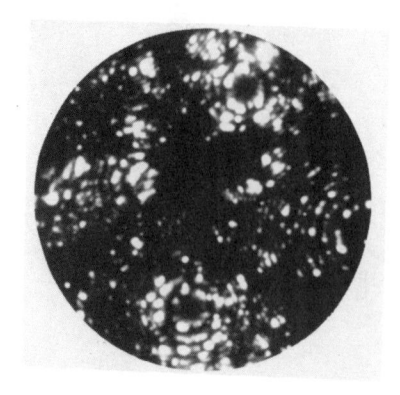

Fig. 3(a) Neon field-ion micrograph of an Alnico-2 magnet alloy (7.3kV). The large dark area in the centre of the field of view is the α_2 phase.

~5nm

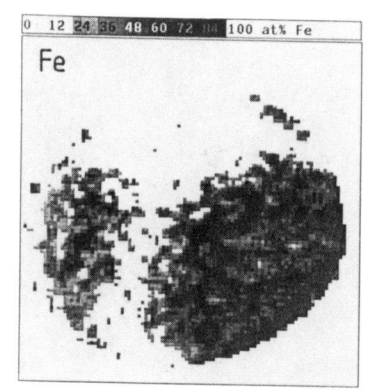

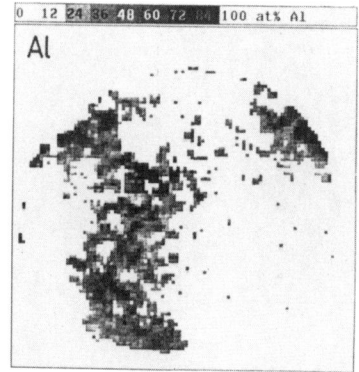

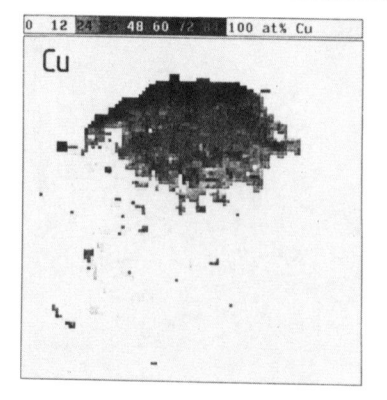

Fig. 3(b) Element-separated POSAP analysis of the central part of the region illustrated in Fig. 3(a). The grey scales represent the local concentrations of Fe, Al, and Cu respectively. The existence of an isolated Cu-rich particle within the α_2 phase is evident.

of small Ag additions on ageing behaviour in the Al-Cu-Mg system might have been attributed to segregation of this species to the interphase interface, but there is no detectable segregation to the broad faces of the precipitate plates. However, there remains the possibility of segregation to the edges of the plates, where the growth process will be mainly concentrated; this will be further investigated.

(b) Alnico permanent magnet alloys

The coercivity of Alnico permanent magnet alloys is known to be increased by the addition of copper, up to about 7wt.% [10]. However, the distribution of copper within the complex microstructure of these alloys has never previously been determined, and consequently the mechanism of the effect has remained uncertain. Fig. 3a shows a neon field ion micrograph of an Alnico-2 alloy (Fe-17wt.%Ni - 13wt.%Co - 9.5wt.%Al - 6wt.%Cu), which has been heat treated to produce optimum properties (solution treated 1250°C for 30 mins, cooled to 650°C in 4 minutes, tempered at 600°C for 4 hours, and slowly cooled to 400°C). The FIM images show the bright α_1 (Fe-Co rich) phase and the darker α_2 (Al-Ni rich) phase, which form by a spinodal process and are fully interconnected in three dimensions. The elemental maps for Fe, Al and Cu for a selected area of this microstructure are shown in Fig.3(b). The copper atoms are concentrated into discrete particles, ~4nm diameter, which are located within the α_2 phase. It appears that copper partitions preferentially towards the α_2 phase during the spinodal decomposition stage of reaction in this alloy, and subsequently precipitates from the supersaturated α_2 solid solution by a classical nucleation and growth process. These particles had not been detected previously by electron microscopy, because of the complexity of the host microstructure. The copper particles, being non-magnetic, will pin domain boundaries, and this appears to be the dominant mechanism for the additional increase in coercivity which is observed.

ACKNOWLEDGEMENTS

We thank SERC and CEGB Berkeley Nuclear Laboratories for financial support. One of us (Alfred Cerezo) wishes to thank the Royal Society for the award of a research fellowship.

REFERENCES

[1] E.W.Muller, J.A.Panitz and S.B.McLane, Rev. Sci. Instrum. 39, 83 (1968).
[2] J.A.Panitz, Rev. Sci. Instrum. 44, 1034 (1973).
[3] A.Cerezo, T.J.Godfrey and G.D.W.Smith, Rev. Sci. Instrum. 59, 862 (1988).
[4] C.Martin, P.Jelinsky M.Lampton, R.F.Malina and H.O.Anger, Rev. Sci. Instrum. 52, 1067 (1981).
[5] A.Cerezo, T.J.Godfrey, C.R.M.Grovenor, M.G.Hetherington, R.M.Hoyle, J.P.Jakubovics, J.A.Liddle, G.D.W.Smith and G.M.Worrall, J. Microscopy, in press.
[6] A.Cerezo, T.J.Godfrey and G.D.W.Smith, Proc. 35th International Field Emission Symposium, Oak Ridge, 1988, J. de Physique, in press.
[7] J.A.Liddle, A.Cerezo and C.R.M.Grovenor, Proc. 35th International Field Emission Symposium, Oak Ridge, 1988, J. de Physique, in press.
[8] J.A.Liddle, A.G.Norman, A.Cerezo and C.R.M.Grovenor, paper submitted to Appl. Phys. Lett. (November 1988).
[9] I.J.Polmear and M.J.Couper, Metall. Trans. 19A, 1027 (1988).
[10] M.G.Hetherington, A.Cerezo, J.P.Jakubovics and G.D.W.Smith, Proc. International Conference on Magnetism, Paris 1988, in press.

STUDY OF THE LATTICE DYNAMICS OF
IRON NANOCRYSTALS BY MÖSSBAUER SPECTROSCOPY*

J. Childress, A. Levy** and C.L. Chien
Department of Physics and Astronomy
The Johns Hopkins University, Baltimore, MD 21218

ABSTRACT

The vibrational properties of iron particles with nanometer sizes in Fe-SiO$_2$ and Fe-Al$_2$O$_3$ granular materials are studied by Mössbauer spectroscopy. The spectra include contributions from metallic Fe atoms at the core of the particles and apparently from surface and oxidized Fe atoms, which are analyzed separately. The temperature dependences of the recoilless fractions are analyzed using the Debye model. The Mössbauer Debye temperature Θ_M for metallic Fe is found to decrease dramatically with smaller grain sizes, indicative of a "softening" of the phonon spectrum.

INTRODUCTION

When the size of a three-dimensional crystal is reduced sufficiently, it is expected that the increased surface-to-volume ratio will result in static and dynamical crystalline properties significantly different from the bulk properties of that material. For example, the surface may contribute additional surface phonon modes to the the phonon spectrum, while the finite size of the crystal will inhibit the existence of long-wavelength phonons. The average lattice spacing may also change due to stresses and strains at the surface. Additionally, the properties of the particles may be influenced by the nature of the matrix in which they are imbedded. Hence the study of small particle systems can reveal important information about the crystalline state in the limit of small crystal size and about the interaction of such particles with their surroundings[1].

The lattice dynamics of microcrystals have previously been studied by Mössbauer spectroscopy in several systems. Mössbauer spectroscopy is ideally suited for such studies since it directly probes the local environment of the absorbing nucleus without being adversely affected by the macroscopic properties of the material studied. Hence a direct comparison between the behaviors of atoms in bulk material and in small particles is easily achieved. Previous studies of other metallic microcrystals have, in some cases, demonstrated a small decrease in the lattice Debye temperature Θ_D with decreasing crystal size, while others have reported a small increase in Θ_D[1,2,3].

Interestingly, iron particles, which could be studied using the most common Mössbauer isotope ^{57}Fe, have rarely been studied mainly because of the difficulty in controlling the chemical activity of isolated Fe grains. Hayashi et al.[4] studied the Debye temperature of 66 Å Fe particles prepared by evaporation and found no difference from the bulk behavior. We use co-deposition of Fe with an amorphous insulator such as SiO$_2$ or Al$_2$O$_3$ to obtain isolated Fe nanocrystals of fairly uniform size and shape. Control over the deposition conditions allows the fabrication of particles over a wide range of average diameters (10-200Å). In this paper we report on the first part of a study on the lattice dynamics of such a system, in the region of very small particle size (10-25Å).

EXPERIMENTAL METHOD

The samples investigated here are Fe-SiO$_2$ and Fe-Al$_2$O$_3$ granular materials containing 15 to 35 volume % of Fe. The compositions were chosen to ensure that the iron grains would be well isolated and superparamagnetic at room temperature. All the samples were obtained by RF sputtering at room temperature onto Kapton substrates. The sputtering targets were homogeneous mixtures of Fe with SiO$_2$ or Al$_2$O$_3$ made to the desired proportions. A high-rate magnetron sputtering system with base pressure 2x10^{-7} mTorr was used, with 4mTorr of Argon as the sputtering gas. During deposition, the gas pressure as well as the substrate temperature determine the size of the granules. The newly formed iron grains are then buried under SiO$_2$ or Al$_2$O$_3$, which limits their oxidation and final size. All other things equal, it has been found that an Al$_2$O$_3$ matrix yields smaller grain sizes. The granular nature of the samples obtained was confirmed by transmission electron microscopy (TEM). Inspection of the TEM pictures (Fig.1) revealed all samples to have a narrow distribution of grain sizes, with mean diameters 15Å for Fe$_{15vol\%}$-SiO$_2$, 25Å for Fe$_{30vol\%}$-SiO$_2$, 14Å for Fe$_{18vol\%}$-Al$_2$O$_3$, and 17Å for Fe$_{35vol\%}$-Al$_2$O$_3$. The thickness of the sample films was approximately 2 μm. Several such layers were stacked and incased in boron nitride to make the Mössbauer absorbers. The Mössbauer absorption data obtained was computer-fitted with Lorentzians using a least-squares method, with all the parameters allowed to vary freely, with the exception that doublet peaks were required to be symmetrical.

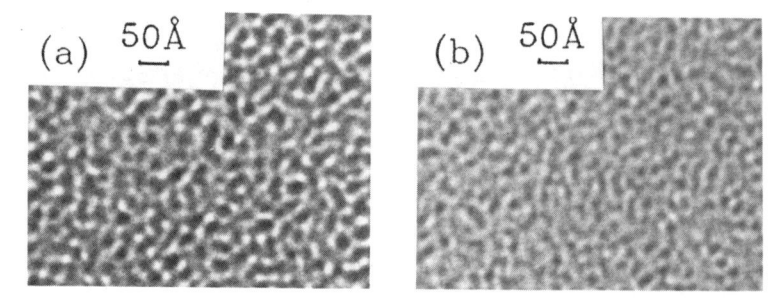

Fig. 1 TEM micrographs of (a) Fe$_{30vol\%}$-SiO$_2$ (Average diameter $\approx$ 25Å) and (b) Fe$_{18vol\%}$-Al$_2$O$_3$ (Average diameter $\approx$ 14Å).

The dynamical quantity available from a Mössbauer experiment is the mean-square displacement $<x^2>$, which is related to the fraction f of atoms which participate in the recoilless absorption of gamma-rays by:

$$f = \exp\left(-\frac{E_\gamma^2 <x^2>}{(hc)^2}\right)$$

where E_γ is the gamma-ray energy, h is Planck's constant divided by 2π, and c is the speed of light. The Debye model can be used to relate $<x^2>$ to the temperature of the solid, and f is proportional to the area A under the absorption peaks. In the high-temperature range ($T > \frac{\Theta_D}{2}$), this leads to: [5]

$$\frac{d(\ln A)}{dT} = \frac{d(\ln f)}{dT} = \frac{-3E_\gamma^2}{Mc^2 k_B \Theta_M^2} \qquad (T > \frac{\Theta_M}{2})$$

where M is the mass of the Mössbauer atom, and k_B is Boltzmann's constant. Θ_M is the Mössbauer Debye temperature, which has been substituted for the usual Debye temperature Θ_D because of the frequent difference in these quantities when Θ_D is measured by other means such as specific heat. Using the correct parameters for a ^{57}Fe Mössbauer experiment, one finds:

$$\Theta_M \approx -11.6 \left(\frac{d\ln A}{dT}\right)^{-\frac{1}{2}} \quad (T > \frac{\Theta_M}{2})$$

Hence Θ_M can be directly obtained from the slope of a lnA vs. T curve, without the painstaking corrections to the background counts that are needed for an accurate determination of f.

RESULTS AND DISCUSSION

The Mössbauer spectrum obtained for $Fe_{15vol\%}$–SiO_2 at 4.2 K is shown on Fig. 2. Fe grains of such small sizes contain a single magnetic domain, and exhibit superparamagnetic behavior. Hence these samples will be ferromagnetic up to their blocking temperature T_B (around 100 K for our samples). Indeed at 4.2 K the absorption spectrum is split into six broad lines, with a correspondingly broad distribution of hyperfine fields. Such a distribution of hyperfine fields is to be expected in ultrafine particles. The metallic Fe atoms near the center of the grain have a hyperfine field of about 345 kOe, while the Fe atoms at or near the grain surface have higher hyperfine fields because they are either oxidized or highly perturbed.

When the spectrum is taken above the superparamagnetic blocking temperature, the magnetic hyperfine splitting vanishes and the metallic Fe atoms near the center of the grain will gives rise to a single unsplit Mössbauer absorption line. Oxidized Fe, however, will be subjected to an electric quadrupole interaction and should give rise to two lines arranged in a doublet, as would metallic Fe atoms near the surface, where the lattice may lose its cubic symmetry and give rise to a non-zero electric field gradient. The actual spectra for the $Fe_{15vol\%}$–SiO_2 and $Fe_{35vol\%}$–Al_2O_3 samples at 300 K are shown on figs. 3a and 3b, respectively. Such spectra were taken between 300 and 500 K, a region where the high-temperature limit approximation is valid. The single line corresponding to metallic Fe is present as expected, and it is found that the existence of two quadrupole sites is needed to obtain accurate and consistent fits at various temperatures. A smooth and constant variation of the fit parameters as the temperature was increased served as a check of the consistency in the analysis. It is worth noting that similar contributions to the spectra were found for all samples, regardless of the specific matrix composition. The isomer shift and quadrupole splitting values for the doublet peaks (given by the position and separation of the lines) are typical of what is expected in microcrystals from Fe^{3+} (smaller shift and smaller splitting) and Fe^{2+} (larger shift and larger splitting)[6]. Hence a certain amount of oxidation is present in all samples. A more detailed analysis of these surface iron sites will appear in a forthcoming paper.

To study the lattice dynamics of metallic Fe, the area under the fitted singlet line was computed separately at each temperature and normalized to the total baseline count. Fig.4 shows a plot of the logarithm of these areas versus temperature, as well as the result of similar measurements made on the six-line spectrum of a bulk Fe film. As expected, the data can be fitted to straight lines, and the slope obtained by a least-squares method is then used to compute the Mössbauer Debye temperature Θ_M at that site, as explained previously. Θ_M for bulk Fe is found to be 388±20K, in good accord

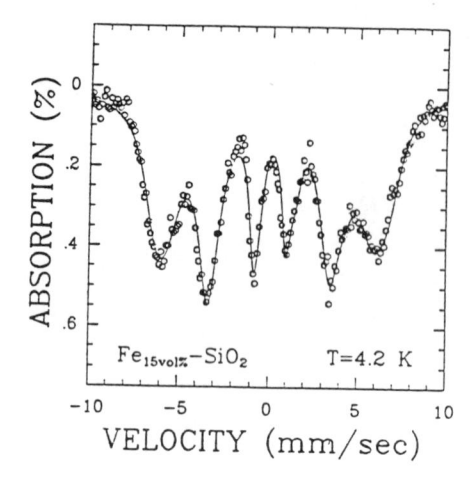

Fig. 2 Mössbauer spectrum of Fe$_{15vol\%}$–SiO$_2$ at 4.2 K.

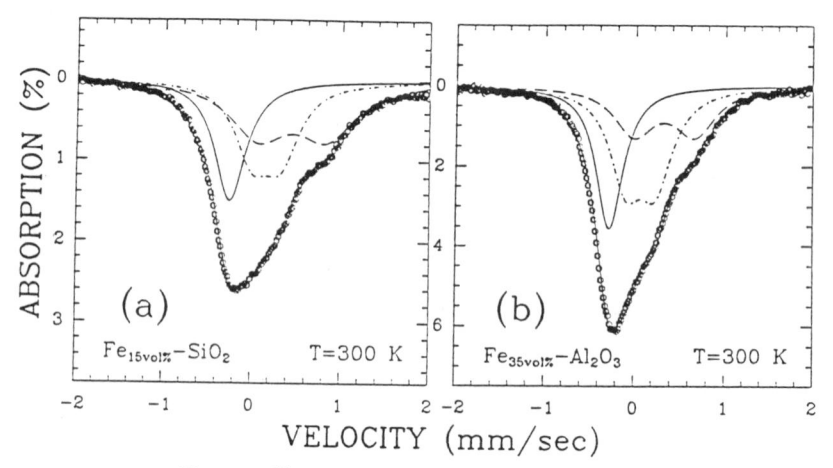

Fig. 3 Mössbauer spectrum of (a) Fe$_{15vol\%}$–SiO$_2$ and (b) Fe$_{35vol\%}$–Al$_2$O$_3$ at 300K.

with previously published results[7]. However, Θ_M for metallic Fe sharply decreases with decreasing particle size to 344±16K and 259±18K for 25Å and 15Å particles in SiO$_2$, respectively, and to 264±16K and 251±6K for 17Å and 14Å particles in Al$_2$O$_3$, respectively. Proportionally, this is the largest decline of characteristic lattice temperature observed in microcrystals by the Mössbauer effect, with a 35% change from the bulk value for the smallest particle size. A decrease in Θ_M corresponds to a "softening" of the phonon spectrum, i.e., to a shift of the spectrum towards longer wavelengths.

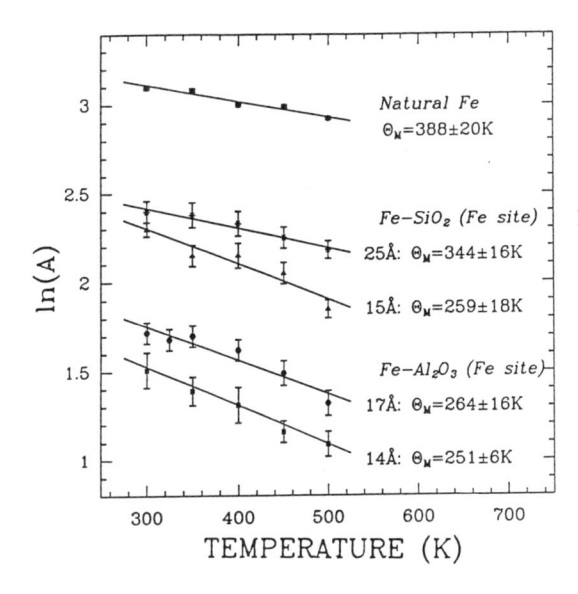

Fig. 4 Logarithm of absorption peak area versus temperature for bulk natural Fe and for metallic Fe in granular Fe-SiO$_2$ and Fe-Al$_2$O$_3$ materials with different average particle sizes. The Mössbauer Debye temperatures Θ_M are computed from the slopes of the fitted lines.

Several things can be mentioned about the possible origins of such change in the dynamics of metallic iron in granular materials. First, lattice relaxation is unlikely to be the major mechanism involved, since an enormous increase of more than 6% in the lattice spacing would be needed to explain the observed effect. Second, the value of Θ_M seems to depend only on the granular size. The two samples with SiO$_2$, for example, have very different grain sizes and accordingly different Θ_M's, and the two samples with Al$_2$O$_3$ have similar grain sizes and also similar values for Θ_M. If the effect was simply due to the dilution of iron into a medium with "softer" dynamical properties, then Θ_M for Fe$_{35vol\%}$–Al$_2$O$_3$ (17Å particles) should be at least equal if not larger than for Fe$_{30vol\%}$–SiO$_2$ (25Å particles), since the Debye temperature of Al$_2$O$_3$ is typically larger than that of SiO$_2$. Therefore the decrease in Θ_M seems to be the result of finite size effects intrinsic to Fe nanocrystals bound in a matrix rather than a property of the particular granular composite studied.

REFERENCES

* This work supported by Office of Naval Research Contract No. N00014-85-K-0175.
** Present address: LCMB-Gerontology Research Center, Francis Scott Key Medical Center, Baltimore, MD 21224.

1. See, e.g., D. Schroeer, Mössbauer Effect Methodology, vol.5, p.141 (Plenum Press, New York 1970) and references therein.
2. M.P.A Viegers and J.M. Trooster, Phys. Rev. B 15, 72 (1977).
3. S. Akselrod, M. Pasternak and S. Bukshpan, Phys. Rev. B 11, 1040 (1975).
4. M. Hayashi, I. Tamura, Y. Fukano and S. Kanemaki, Surface Science 106, 453 (1981).

5. R.H. Herber, Chemical Mössbauer Spectroscopy p.199, edited by R.H. Herber (Plenum Press, New York, 1984) and references therein.
6. See, e.g., Mössbauer Spectroscopy, edited by D.P.E. Dickson and F.J. Berry (Cambridge University Press, 1986).
7. L.D. Lafleur and C. Goodman, Phys. Rev. B 4, 2915 (1971).

MICROSTRUCTURE AND MAGNETO-OPTICAL PROPERTIES OF
TbFeCo FILMS PREPARED BY FACING TARGETS SPUTTERING

H. ITO, T. HIRATA AND M. NAOE
Department of Physical Electronics, Tokyo Institute of Technology, 2-12-1 O-okayama, Meguro-ku, Tokyo 152, Japan

ABSTRACT

The microstructure of TbFeCo amorphous films deposited by a facing targets type of sputtering (FTS) method varied with the change of deposition rate Rd. The films deposited at Rd of about 100-200 nm/min did not present any distinguishable columnar structure while the ones at Rd of about 50 nm/min revealed obvious columnar structure. The films with columnless structure exhibited Kerr rotation angle θk of 0.37 deg. of which the value was higher than that of the films with columnar structure over the wide range of Tb content. The torque measurement implied that the presence of columnar structure was not mainly necessary for the occurrence of large perpendicular magnetic anisotropy. Under satisfactory confinement of plasma during sputtering, the dense, uniform and columnless TbFeCo films can be obtained on plasma-free substrate even at Rd as high as 200 nm/min, without destroying the excellent magnetic properties and magneto-optical characteristics.

INTRODUCTION

Rare earth-transition metal(RE-TM) amorphous thin films with perpendicular magnetic anisotropy have attracted very strong attention as one of promising materials for achieving magneto-optical media with high-density and erasability. It is well known that the magnetic properties and magneto-optical characteristics of these films are closely related to microstructure which depends strongly on deposition parameters such as argon gas pressure P_{Ar}[1]-[5], substrate bias voltage [1],[5]-[7],[8] and substrate temperature[9].
The most essential and important problems for widely practical application are to improve the carrier to noise (C/N) ratio, write/erase sensitivity and environmental stability of these media. So, many attempts have been carried out to improve these practical characteristics. For example, improvements of the aging characteristics of RE-TM amorphous films have been achieved by using proper protective layers and adding appropriate other metals. Another way of improving their long-term stability and the magneto-optical characteristics of the films is to optimize microstructures[5][10].
In order to achieve these purpose, however, the intrinsic relationships among preparation condition, microstructure, magnetic properties and magneto-optical characteristics have to be understood. There are, however, few report of the relationships as for the TbFeCo films. We have prepared the dense, uniform and columnless TbFeCo films by using the FTS method which can form the RE-TM films at high deposition rate on the plasma-free substrate. Consequently, these films exhibited high C/N ratio and corrosion resistance[11].
In this paper, the microstructure, magnetic properties and magneto-optical characteristics of the TbFeCo films will be described.

EXPERIMENTAL PROCEDURES

The films with composition of $Tb_x(Fe_{100-y}Co_y)_{100-x}$:x=14-24 and y=9.7 and thickness in the range of 50-120 nm have been prepared by the Facing

Targets Sputtering (FTS) method[12]. The FTS apparatus used in this study was same as one as described in previous papers [11][13]. The specimen films were deposited on substrates of glass slides and PMMA plates at deposition rate Rd of about 50-200 nm/min. The back ground gas pressure P_B was below 7×10^{-4} mTorr, and the argon pressure P_{Ar} during deposition was in the range of 1.1-6.0 mTorr. The thickness and composition were determined and analyzed by x-ray fluorescence method.

Microstructure of the films was observed by means of a high-voltage transmission electron microscope(TEM): JEM-2000FX with acceleration voltage of 200 kV and resolution length of 0.14 nm. For observation of their cross-section, the specimen films were cut by using an ultra-microtome equipped with a diamond knife.

Kerr rotation angle θ_k and reflectivity R were determined at a wavelength of 830 nm. R of the films used in this study were 43-47 %. Coercivity Hc was measured by Kerr hysteresis loop tracer installed with a He-Ne laser(633 nm). These measurements were carried out through the substrate at room temperature. Saturation magnetization Ms and perpendicular magnetic anisotropy constant Ku were determined on the M-H hysteresis loops and torque curves by using a vibrating sample magnetometer (VSM) and a high sensitive torque magnetometer up to a field of 15.0 kOe, respectively.

RESULTS AND DISCUSSION

A. Microstructure

Table 1 shows the sputtering conditions of the TbFeCo films for TEM observations.

Table 1. Sputtering conditions of films for TEM observations.

Specimen film	A	B	C
P_B(Torr)	1×10^{-6}	7×10^{-7}	5×10^{-7}
P_{Ar}(mTorr)	5.0	3.6	3.6
Rd (nm/min)	50	100	200

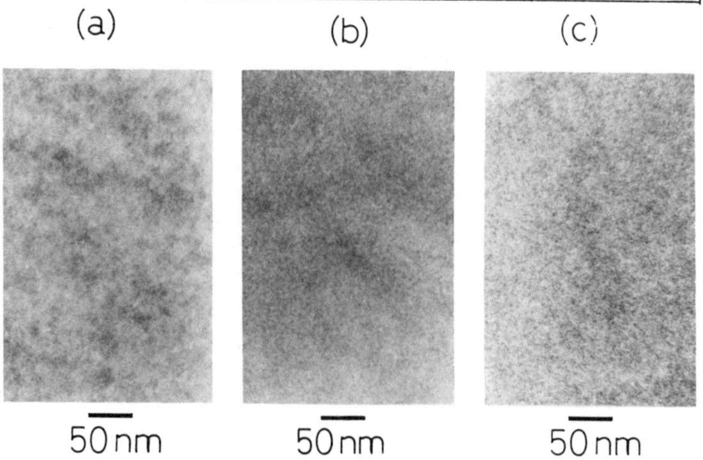

Figure 1 Bright field images at Rd of (a)50,(b)100 and (c)200 nm/min.

Bright field TEM micrographs of the TbFeCo films are shown in Figure 1, where (a), (b) and (c) represent the deposition rate Rd of 50, 100 and 200 nm/min, that is, the specimen films A, B and C deposited under sputtering conditions as listed in Table 1, respectively. The film A at Rd of 50 nm/min presented the obvious texture composed of round grains surrounded by the low-density network which was found in the bias-sputtered GdCo films[14]. On the other hand, the films B and C deposited at Rd of 100 and 200 nm/min, respectively, did not present any texture as observed in the film A.

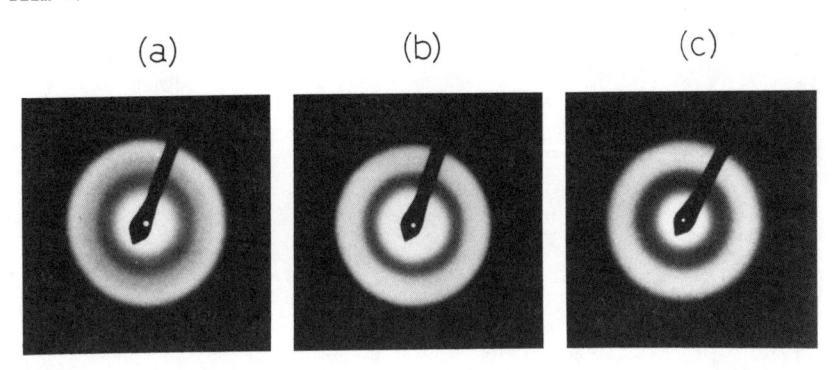

Figure 2 Selected-area electron diffraction patterns at Rd of (a)50, (b)100 and (c)200 nm/min.

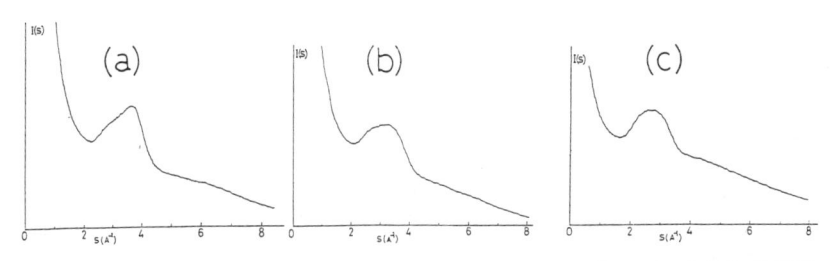

Figure 3 Microdensitometer traces of SAED patterns at Rd of (a)50, (b)100 and (c)200 nm/min.

The selected-area electron diffraction (SAED) patterns were taken from the region of diameter as small as 400 nm and were broad halo as shown in Figure 2. This indicates that all specimen films are amorphous. Figure 3 shows the traces of SAED patterns by the microdensitometer. The film at Rd of 50 nm/min showed a sharp peak at scattering vector $s(=4\pi \sin\theta/\lambda)$ of 3.6 Å^{-1}. Such a peak was also observed in the GdCo films[14]. On the other hand, the films at Rd of 100 and 200 nm/min showed the relatively broad peaks. The existence of amorphous or microcrystalline TbO_n (n=1.5-1.8) formed by a preferential oxidation of Tb was not distinctly confirmed, because the peaks corresponding to TbO_n (n=1.5-1.8) were not clearly observed at s of 2.09-2.02 Å^{-1} [15].

Figure 4 shows high resolution TEM micrographs of cross section of specimen films about 100 nm thick. The cross-sectional images of the film at Rd of 50 nm/min revealed the clear and uniform columnar structure composed of fine grains with mean diameter of about 8 nm and random orientation. The axes of columnar grains were nearly perpendicular to the film plane. On the other hand, the cross-sectional images of the films at Rd of 100 and 200 nm/min showed the columnless structure and they were composed of only very fine and nearly uniform grains as seen in these figure.

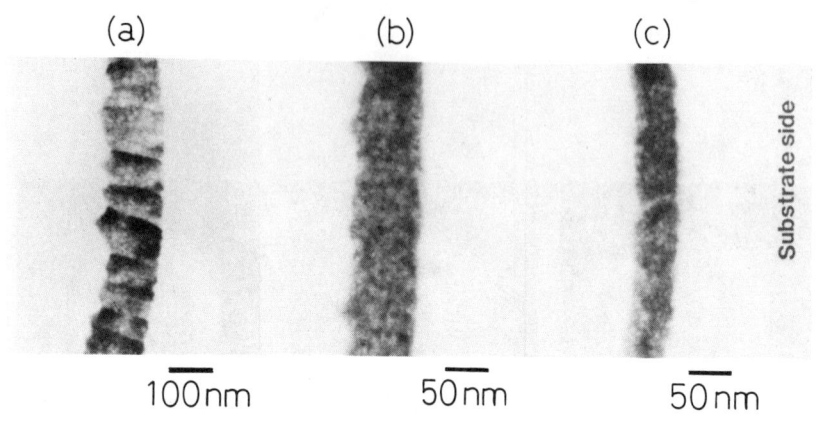

Figure 4 Cross-sectional TEM images at Rd of (a)50,(b)100 and (c)200 nm/min.

B. Magnetic and magneto-optical properties

Fig. 5 shows the dependences of Kerr rotation angle θ_k on Tb content in the TbFeCo films, which were deposited at constant P_{Ar} of 3.6 mTorr and at various Rd in the range of 50-200 nm/min. With increase of Tb content, θ_k at Rd of 50 nm/min significantly decreased, while that at Rd of 100 and 200 nm/min slightly decreased. This imply that the magneto-optical characteristics may correlate closely with the microstructure such as size and axis orientation of columnar grains.

The perpendicular magnetic anisotropy constant Ku can be derived from the relationship between $(L/H)^2$ and L.[16], which is induced by applying the magnetic field H in the direction of 45 deg. with respect to the film plane($Ku=K + 2\pi Ms^2$). Where L represents the torque amplitude. Figure 6 shows an example of the relationship, where the specimen film was deposited at Rd of 100 nm/min and has Ku of 3.6×10^6 erg/cm^3. Its value was approximately equal to that at Rd of 50 nm/min. This indicates that the perpendicular magnetic anisotropy in the TbFeCo amorphous films may mainly not correlate so closely with their microstructure.

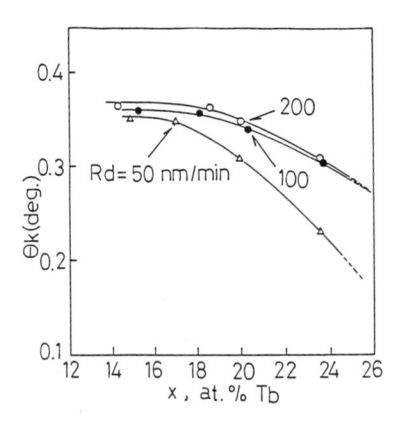

Fig. 5 Dependence of Kerr rotation angle θ_k on Tb content for various Rd.

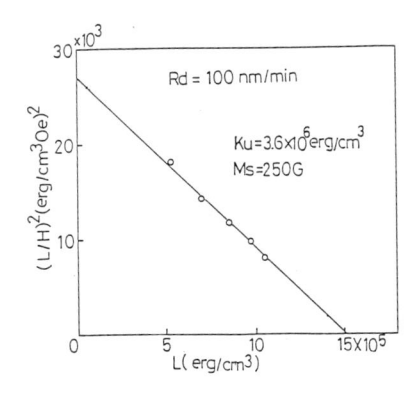

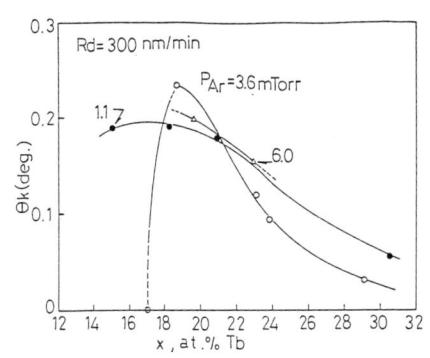

Fig. 6 $(L/H)^2$ versus L characteristic at Rd of 100 nm/min. (Thickness and volume of specimen film were 120 nm and 1.0×10^{-5} cm^3, respectively.)

Fig. 7 Dependence of Kerr rotation angle θk on Tb content for various P_{Ar}.

Figure 7 shows the Tb content dependences of θk at Rd of 300 nm/min and various P_{Ar} in the range of 1.1-6.0 mTorr. This figure indicates that the region of Tb content where the effective perpendicular magnetic anisotropy K is positive expanded considerably to the value below 17 at. % with decrease of P_{Ar}, and in addition, θk at P_{Ar} of 1.1 mTorr is as relatively large as about 0.2 deg. It has been found that the dependences at Rd of 300 nm/min are more drastical than those at Rd of 100 and 200 nm/min. This may result from the facts that the films were damaged by bombardment of negative transition metal iones and the substrates were heated by irradiation of high-energy r-electrons from targets due to unsatisfactory confinement of plasma during sputtering. These results suggest that the perfect confinement of plasma by sufficiently strong magnetic field and the film deposition at lower P_{Ar} are necessary for improving magneto-optical characteristics of the films deposited at Rd as high as 300 nm/min.

CONCLUSION

The TbFeCo amorphous films with magnetization perpendicular to the film plane have been prepared by using the FTS apparatus, and their microstructure, magnetic properties and magneto-optical characteristics were investigated. The distinguishable columnar structure was not observed in the films deposited at higher Rd of 100 and 200 nm/min, while the obvious columnar structure was observed in the films deposited at lower Rd of about 50 nm/min. The columnless films were very dense and composed of uniform and fine grains of about 8 nm diameter throughout the whole film. The Kerr rotation angle of columnless films was about 0.37 deg. and was higher than that of columnar films over the wide range of Tb content. Both columnless films and columnar ones exhibited the same perpendicular magnetic anisotropy constant Ku of about 3.6×10^6 erg/cm^3. This may imply that the Ku of the TbFeCo amorphous films does not correlate so closely with their microstructure.

ACKNOWLEDGMENT

We would like to thank Mr.T.Kudo for TEM and SAED observations and Mr.N.Onagi for x-ray fluorescence analysis, Corporate Research and Developement Lab., Pioneer Electronic Corp.

REFERENCES

[1] M.Ohkoshi, M.Harada, Y.Tokunaga, S.Honda, and T.Kusuda, IEEE Trans.Magn., MAG-21, 1635 (1985)
[2] M.Hong, E.M.Gyorgy, R.B.van Dover, S.Nakahara, D.D.Bacon and P.K.Gallagher, J. Appl. Phys. 59, 551 (1986)
[3] T.Suzuki, H.Ichinose and E.Aoyagi, Jpn. J. Appl. Phys. 23, 585 (1984)
[4] J-W.Lee, H-P.D.Shieh, M.H.Kryder, and D.E.Laughlin, J. Appl. Phys. 63, 3624 (1988)
[5] H-P.D.Shieh, M.Hong and S. Nakahara, J. Appl. Phys. 63, 3627 (1988)
[6] Yasugi, S.Honda, M.Ohkoshi and T.Kusuda, J. Appl. Phys. 52, 229 (1981)
[7] Y.Togami, N.Saito and K.Okamoto, J. Appl. Phys. 60, 3691 (1986)
[8] T.Suzuki, J. Mag. Mag. Mat. 35, 232 (1983)
[9] T.Takeno, M.Suwabe, T.Sakurai and K.Goto, Jpn. J. Appl. Phys. 25, L657 (1986)
[10] S.Nakahara, M.Hong., R.B.van Dover, E.M.Gorgy and D.D.Bacon, J. Vac. Sci. Technol. A4, 543 (1986)
[11] M.Naoe, N.Kitamura and H.Ito, J. Appl. Phys. 63, 3850 (1988)
[12] M.Naoe, S.I.Yamanaka, and Y.Hoshi, IEEE Trans. Magn., MAG-16,646 (1080)
[13] H.Ito, T.Hirata, N.Kitamura and M.Naoe. J. Magn. Soc. Jpn.Vol.11 Supplement No.S1, 225 (1987)
[14] H.J.Leamy and G.Dirks, J. Appl. Phys. 50, 2871 (1979)
[15] C.D.Wright, P.J.Grundy and E.T.M.Lacey, IEEE Trans. Magn. MAG-23, 162 (1987)
[16] H.Miyajima, K.Sato and T.Mizoguchi, J. Appl. Phys. 47, 4669 (1976)

STUDY OF MULTICOMPONENT MAGNETIC NANOSTRUCTURES WITH DIGITAL IMAGE PROCESSING TECHNIQUE.

A.P. VALANJU[1], I.S. JEONG[2], D.Y. KIM[1], AND R.M. WALSER[1,2,3]
[1] Department of Electrical and Computer Engineering, University of Texas, Austin, Texas 78712
[2] Center for Materials Science and Engineering, University of Texas, Austin, Texas 78712
[3] J. H. Herring Centennial Professor in Engineering

ABSTRACT

Previously, we used Digital Image Processing (DIP) to explore the relationships between the growth morphologies of sputtered two-phase nanostructures and their soft magnetic properties [1]. In this work we extended the application of DIP to analyse the effects of deposition parameters and annealing conditions on their soft magnetic properties including disaccommodation.

Magnetically soft, amorphous $Co_{61}B_{39}$ thin films, exhibiting a two-phase structure, were deposited by sequentially co-sputtering cobalt and boron. We digitized TEM micrographs of these thin films, prepared under different deposition conditions, and subjected to various post deposition processing. Digital Fourier transforms of the TEM micrographs were studied for evidence that the film anisotropy could be correlated with morphological order arising from long and short range interactions between particles over distances of $\approx 0.2 - 1.0$ nm.

Our qualitative studies showed that important changes in soft magnetic properties were associated with changes in the two phase morphologies. We determined, for example, that specific morphological changes were associated with the reduction in the magnetic anisotropy produced by annealing. In general, decreases in anisotropy were associated with increased isotropy in Fourier space. The largest reductions and circular symmetric 2D Fourier transforms were produced by rotating field annealing.

INTRODUCTION

Evaporated and sputtered thin films develop anisotropic columnar nanostructures due to self shadowing effects. The nanoscale features in the morphology of these films can have important effects on the macroscopic magnetic film anisotropy. Existing models of columnar growth associate the magnetic anisotropy with the ellipticity and size of the individual columns [2,3]. The alignment of the columns is implicitly assumed in these models. In previous work, we showed that the ellipticity and diameter of the columns can not account for the observed anisotropy of $Co_{61}B_{39}$ films, although there is a possible correlation with their alignment [1]. Films with this composition were chosen for detailed morphological study because they have interesting magnetic properties [4,5], and because they have a large electron contrast in TEM that is useful for image analysis.

These studies indicated that short range exchange interactions, and/or longer ranged dipolar interactions between nanoscale morphological features may contribute to the in-plane film anisotropy. The possibility of nanoscale morphological order in amorphous thin films has received only limited previous investigation, and there appear to have been no studies of morphological order involving Fourier components in the range important to macroscopic, in-plane, magnetic anisotropy. This correlation is important in films with an in-plane anisotropy, especially when one is concerned with understanding the small residual contributions in films in which vanishing small anisotropies are desired. For inter-particle exchange and dipolar interactions, these are estimated to involve Fourier wavelengths in the 0.1-10 nm. Small angle electron scattering (SAES) can be used to investigate wavelenghs in the 1.0-10 nm range. These studies have revealed the existence of a long range order correlated with the columnar structure of films exhibiting perpendicular anisotropy [6,7]. In research closely related to the present work, Yudin et. al. [8~10] detected morphological anisotropy in magnetic thin films using optical Fraunhofer patterns. This technique cannot, however, resolve morphological order involving dimensions less than approximately 10 nm.

In this work we used digital Fourier processing of the TEM micrographs to qualitatively examine the relationship of morphological and in-plane magnetic film anisotropies. This technique permitted the investigation of morphological anisotropies involving Fourier wavelengths in the range thought to be important (0.1~1.0 nm) to this issue. We conducted systematic studies of the effects of deposition parameters and post-deposition processing on the 2D Fourier transforms of the corresponding micrographs. This appears to be the first research along these lines and the results indicate that further, possibly quantitative, study is justified.

EXPERIMENTAL PROCEDURE

We prepared amorphous films of $Co_{61}B_{39}$ by RF sequential co-sputtering onto water cooled substrates rotating at 2 rpm in technical vacuum (10^{-7} Torr backbround). An argon sputter gas pressure of 10 mTorr was used. Samples were made with RF substrate bias of 0V and -60V. The effect of beam cut-off angle Ω was controlled by placing the substrate inside pyrex cylinders of appropriate dimensions. Some films were deposited by simultaneously applying a DC magnetic field of 20 Oe during deposition. The films were deposited on 1" diameter silicon substrates for magnetic measurements. The TEM samples were prepared at the same time by direct deposition of $Co_{61}B_{39}$ films on carbon backed copper grids. The average composition was controlled by varying the target voltage and was analysed by Inductively Coupled Plasma Atomic Emission Spectroscopy. The Co-B composition, and Ar and O impurity concentrations were depth profiled by Auger Electron Spectroscopy. Some as-deposited films were annealed in an inert environment of Ar by a rotating field annealing (RFA) technique at 275° to 300° C in a 1.8 KOe magnetic field. Other films were thermally annealed (TA) in Ar at 300° C. In order to study the effect of annealing and aging, magnetic and morphological data was obtained on samples annealed by RFA.

A Jeol JEM 200CX TEM with underfocused phase current was utilized for making micrographs of nanostructure morphologies. Statistical measurements and Fourier spectra were obtained with a Kontron Electronics IPS image processing system. Magnetic parameters were measured with a B-H loop tracer. The coercivity was obtained from the hysteresis loop along the easy axis which was taken as the direction of minimum coercivity. The anisotropy was taken as the intersection of the hard axis loop with the saturation magnetization at a small drive field.

RESULTS AND DISCUSSION

The films were deposited in a single pumpdown to ensure an identical composition. Substrates placed in cylinders of different dimensions received different amounts of particle flux and were of different thicknesses. By analyzing the combined effects of the beam cut-off angle Ω and thickness on magnetic properties, our previous work [1] established that the magnetic properties were dominated more by the effect of Ω and not by the film thickness.

Electron diffraction patterns indicated that the $Co_{61}B_{39}$ films were amorphous and hence lacked long range order related to atomic scale structure. The hysteresis loops of these films revealed a macroscopic, in-plane, uniaxial magnetic anisotropy. Annealing studies indicated that this anisotropy was not associated with intra-particle, atomic scale spin orientations, nor with long range strains in the film. As indicated above, the magnetic anisotropy could not be conclusively associated with the shape anisotropy of nanoscale morphological film features. Accordingly we studied the possible role of inter-particle interactions, the presence of which might be revealed in the anisotropy of the in-plane film morphology. Since it is difficult, if not impossible, to directly detect this anisotropy in the micrographs (Figure 1), we transformed the morphological information to Fourier space (Figure 2). Various n-fold symmetry and inversion axes in the 2D Fourier transforms were noticeable. As expected, the diameter of the central dark ring was inversely proportional to the size of the largest structures in the micrographs. The morphological anisotropy was most evident at high frequencies corresponding to distances of approximately 1 nm in real space.

To associate the anisotropy in the high spatial frequency region of the Fourier transforms with specific features of the nanostructure in the micrographs, we separated (spatially filtered) the most intense (bright) regions in the original micrographs and constructed binary images from the filtered micrographs. Fourier transforms of the binary images closely resembled those of the original micrographs (Figure 3). Thus the "gaps" (brightest regions) separating the "particles"

175

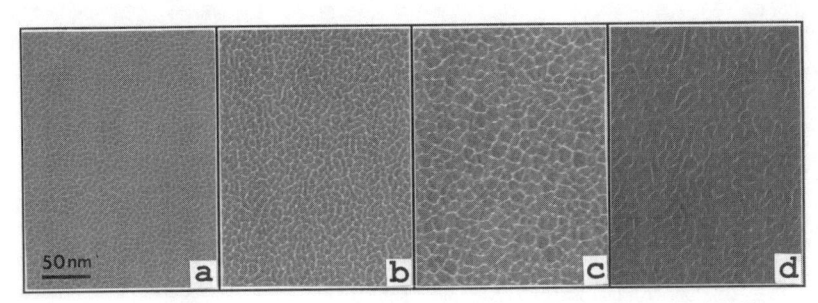

Figure 1: TEM micrographs and magnetic data (coercivity H_c and anisotropy H_k) for $Co_{61}B_{39}$ under different beam cut-off angles (Ω) and substrate bias. (a) $\Omega=10°$, bias=0V, $H_c = 8$ Oe, $H_k = *$, (b) $\Omega=10°$, bias=-60V, $H_c = 14.9$ Oe, $H_k = *$, (c) $\Omega=60°$, bias=0V, $H_c = 36.2$ Oe, $H_k = 85$ Oe, (d) $\Omega=60°$, bias=-60V, $H_c = 34.4$ Oe, $H_k = 50$ Oe. * denotes small rotatable anisotropy.

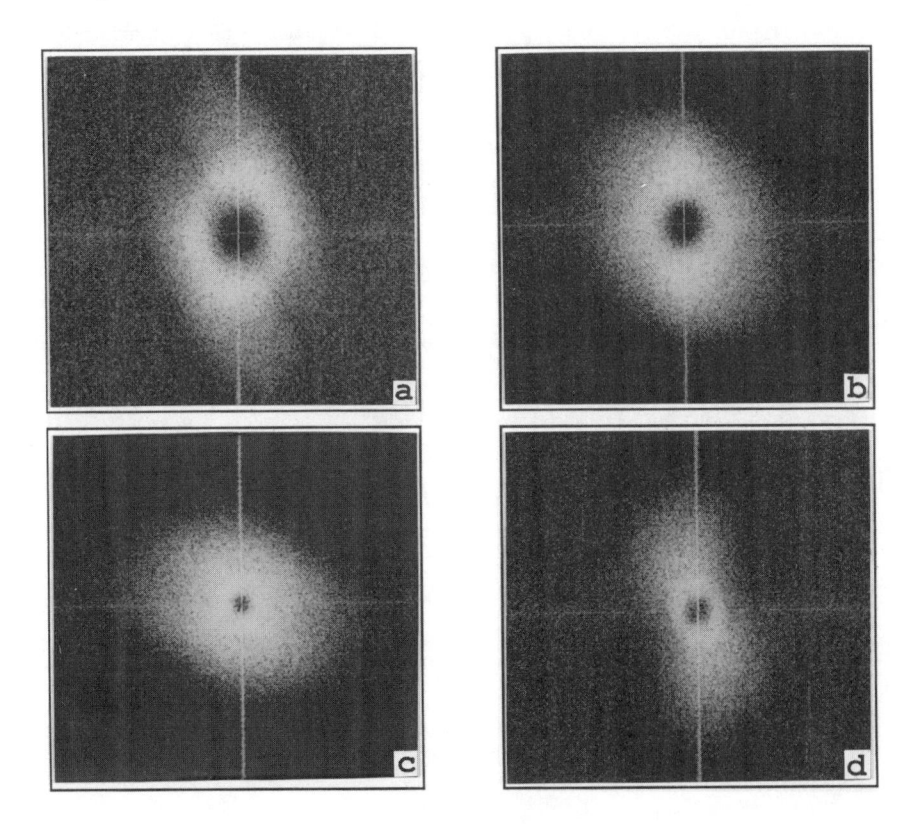

Figure 2: 2D Digital Fourier transforms of micrographs in figure 1.
(a) $\Omega=10°$, bias=0V, (b) $\Omega=10°$, bias=-60V, (c) $\Omega=60°$, bias=0V, d) $\Omega=60°$, bias=-60V.

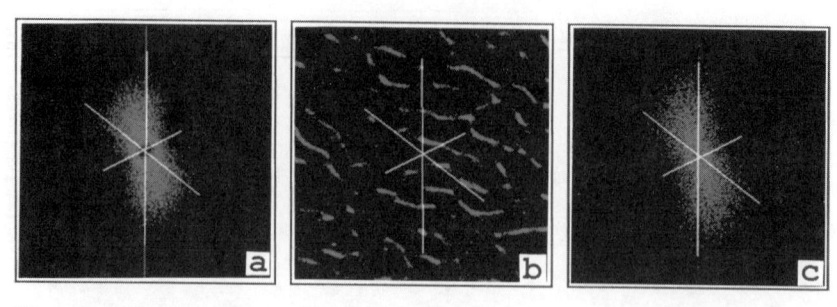

Figure 3: (a) Fourier transform of TEM micrograph 1d, (b) binary image created from 1d, (c) Fourier transform of 3b.

(darkest regions) in the micrographs appeared to be the dominant features determining the symmetries in the micrographs.

In order to study the origin of magnetic anisotropy in $Co_{61}B_{39}$ films, we carried out RFA and TA annealing experiments. The magnetic anisotropy reduced with annealing, although a larger reduction was achieved with RFA. Also RFA substantially increased the permeability. We then compared morphologies of RFA, TA and as-deposited films by comparing their 2D Fourier power spectra (Figure 4). The results indicated that the anisotropy of annealed films was associated with the degree of isotropy in their 2D Fourier spectra. While the Fourier transform of the TA film resembled that of the as-deposited film, that of the RFA film was essentially isotropic.

It is well known that, after a characteristic time that depends on their composition and processing conditions, the properties of annealed, amorphous thin films relax toward their magnetically hard as-deposited state. The 2D Fourier transforms of annealed and partially relaxed films were examined to determine if the relaxation was correlated with morphological changes. We compared the 2D Fourier transforms of as-deposited, RFA annealed, and aged films (Figure 5). As seen earlier, the RFA film exhibited an isotropic distribution in Fourier space. However, after aging for 55 days, the Fourier transform of this film exhibited an anisotropic distribution.

CONCLUSION

In this study we qualitatively demonstrated the usefulness of processing the digital Fourier transforms of TEM micrographs to detect morphological anisotropy. Apart from the convenience of use, this technique offers several other advantages. The resolution in Fourier space is set by an adjustable sampling pixel size and not by external parameters like the wavelength and angle of incident light etc. Also the low frequency features are not obscured by the direct beam as in TED and SAES. The ease of filtering in the frequency and spatial domains makes this techniques especially useful. The electron diffraction spectrum resolves periodicities on atomic scale (0.1 ~ 1 nm), whereas small angle scattering and laser Fraunhofer diffraction techniques resolve structures on 10 to 100 nm scale. The digital technique reported here was found to be useful for studying structures on intermediate scales (0.2 ~ 10 nm) that appear to be important in determining the anisotropy of films with macroscopic anisotropies involving nanoscale, interparticle interactions.

ACKNOWLEDGEMENT

This work was sponsored by the Air Force Office for Scientific Research under Contract No. F49620-87-C-0067

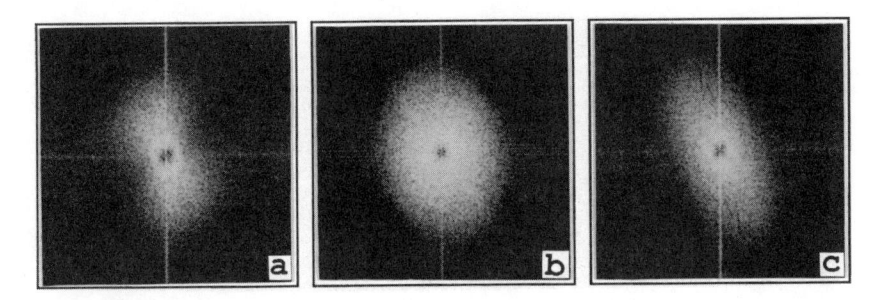

Figure 4: Effect of annealing on morphology and magnetic properties (coercivity H_c, anisotropy H_k and relative permeability μ_i) of $Co_{61}B_{39}$ films deposited with $\Omega=60°$, bias=-60V. Digital Fourier transform of films (a) as deposited, $H_c = 3\text{-}4$ Oe, $H_k = 7\text{-}8$ Oe, $\mu_i = 0.2$, (b) after rotating field annealing, $H_c = 0.02$ Oe, $H_k =$ isotropic, $\mu_i = 7$, and (c) after thermal annealing, $H_c = 0.6$ Oe, $H_k = 1\text{-}2$ Oe, $\mu_i = 2$.

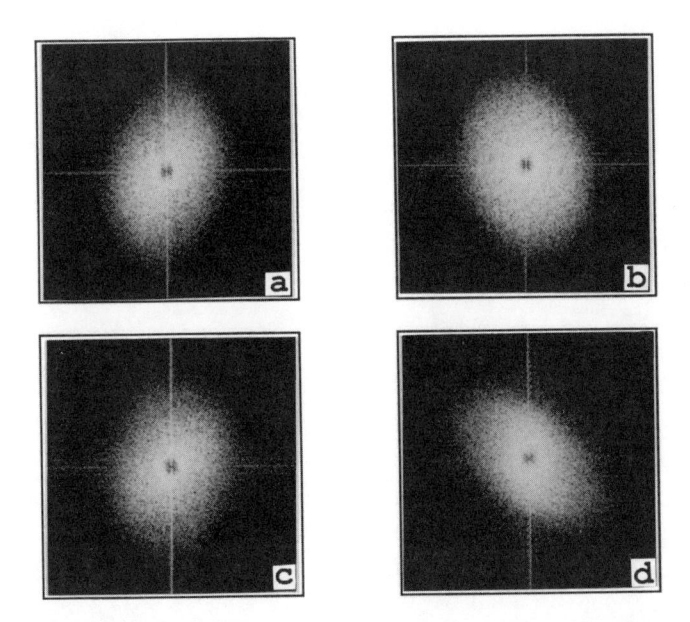

Figure 5: Effect of aging on morphology and magnetic properties (coercivity H_c, anisotropy H_k, relative permeability μ_i, and disaccommodation DA) of $Co_{61}B_{39}$ films deposited with $\Omega=60°$, bias=-60V, and DC magnetic field 20 Oe. Digital Fourier transform of films (a) as deposited, $H_c = 0.5$ Oe, $H_k = 8\text{-}12$ Oe, $\mu_i = 0.3$, (b) after rotating field annealing, $H_c = 0.35$ Oe, $H_k = 1.5$ Oe, $\mu_i = 6$, (c) 25 days after annealing, $H_c = 0.3$ Oe, $H_k = 2$ Oe, $\mu_i = 5$, DA = 20%, and (d) 55 days after annealing, $H_c = 0.3$ Oe, $H_k = 2$ Oe, $\mu_i = 4$, DA = 30%.

REFERENCES

1. A.P. Valanju, I.S. Jeong, D.Y. Kim, and R.M. Walser, J. Appl. Phys. 64, 5443 (1988).
2. W. Metzdorf and H.E. Wiehl, Phys. Stat. Sol. 17, 285 (1966).
3. H.N. Oredson and E.J. Torok, J. Appl. Phys. 36, 950 (1965).
4. D.Y. Kim and R.M. Walser, J. Appl. Phys. 64, 5676 (1988).
5. I.S. Jeong, A.P. Valanju, and R.M. Walser, J. Appl. Phys. 64, 5679 (1988).
6. R.H. Wade and J. Wilcox, Appl. Phys. Lett. 8, 7 (1966).
7. H.J. Leamy and A.G. Dirks, J. Appl. Phys. 49, 3430 (1978).
8. V.V Yudin et. al., Sov. Phys. Solid State 24, 250 (1982).
9. G.P. Timanova et. al., Phys. Met. Metall. 44, 63 (1979).
10. G.P. Timanova et. al., Phys. Metal. Metall. 47, 55 (1980).

MAGNETIC PROPERTIES OF IRON/SILICA GEL NANOCOMPOSITES

Robert D. Shull, Joseph J. Ritter, Alexander J. Shapiro, Lydon J. Swartzendruber, and Lawrence H. Bennett, Institute for Materials Science and Engineering, National Institute of Standards and Technology, Gaithersburg, MD 20899.

ABSTRACT

Homogeneous gelled composites of iron and silica containing 5-30 wt. % Fe have been prepared by low temperature polymerization of aqueous solutions of ferric nitrate, tetraethoxysilane, and ethanol (with an HF catalyst). X-ray diffraction data, characterized by the presence of a diffuse scattering peak centered at 2θ≈24 degrees and the absence of any strong Bragg scattering from the iron-containing regions, indicates that these bulk materials are comprised of nanometer-sized regions of iron compounds embedded in a silica gel matrix. Scanning electron microscopy observations show that this matrix is characterized by the presence of many interconnected pores and that the size of these pores is related to the particle size of the Fe-containing regions. The paramagnetic nature of these materials at room temperature, as well as the small size of the iron-containing regions, is indicated by the appearance in many of the samples of only a high intensity central doublet in the ^{57}Fe Mössbauer spectra. The Mössbauer effect data demonstrates that the form of the iron can be changed by a subsequent treatment in an atmosphere of ammonia or hydrogen at elevated temperatures: for a 10 wt. % Fe sample treated with ammonia, only a central doublet was observed but with a much larger quadrupole splitting and isomer shift. Both of these subsequently treated materials became superparamagnetic at room temperature. In addition, magnetic susceptibility measurements indicate that the hydrogen treated material becomes a spin glass at low temperatures.

INTRODUCTION

When an immiscible metal and non-metal are co-sputtered onto a suitable substrate, they can form a material which is an intimate mixture of the two constituent phases with the size scale of these phases being on the order of nanometers. This material, sometimes referred to as a granular metal [1], possesses unusual properties which are highly composition dependent, but which are related to those of the constituent phases. For the development of these unusual properties, the nanometer particle size is important. A chemical method for creating small metal particles in a non-metallic matrix has been described by Bilisoly [2], Roy [3], and Pope [4]. This latter method would be especially advantageous for the preparation of materials which are stable only at low temperatures. This study was initiated to study the feasibility of using a sol gel process to create small particle magnetic systems. For this purpose the Fe + silica gel system was chosen because of its similarity to the Fe+SiO$_2$ granular metal system previously studied [5] and which was found to possess superparamagnetic behavior.

EXPERIMENTAL PROCEDURE

Iron and silica gel nanocomposites containing 10-30 % Fe (all iron contents are given in weight percents) were prepared by adding the appropriate amounts of an aqueous solution of ferric nitrate to a 50% solution (by volume) of tetraethoxysilane and ethanol, with a few drops of aqueous HF added as a catalyst. Gel times varied from 12 hours for the 10 % Fe samples to ~6 days for the 30 % Fe samples. After slowly air drying in loosely

covered dishes for ~2 weeks, the gels formed as a hard brittle solid. One
series of samples, series A, was prepared without regard to the water
content of the solutions, under the assumption that the water evaporates.
Series B samples, however, were prepared keeping the water content of the
solutions constant as the Fe content was varied. Sample notation (e.g. 10A
and 11B) herein indicates both the Fe content and preparation series.
  X-ray diffraction and magnetic susceptibility measurements were
performed on powders of these solids. Mössbauer spectra were measured on
compacts of these sample powders mixed with phenolic. Magnetization
measurements were performed as a function of magnetic field (-10 kOe < H <
+10 kOe) and temperature (10 K < T < 300 K) using a vibrating sample
magnetometer. The Mössbauer effect measurements were performed using the
sample as the absorber and a 0.5 mCi ^{57}Co in Rh source. Velocity
calibration of the Mössbauer equipment was performed using Fe_2O_3, and zero
velocity is the center of a pure Fe spectrum. Scanning electron microscope
(SEM) observations were performed at either 15 or 25 kV on freshly fractured
surfaces of the hardened gel. Prior to observation, a thin Au coating was
sputtered on top of the specimen in order to eliminate charging effects.

RESULTS AND DISCUSSION

All of the specimens possessed x-ray diffraction patterns characterized by
the presence of two broad scattering bands (centered at ~24 and ~47 degrees
2θ) similar to that observed for amorphous materials. The low temperature
Mössbauer data presented later shows that the Fe-containing regions in these
samples, if crystalline, are too small to give intense Bragg diffraction
peaks.
  The room temperature Mössbauer patterns measured for the air dried
iron/silica gel nanocomposites were of the two basic types shown in figure
1. Either they possessed only a strong central doublet with a small 0.4
mm/sec isomer shift or they possessed this same central doublet in
combination with a broad multiple line spectrum extending to high velocities
indicating the presence of magnetically ordered material.

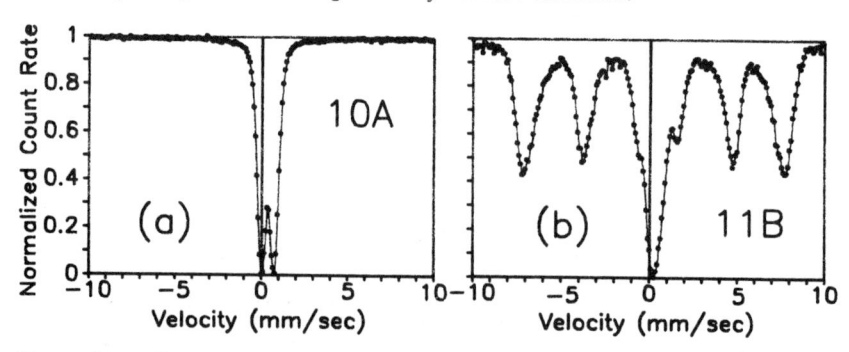

Figure 1. Room temperature Mössbauer patterns for (a) sample 10A and (b) sample 11B.

The magnetic hyperfine field ($\underline{H}$) for this multiple line spectrum, 453 kOe,
is smaller than the 517 kOe field of Fe_2O_3, larger than the 330 kOe field
for pure iron, but close to the 450 kOe field commonly observed for iron
compounds in which the iron occurs as Fe^{2+}. The central doublet found in
all the room temperature spectra comes from very small magnetic regions
(estimated diameters<200 Å) above their blocking temperatures similar to
that reported by Kundig [6] for submicron particles of Fe_2O_3 and by Shull
[7] for nanocomposites of $Ag+Fe_3O_4$. This particle size effect is shown very

dramatically in the sequence of Mössbauer spectra shown in figure 2 for specimen 10A measured at reduced temperatures. At 45 K the spectra shown in figure 1a begins to show a magnetic component. By 20 K, the magnetic component is well developed; and the 4.2 K spectrum shows that the magnetic component grows at the expense of the high temperature central doublet. The positions of the multiple line spectrum at 4.2 K ($\underline{H}{\approx}470$ kOe) in this sample are very close to those observed in the multiple line spectra measured at room temperature for those samples possessing large Fe-containing particles. Figures 3 and 4 show that the magnetic hyperfine field observation in sample 10A at low temperatures is not due to the presence of a ferromagnetic transition. This 10 % Fe sample possesses a linear field dependence at both room temperature and 10 K indicating it is paramagnetic (on the time scale, ~1 sec, of the magnetization measurement) in this temperature region, and there is no discontinuity in the temperature dependence of the magnetization (at 2 kOe field) down to 10 K. The magnetic hyperfine field component in the Mössbauer spectra of figure 2 reflects the appearance at low temperatures of cooperative magnetic behavior in sample 10A, but only on a very short time scale ($\leq 10^{-7}$ sec, ~the Larmor precession period of the ^{57}Fe nuclei).

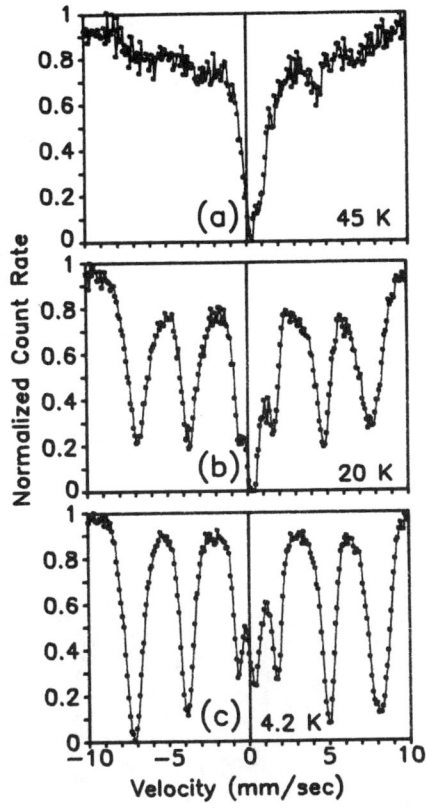

Figure 2. Mössbauer patterns measured for the 10 % Fe sample 10A at (a) 45 K, (b) 20 K, and (c) 4.2 K showing the effect of small particles in this system.

Figure 3. Magnetization vs. applied field data for sample 10A measured as the field was cycled between +10 kOe and -10 kOe at room temperature (open symbols) and 10K (filled symbols).

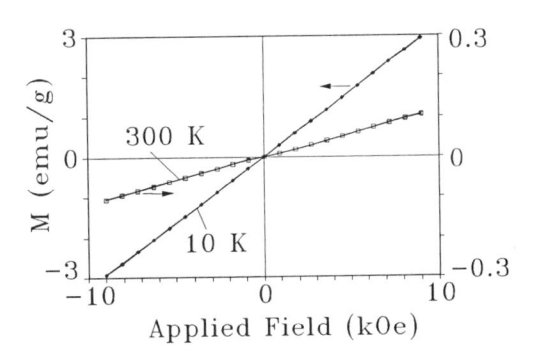

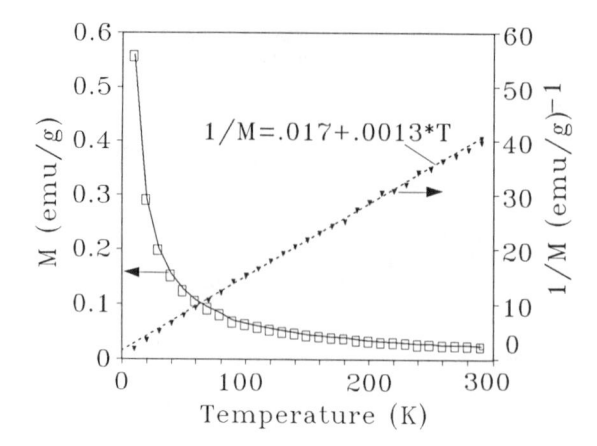

Figure 4. Magnetization, M, (open symbols) and reciprocal magnetization (filled symbols) vs. temperature data for sample 10A measured during cooling in a 2 kOe applied field. The dashed line is a least squares fit of the T>50 K data to a Curie-Weiss law. Note that all the 1/M data points fall below this line for T<60 K.

Figure 4 shows that small negative deviations from Curie Weiss behavior do occur around 60 K, and the possibility of an antiferromagnetic transition is suggested by the intercept at ~-13 K of the extrapolated high temperature behavior. Further magnetization measurements at lower applied fields in addition to zero-field-cooling measurements will clarify this possibility.

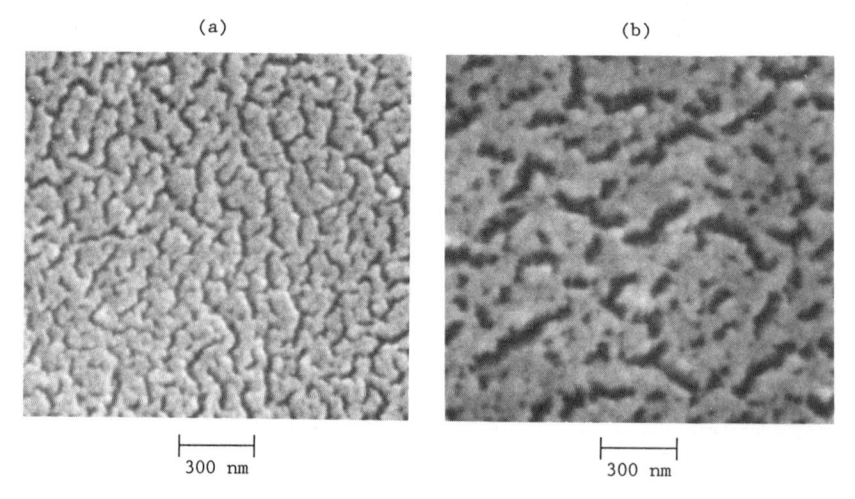

Figure 5. Scanning electron micrographs of samples (a) 10A and (b) 11B taken at 50,000x showing the interconnected network of pores (dark) existing throughout the matrix (light) of these materials.

Series A samples containing 25 and 30 % Fe possessed identical room temperature Mössbauer spectra to that shown for specimen 10A in figure 1a, indicating a small particle size for the Fe-containing regions in these samples. However, larger particle sizes were found to be present in specimen 18A as it possessed a spectra similar to that shown in figure 1b. For series B samples, 11B and 18B possessed room temperature spectra containing a magnetically ordered component while that measured for 25B only contained the large central doublet. Consequently, the Fe and H_2O contents of the preparation solutions are not the only parameters affecting the particle size of the iron in these nanocomposites. A correllation, however, was observed between the morphology of the samples and their high temperature Mössbauer patterns. All the materials prepared here possessed an interconnected network of pores throughout the matrix as shown in figure 5. However, those materials which possessed large (≥45 nm) pore diameters as shown in figure 5b for sample 11B also possessed Mössbauer spectra with a magnetically ordered component visible at room temperature. If the Fe-containing regions are localized in the pore areas, one might expect their sizes to be proportional to the amount of cation (Fe) in the pores during the drying process and, consequently, to the pore volume. This suggestion was also made earlier by Roy [3] in connection with Ag/silica gel nanocomposites. Note from figure 5 that even for those materials with large pore size (and therefore possessing "large" iron particle sizes), the particle sizes of the Fe-regions are still too small (≤20 nm) to be imaged in the conventional SEM.

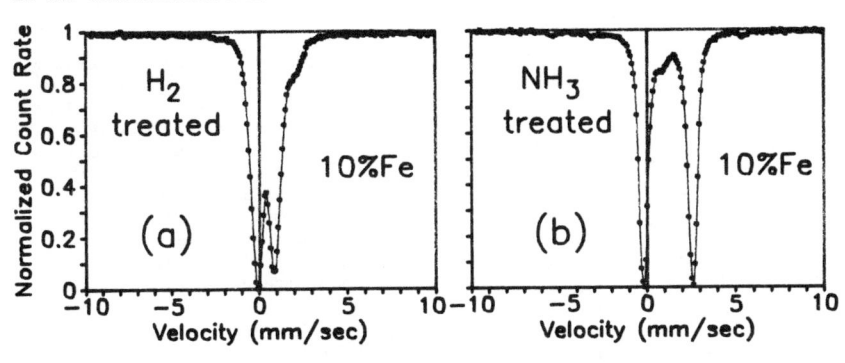

Figure 6. Room temperature Mössbauer patterns measured for sample 10A following treatments in (a) hydrogen and (b) ammonia atmospheres.

Evidence was found for localizing the Fe-containing regions in the pore areas by subjecting separate pieces of the dried nanocomposite 10A to treatments in 1 atmosphere of hydrogen gas (378 C for 20 hours) and 1 atmosphere of NH_3 gas (460 C for 1 hour following the previous hydrogen treatment). Due to the interconnection of the pore areas, any Fe-regions located in these areas would be likely to change their form during these treatments. The Mössbauer spectra for these altered samples, displayed in figure 6, indeed show distinct changes in the form of the iron in these nanocomposites. The hydrogen treatment resulted in a pattern with a slightly broadened strong central doublet with an admixture of a less intense peak having a greater isomer shift; the NH_3 treated sample resulted in a material primarily possessing a much larger isomer shift of 1.4 mm/sec. Interestingly, at room temperature both of these altered materials possessed increased magnetization values (by factors of 2 and 3 respectively for the NH_3 and H_2 treated specimens) and changed their magnetic state from paramagnetic to superparamagnetic (see figure 7). Even though the Mössbauer pattern for the H_2-treated sample was not remarkably different from the untreated sample 10A (figure 1a), its magnetic state is quite different. In fact, figure 8 shows the H_2-treated 10 % Fe sample possesses thermomagnetic history effects at temperatures below about 30 K. These effects combined with the observation of a displaced hysteresis loop along the field axis at low temperatures are characteristic of spin glass magnetic behavior [8]. This magnetic behavior results when the magnetic spins on the Fe become frozen in their high temperature orientation when cooled to below their freezing temperature (which in this case is ~30 K). Because this is such an unusual magnetic state, it will be discussed at greater length elsewhere.

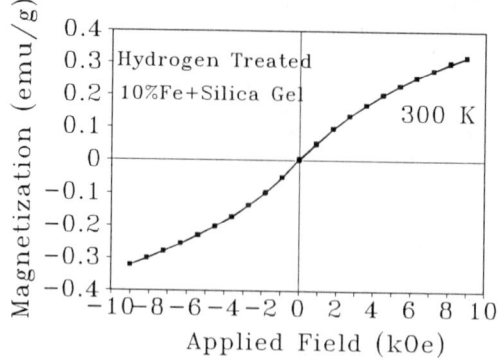

Figure 7. Magnetization vs. applied field data for sample 10A at 300 K following a treatment in hydrogen gas. The room temperature data measured for the NH_3-treated material looked similar, but with a maximum magnetization at 9 kOe of 0.2 emu/g.

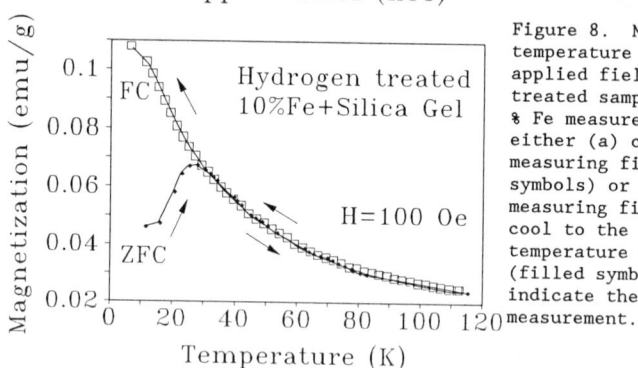

Figure 8. Magnetization vs. temperature data at 100 Oe applied field for the H_2-treated sample containing 10 % Fe measured as it was either (a) cooled in the measuring field (open symbols) or (b) warmed in the measuring field following a cool to the lowest temperature in zero field (filled symbols). Arrows indicate the direction of measurement.

CONCLUSIONS

Nanocomposites of iron and silica gel can be prepared over a composition range of 10-30 % Fe by a sol gel process. The Fe-containing regions are localized primarily in the pore areas and their particle size is found to be a function of pore size. Hydrogen and ammonia gas treatments of a 10 % Fe nanocomposite increased its magnetization and created a superparamagnetic state, resulting in one case in the formation of a low temperature magnetic spin glass.

REFERENCES

[1] B. Abeles, Appl. Solid State Sci. 6, 1 (1976).
[2] J. P. Bilisoly, U.S. Patent No. 2496265 (Feb. 7, 1950).
[3] R. A. Roy and R. Roy, Mat. Res. Bull. 19, 169 (1984).
[4] E. J. A. Pope and J. D. Mackenzie, J. Non-Crys. Sol. 87, 185 (1986).
[5] Gang Xiao, S. H. Liou, A. Levy, J.N. Taylor, and C. L. Chien, Phys. Rev. B 34, 7573 (1986).
[6] W. Kundig, H. Bommel, G. Constabaris, and R. Lindquist, Phys. Rev. 142, 327 (1966).
[7] R. D. Shull, U. Atzmony, A. J. Shapiro, L. J. Swartzendruber, L. H. Bennett, W. J. Green, and K. Moorjani, J. Appl. Phys. 63, 4261 (1988).
[8] A. K. Mukhopadhyay, R. D. Shull, and P. A. Beck, J. Less Common Metals 43, 69 (1975).

COERCIVITY IN GRANULAR Fe - Al$_2$O$_3$[*]

F.H. Streitz and C.L. Chien
The Johns Hopkins University
Department of Physics and Astronomy
Baltimore, Maryland 21218

ABSTRACT

We present the results of a comparison between the magnetic properties of granular Fe–Al$_2$O$_3$ and Fe–SiO$_2$. Granular Fe-SiO$_2$ was found earlier to display an anomalous increase in coercivity which appears to scale with the Fe grain size. The Fe-Al$_2$O$_3$ samples were prepared under similar conditions and their magnetic properties determined. The coercivity in granular Fe–Al$_2$O$_3$ is also seen to increase with increasing grain size, showing that the effect is not unique to the Fe-SiO$_2$ system. A major difference between granular Fe–Al$_2$O$_3$ and Fe–SiO$_2$ is the size of the Fe grains, which are smaller in Al$_2$O$_3$.

INTRODUCTION

A granular metal is an artificial material composed of nanocrystals of metal imbedded in an amorphous insulator. Depending on the type and amount of metal which is imbedded, the granular composite displays a rich variety of physical properties owing to their unique nanostructure[1]. In the case of magnetic granular solids, the small grains are single domain particles, (i.e., they are so small that the formation of multiple magnetic domain structures is energetically prohibitive). An assembly of single domain particles exhibit many intersting magnetic behaviors. Inherent to all small magnetic particles is the phenomenon of superparamagnetic behavior. Above a well defined temperature, known as the blocking temperature (T$_B$), the moments of the particles will be thermally relaxed, allowing them to fluctuate in unison. The assembly will behave as a paramagnet[2,3]. The value of T$_B$ is well known to be proportional to the size of the particle.

The coercive behavior of an assembly of single domain particles has long been of interest, because of the predicted high coercivity that could be obtained in such a system[4]. Recent work on the magnetic properties of granular materials has revealed an unusual enhancement of the coercivity in the Fe$_x$(SiO$_2$)$_{100-x}$ system[5]. The coercivity measured at 2 K is reported to increase from a nominal value of 500 Oe for x $\sim$ 30% to approximately 2500 Oe at x $\sim$ 50%. All compositions x are given as volume percents. The increase in coercivity has been found to be related to the increase in grain size as x approaches x$_c$ (the percolation threshold). Furthermore, measurements made on samples with larger grain sizes but identical volume fractions also revealed enhanced coercivities[6].

We present in this paper some of our results of magnetic measurements made on the $Fe_x(Al_2O_3)_{100-x}$ granular system. Al_2O_3 was chosen as the insulating matrix in order to determine what role, if any, the matrix is playing in the enhanced coercivity effect in particular, and the formation of Fe grains in general. We show that the $Fe-Al_2O_3$ system also displays an increase in coercivity with grain size, but the results are not as dramatic as in the $Fe-SiO_2$ case. Furthermore, the sizes of Fe grains produced are significantly smaller in the Al_2O_3 matrix than in SiO_2.

SAMPLE PREPARATION AND CHARACTERIZATION

All of the samples discussed here were prepared by rf sputtering onto controlled temperature substrates from composite targets of appropriate mixtures of pure Fe and Al_2O_3. Sputtering was performed in 4 mT of Argon at a power of 100 W, resulting in a deposition rate of ~ 1 μm/hr. Magnetic and x-ray measurements were performed on samples which were 2-3 μm thick sputtered on Corning 7059 glass while electron microscopy was facilitated by sputtering films 100 - 200 Å thick directly onto carbon coated copper transmission electron microscope (TEM) grids. The samples were sputtered at temperatures from 100-500 °C.

X-ray diffraction was used to ascertain the structure of the resulting samples. Fig. 1 shows the x-ray spectrum taken on a Philips APD 3720 automated diffractometer. The broad peak at ~ 25° is the primary feature of the amorphous Al_2O_3, while the other peaks can be indexed to bcc Fe as shown. Note that the iron peaks are significantly broadened due the small crystallite size (20-50 Å).

A Philips EM420 transmission electron microscope was used to determine the mean grain sizes in the sputtered samples. Fig. 2 shows TEM micrographs for $Fe_{45}(Al_2O_3)_{55}$ sputtered at two substrate temperatures (175 °C and 450 °C). The grain size in these samples was determined by digital analysis of the micrographs to be 22 Å and 31 Å, respectively. Similar micrographs were obtained for the other samples.

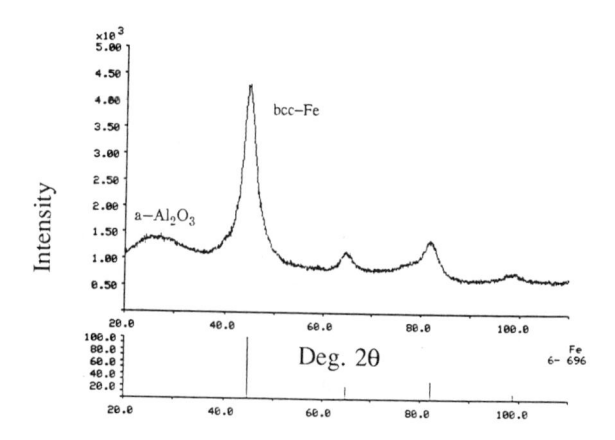

Fig. 1.- $\theta-2\theta$ x-ray diffraction pattern obtained from a granular sample of $Fe-Al_2O_3$.

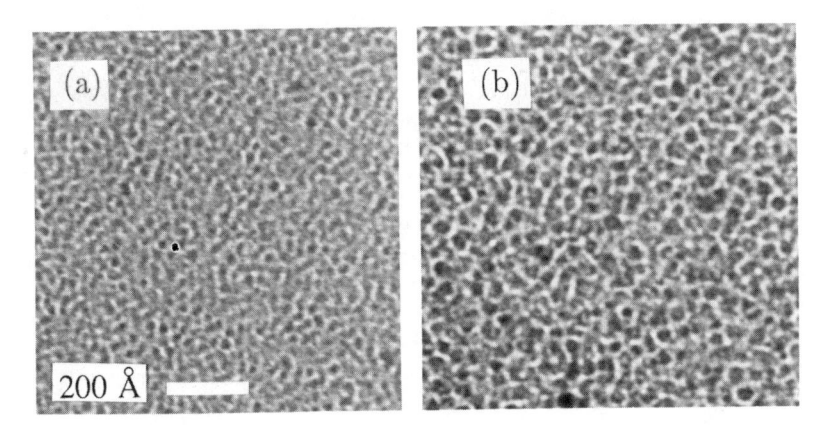

Fig. 2.- TEM micrographs of $Fe_{45}(Al_2O_3)_{55}$ sputtered at a substrate temperature $T_s = 175\,°C$ (a) and $450\,°C$ (b). The mean grain size determined in (a) is 22 Å and in (b) is 31 Å.

The magnetic properties of the samples were studied using an SHE SQUID magnetometer in the field range 0-50 kOe and 2-400 K. To determine the coercivity, the sample is cooled to ~ 2 K and saturated in the maximum field of 50 kOe. A reversing field is then applied until the coercive field is found. The blocking temperature is determined by cooling the sample in zero field and measuring its response to a 5 Oe applied field as a function of temperature. Fig. 5 displays a typical M vs. T curve thus obtained, with T_B indicated.

DISCUSSION

Analysis of the TEM micrographs reveals that the grains formed in these materials are roughly spherical in shape and narrowly distributed about some well defined mean grain size. (See Fig. 2). The primary method which we used to control the mean grain sizes is the variation of substrate temperature during deposition, which has been found to be effective in Fe—SiO_2[6]. The dependence of mean grain size $<D>$ with substrate temperature T_s is shown in Fig. 3. As expected, larger $<D>$ were obtained at higher T_s but the dependence is seen to be much weaker than in Fe—SiO_2. This demonstrates that the matrix has a pronounced effect on the size of the granules; a-SiO_2 accomodates much larger Fe grains than does a-Al_2O_3.

The iron grains in these granular materials are small enough to be single domain particles. The coercivity of an assembly of single domain particles was considered in the classic paper by Stoner and Wohlfarth[7], in which they showed that the coercivity at 0 K is given by

$$H_c \approx \frac{2K_{eff}}{M_s}, \tag{1}$$

where K_{eff} is the effective volume anisotropy and M_s is the saturation magnetization.

As neither M_s nor K_{eff} are expected to vary with the size of the grain[8], the zero temperature coercivity of the material should be size independent.

Fig. 4 depicts the variation of coercivity H_c with $<D>$ for both the Fe-Al$_2$O$_3$ and Fe-SiO$_2$ systems. In neither system is H_c a constant, as suggested by Eqn. (1). Instead, H_c shows an approximately linear increase with grain size. This increase must be considered a general feature of granular Fe systems, although the values of H_c obtained in each system will be matrix dependent. Fig. 4 shows that for the same $<D>$, the value of H_c for Fe-Al$_2$O$_3$ is less than that for Fe-SiO$_2$.

From measurements of the blocking temperature T_B the effective volume anisotropy constant K_{eff} can be determined. From a zero field cooled state, the magnetization increases with increasing temperature as the moments become "unfrozen" before decreasing paramagnetically. The blocking temperature is taken as that temperature for which the zero field cooled branch of the magnetization curve is at a maximum, as shown in Fig. 5. Assuming a simple Arrhenius relaxation law

$$T_B = \frac{K_{eff}V}{\ln(\tau_{SQUID}/\tau_0)}, \qquad (2)$$

K_{eff} can be determined from measured values of T_B and V, the particle volume. The measured values of T_B are shown in Fig. 6 as a function of D^3. The linear dependence of T_B on D^3 indicates that K_{eff} is approximately a constant, independent of grain size. From Fig. 6, a value of $K_{eff} = 3 \times 10^7$ erg/cm^3 has been found. A similar value of 10^7 erg/cm^3 has also been found in granular Fe–SiO$_2$[5]. These values are much larger than the magnetocrystalline anisotropy of bulk Fe ($\sim 5 \times 10^5$ erg/cm^3) indicating either that the ultrafine Fe granules take on very large magnetocrystalline anisotropy, or (more likely) that K_{eff} is dominated by other contributions such as surface and lattice strains (which have been neglected in many theoretical treatments). Finally, the observed dependence on grain size of H_c in these Fe granular systems can not be described by the simple relation of Eqn. (1), as there is no evidence for a size dependent K_{eff} or M_s.

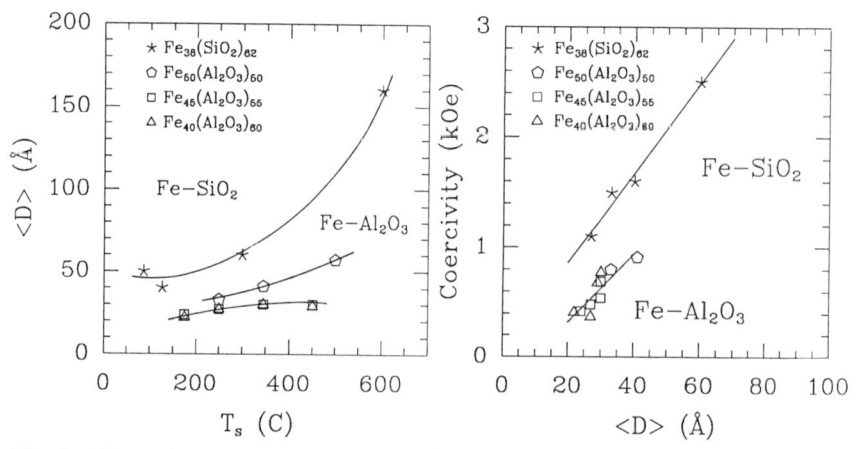

Fig. 3.- Observed mean grain size versus deposition substrate temperature for granular Fe–Al$_2$O$_3$ and Fe–SiO$_2$.

Fig. 4.- The variation of coercivity with mean grain size for granular Fe–Al$_2$O$_3$ and Fe–SiO$_2$.

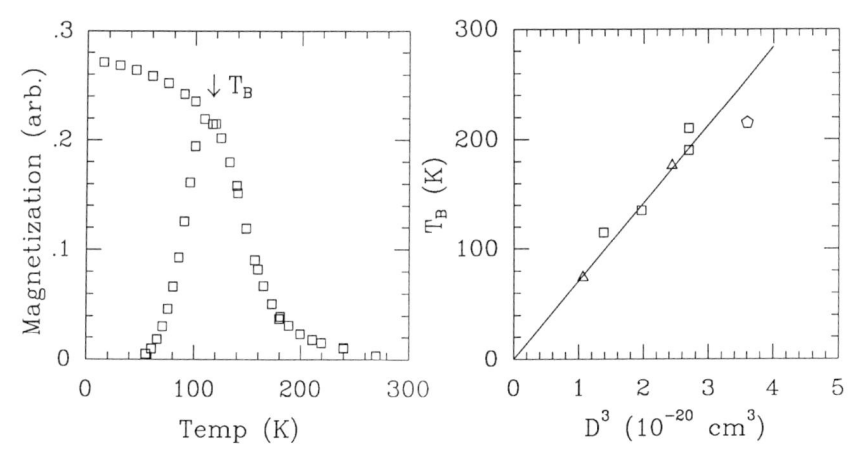

Fig. 5.- Typical magnetization curve for an assembly of single domain particles. The peak in the magnetization curve is taken to be T_B, the blocking temperature.

Fig. 6.- Blocking temperature T_B plotted against D^3. The solid line is corresponds (see text) to a K_{eff} of $\sim 3 \times 10^7$ erg/cm^3 for all samples.

CONCLUSIONS

We have studied the magnetic properties of rf sputtered Fe–Al$_2$O$_3$ granular metals and compared them to similarily prepared granular Fe–SiO$_2$. Both materials display an increase in coercivity with increasing grain size The grain size dependence of coercivity is thus a more general feature of granular Fe systems. The matrix has a pronounced effect on the size of the Fe granules which are grown. The grain sizes in granular Fe-Al$_2$O$_3$ are not as large as those obtained under identical conditions for Fe-SiO$_2$; in other respects the morphologies of the two systems are similar. The value for anisotropy energy density for Fe-Al$_2$O$_3$ is very large ($\sim 10^7$ erg/cm^3) and comparable to that for Fe-SiO$_2$.

REFERENCES

[*] Work supported by ONR Contract No. N00014-85-K-0175.

1. B. Abeles, P. Sheng, M.D. Couts, and Y. Arie, Adv. Phys. **24**, 407 (1975).
2. C.P. Bean and J.D. Livingston, J. Appl. Phys. **30**, 120S (1959).
3. L. Néel, Compt. rend. **228**, 664 (1949)
4. see e.g., A.H. Morrish, *The Physical Principles of Magnetism*, Krieger Publishing, Malabar, Florida (1983).
5. Gang Xiao and C.L. Chien, Appl. Phys. Lett. **51**, 1280 (1987).
6. S.H. Liou and C.L. Chien, Appl. Phys. Lett. **32**, 512, (1988).
7. E.C. Stoner and E.P. Wohlfarth, Phil. Trans. Roy. Soc. **A-240**, 599 (1948).
8. F.E. Luborsky, J. Appl. Phys. **29**, 309 (1958).

ON THE COERCIVITY OF GRANULAR Fe-SiO$_2$ FILMS*

S.H. LIOU[a], C.H. CHEN[b], H.S. CHEN[b], A.R. KORTAN[b] AND C.L. CHIEN[c]

(a) Department of Physics and Astronomy
University of Nebraska-Lincoln, Lincoln, NE 68588-0111

(b) AT&T Bell Laboratories, Murray Hill, NJ 07974

(c) Department of Physics and Astronomy
The Johns Hopkins University, Baltimore, MD 21218

ABSTRACT

The coercivity of granular Fe embedded inside an SiO$_2$ matrix was as high as 3 kOe at 6K, and 1.1 kOe at 300K. In this study, we observed a linear temperature (T) dependence of the coercivity for the samples prepared at a high substrate temperature (773K), and a $T^{1/2}$ dependence of the coercivity for the sample prepared at a low substrate temperature (473K). This indicates that the microstructures of films prepared at different substrate temperatures are not the same. This phenomenon can be explained if we assume that there are interconnections between particles for the sample prepared at a high substrate temperature. We looked for evidence of interconnections between particles with transmission electron microscopy (TEM).

INTRODUCTION

Granular metal films are composite materials consisting of ultrafine metal particles (with size 20 ∼ 150 Å) dispersed in an insulating medium. These artifical structured materials have been studied since the late 1960's by B. Abeles et al.[1]. Due to the large surface-to-volume ratio and the size of small particles, there are many unusual electronic and optical properties[2,3]. This type of film offers other attractive features which are relevant to applications. Because the insulating components are materials such as SiO$_2$ and Al$_2$O$_3$, granular metals are chemically stable and corrosion resistant. They are expected to achieve a high degree of thermal stability. By controlling the microstructure of the film (such as the size of particles, volume fraction of metal etc.), many desired properties can be tailored through processing conditions. Furthermore, since sputtering is the most effective method of making granular metal films, the fabrication, dispersion, and protection of the ultrafine granules, as well as coating desired surfaces suitable for device applications, can be achieved in a single process.

Recently, we observed that the coercivity (H$_c$) of the granular Fe embedded inside the SiO$_2$ matrix was as high as 3 kOe measured at 6K and 1.1 kOe measured at 300K, and the magnetization was 150 emu/g (which is about twice that of γ-Fe$_2$O$_3$ and Co-Cr). These characteristics compare very favorably with many existing media for magnetic recording [4].

In this paper, we will discuss the temperature dependence of the coercivity in granular Fe$_{75}$(SiO$_2$)$_{25}$ (42 volume % of Fe) films and their microstructures.

FILM PREPARATION AND CHARACTERIZATIONS

The granular Fe-(SiO$_2$) films were prepared by rf magnetron sputtering with a composite target. The target consists of homogeneous mixtures of pure Fe and SiO$_2$ powder (75 atomic % of Fe and 25 atomic % of SiO$_2$). Sputtering was performed in 4 mtorr of Argon gas. Films were deposited on glass substrates and carbon grids, the temperatures of which were varied from 77K to 875K. Typical film thickness was about 2μm for the magnetic study and 500 Å for the transmission electron microscopy (TEM) study. The magnetic properties were measured by a SQUID magnetometer with a field range of 0 - 50 kOe, and a temperature range of 2 - 400K.

RESULTS AND DISCUSSIONS

As shown in Fig. 1, H$_c$ measured at 6K increased from 1.5 kOe to about 3 kOe, as substrate temperature (T$_s$) during depositions increased from 300 to 775K. The value of H$_c$ measured at 300K was considerably reduced. In this region, the value of H$_c$ for the sample prepared at low T$_s$ was changing more rapidly with temperature than that at the high T$_s$. A maximum of H$_c$ measured at both 6 and 300K was achieved with T$_s$=775K. Further increasing T$_s$ beyond 775K caused the H$_c$ of samples measured at both 6K and 300K to decrease precipitously. In order to understand the mechanism of coercivity in Fe-SiO$_2$ granular films, we performed a coercivity versus temperature study. We found that the temperature dependence of the coercivity was not the same in films prepared at different substrate temperatures. The coercivity versus the temperature of films prepared at a low substrate temperature (T$_s$=473K) is shown in Fig. 2(a). H$_c$ decreased drastically as the temperature increased. This is because of thermal effects for the single domain particles.

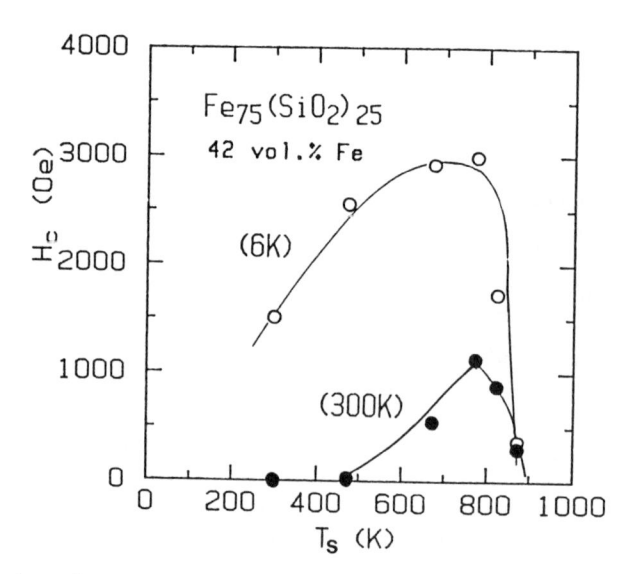

Fig.1 Variations of coercivity of Fe$_{75}$(SiO$_2$)$_{25}$ (42 volume % of Fe) films measured at 6K and 300K as a function of substrate temperature.

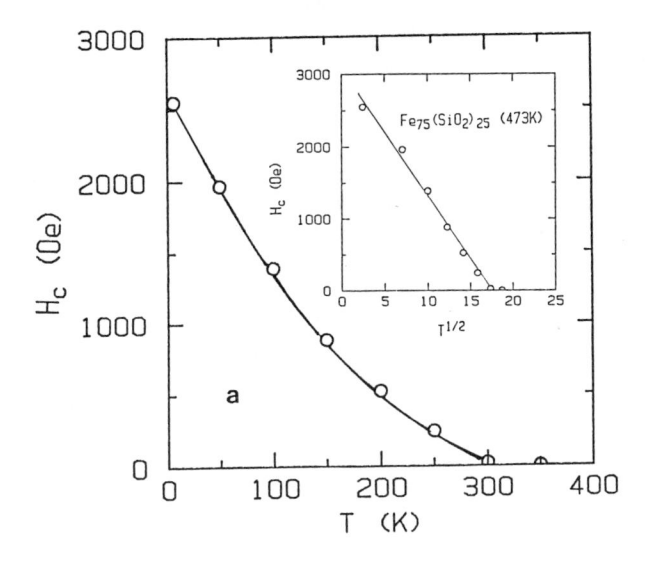

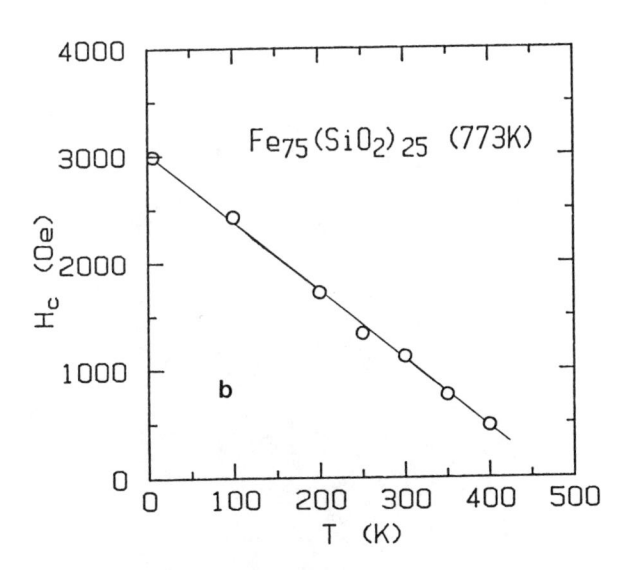

Fig.2 (a) The coercivity versus temperature of films prepared at low substrate temperature (T_s=473K). The insert is a plot of the coercivity versus $T^{1/2}$. (b) The coercivity versus the temperature of films prepared at high substrate temperature (T_s=773K).

According to Kneller and Luborsky [5] the temperature dependence of the coercivity for non-interacting, uniform, and single domain particles is

$$H_c = H_{ci}[1 - (\frac{T}{T_B})^{1/2}], \qquad (1)$$

where H_{ci} is the coercivity at 0 K and T_B is the blocking temperature, above which the superparamagnetism occurs. The coercivity of film at a temperature above T_B is zero.

The insert of Fig. 2(a) shows the plot of H_c as a function of $T^{1/2}$. The $T^{1/2}$ dependence as predicted by eq.(1) is clearly observed. However, for the sample prepared at a high substrate temperature ($T_s = 773K$), we observed a linear dependence of the coercivity which indicated that the particles in this film are no longer non-interacting uniform single domains. The linear dependence of coercivity has been observed in acicular γ-Fe_2O_3 particles [6]. In the system of γ-Fe_2O_3, the coercivity depends upon both shape and crystal field contributions which can be expressed as,

$$H_c = \alpha M + \beta(\frac{K_c}{M}) \qquad (2)$$

where α is the shape factor, β is a constant, K_c is crystal anisotropy, and M is saturated magnetization.

If we assume the same mechanism which also dominated in the Fe-SiO_2 granular films prepared at $T_s = 773K$, we should add the contribution of shape anisotropy to explain the linear behavior of the coercivity (i.e. there is a possibility of an interconnection of particles forming a chain-like behavior in the films). We looked for evidence of interconnections between particles with transmission electron microscopy (TEM).

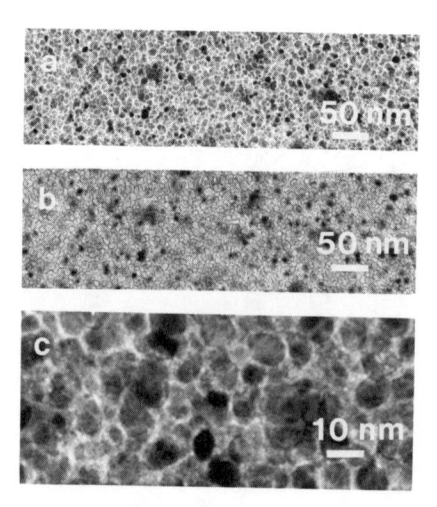

Fig.3 TEM micrographs of $Fe_{75}(SiO_2)_{25}$ (42 volume % of Fe), about 500 Å thick, sputtered at substrate temperatures of (a) 473K, (b) 473K, and (c) 773K. Where (a) is underfocaussed and (b) is overfocussed. They are from the same area.

As shown in Fig. 3(a) and 3(b), the sample deposited at $T_s = 473$K contains granules of about 45 Å. The particles are nearly equiaxial with small aspect ratios. Fig.3(a) and 3(b) are from the same area. The difference is that Fig 3(a) is underfocaussed and Fig 3(b) is overfocussed. We note the reverse of contrast of the thin layer that surround each grains. This suggests that the thin layer has a lower mass density than the grain. These grains are likely to be well separated. The size of the granules increases with increasing T_s. As shown in Figure 3(c), in the films deposited at $T_s = 773$K, the granule size was about 100 Å. The particles were still nearly equiaxial with small aspect ratios. The clustering of particles is seen in the micrograph. However, we are not sure that this interconnection of particles could lead to a chain-like structure in this granule film. The clusterings are even more pronounced in the film prepared with $T_s > 775$K. This massive interconnection between particles may cause the formation of a multidomain, thus reducing H_c measured both at 6K and 300K, as seen in Fig. 1.

Finally, we should mention some possible causes for the large coercivity measured at 6K. It is well known that for single-domain equiaxial Fe particles, H_c is of the order of $\beta K/M$, where K includes all contributions (magnetocrystalline, stress, surface, etc.) to the magnetic anisotropy. If one makes the overly simplistic assumption that magnetocrystalline anisotropy is the only contribution, and that its magnitude is the same as that of bulk Fe, one will conclude that H_c should be only about 600 Oe and independent of particle size [7]. It is clear that some contributions, particularly those due to surface and stress, may well be enhanced, because the Fe granules are not free-standing, but bonded in the SiO_2 matrix. Other possible causes for the enhanced H_c are particles with large aspect ratio, or particles forming very long chains. At present, however, we still do not have enough experimental evidence to support any of these possibilities. Further study is needed to understand the high H_c in Fe-SiO_2 films at low temperature.

In summary, we have studied the coercivity of granular Fe-SiO_2 films. We observed a linear T dependence of H_c for the sample prepared at T_s=773K, and a $T^{1/2}$ dependence of H_c at T_s=473K. The $T^{1/2}$ dependence of H_c is due to the existence of a uniform single domain in the sample prepared at low T_s. The linear behavior of H_c versus T could be explained if we assume that there is a chain-like connection of particles. The evidence of inter-connections of particles for a sample prepared with T_s=773K are revealed by TEM micrograph.

REFERENCES

* This work is partly supported by ONR Contract No:N00014-85-K-0175.

[1] B. Abeles, R.W. Cohen, and G.W. Cullen, Phys. Rev. Lett. **17**, 632 (1966).

[2] B. Abeles, P. Sheng, M.D. Couts, and Y. Arie, Adv. Phys. **24**, 407 (1975).

[3] For a review, see J. A. A. J. Perenboom, P. Wyder, and F. Meier, Phys. Rep. **78**, 173 (1981).

[4] S.H. Liou, and C.L. Chien; Appl. Phys. Lett. **52**, 512 (1988).

[5] E.F. Kneller, and F.E. Luborsky; J. Appl. Phys. **34**, 656 (1963).

[6] D.F. Eagle, and J.C. Mallinson; J. Appl. Phys. **38**, 995 (1967).

[7] I.S. Jacobs and C.P. Bean, in Magnetism III, edited by G.T. Rado and H. Suhl (Academic, New York, 1963), p. 275.

PART IV

Multilayers and Superlattices

CRITICAL PHENOMENA IN NANOSCALE MULTILAYER MATERIALS

T. TSAKALAKOS* and A. JANKOWSKI**
*Dept. of Mechanics and Material Science, Rutgers University, P.O. Box 909, Piscataway, N.J. 08854
**Dept. of Chemistry and Materials Science, Lawrence Livermore National Laboratory, P.O. Box 808, L-350, Livermore, CA 94550

ABSTRACT

We present a review of critical phenomena in a variety of nanoscale multilayer materials. Discontinuities of interdiffusitives at low temperatures, "the Supermodulus effect" and the "softening" of elastic constants of fcc-bcc structures, other mechanical properties at critical layer thickness are also presented. Some new calculations of elastic constants of (111) fcc and (110) bcc metallic superlattices show large changes as a function of homogeneous lattice strains. The effect of structural relaxations on a number of electronic and magnetic properties is discussed and a new analytical model is developed to explain the origin of interference effects when interfaces are apart by a few interplanar spacings. This model can predict relaxations at interfaces for a variety of materials including metals, semiconductors and ceramics.

INTRODUCTION

Recently, interfaces and surfaces have attracted a considerable interest among materials scientists in both theoretical and experimental studies [1,2] The discovery of the "supermodulus" effect [3] in fcc-fcc first, and the "softening" of the elastic constant C_{44} in fcc-bcc metallic superlattices [4,5,6,] later have generated an enormous research activity in the scientific community to explain and understand the role of interfaces in macroscopic properties such as mechanical, electronic, magnetic and thermodynamic [1,2,7,8,9].

This interest has been substantially increased by the ongoing effort to push the size of phase in multiphase materials down to nanoscale dimensions [10]. There are three basic concepts in dealing with interfaces in nanoscale dimensions. The first concept is related to the coherency of atoms across the interface. This concept has an enormous impact in our fundamental understanding of materials properties and results usually in elastic stresses within the various phases. The nature of the interface itself, on the other hand, as it relates to the smoothness and the flatness, the amount of disorder and interdiffusion is of fundamental interest to any interfacial property. Finally, the interference effects when interfaces are apart by a few interplanar spacings, have a profound change of the macroscopic and microscopic behavior of materials.

In this paper we attempt to review and explain some of the phenomena which occur at interfaces. We first review the structural properties and characterization techniques of metallic superlattices. We then furnish a comprehensive review of the elastic properties of these materials with emphasis on the "supermodulus effect" and the "softening" of the elastic constant C_{44}. We briefly review most of the existing theories to explain the origin of this anomalous elastic behavior and for the first time we show new results in [111] and [110] habit orientations of fcc and bcc metals based on pseudopotential energy calculations. Finally we develop for the first time a simple model based on microscopic theory which involves the phonon dispersion relations and the chemical Kazanki forces. This model predicts more realistically the structural relaxations and interference of relaxations and atomic redistribution as a function of interface proximity. An attempt is also made to explain a number of critical phenomena such as diffusivity discontinuities, magnetic phase transitions, elastic behavior, and d-spacing anomalous contractions and expansions in nanoscale multilayer materials. Although the results of this study are for multilayer materials some conclusions can be drawn for nanophase particulate or filamentary composites. Moreover, it is also expected that with proper modifications, the theory and conclusions of this study could be extrapolated to semiconductor-semiconductor or metallic-semiconductor interfaces.

EXPERIMENTALS

Most metallic modulated foils are produced by vapor deposition of two or three metals under high vacuum. The evaporants are interrupted by rotating wheel shutters or properly sequenced mechanical shutters above each crucible. The thickness of each layer and the rate of deposition are controlled by separate quartz-crystal thickness monitors or ionization gauges.

The most common substrate used in the deposition of metallic superlattices is mica for production of [111] texture. NaCl single crystals for [100] texture has been utilized together with a number of other substrates such as sapphire, etc.

In most cases the substrate is heated at a specified temperature for each alloy system. Sometimes it is necessary to depositic a metallic underlayer prior to the deposition of the modulated film. These processes are mainly used to produce film with good texture.

Structure of Multilayered Films

Structural studies by means of transmission electron microscopy in modulated films were performed by a number of investigators [11], [12], [13]. These disclosed the existence of two predominant defects: microtwins and double position twins both having a very small volume fraction. The study also revealed a strong texture, with grains oriented along the <111> direction. Nakahara et. al. [12] studies Cu-Ni modulated foils and observed misfit dislocations for longer wavelengths. For short wavelength modulated films both x-ray and TEM experiments indicated that coherent interfaces had been established. Based on these and other structural studies [12] three basic multilayered structures exist as shown in Fig. 1 (a,b,c):

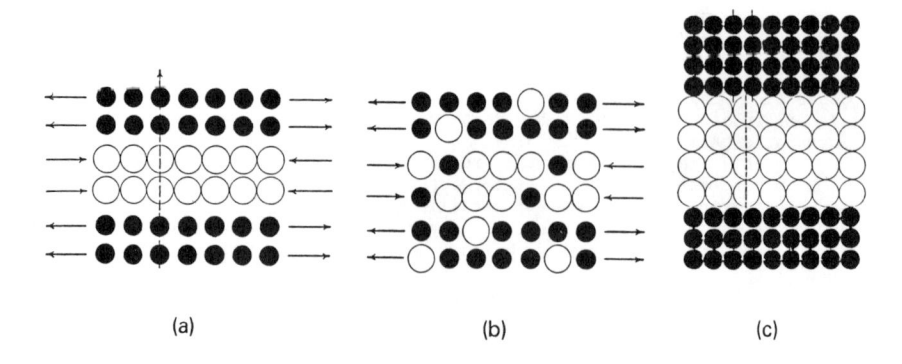

Fig. 1: Multilayered Structures:
(a) coherent superlattice, (b) composition modulated alloys,
(c) incoherent modulated structure.

(a) Superlattices (coherent); (b) Composition Modulated alloys in which coherency is maintained but interdiffusion has taken place during deposition or during subsequent annealing and, finally, c) Incoherent modulated structure in which misfit dislocations appear at the interface. The concept of coherency is extremely essential for nanoscale modulated films and is discussed in some detail below.

In particular, for metallic nanoscale multilayers deposited on mica substrate the grain structure [11] is shown in Fig. 2. In this figure it is shown that all grains are oriented along 111 and that the grains have a 60° symmetry orientation on the plane.

Characterization Model

The x-ray diffraction is routinely used to determine the wavelength, composition and strength of modulation. A standard diffractometer can be used to study the evolution of a dimensional modulation (through isothermal anneals) and quantitative information can be obtained by using a proper diffraction model for the analysis of the diffracted data.

For a sinusoidal composition fluctuation Guinier [14] developed a diffraction model which yields sharp Bragg peaks representative of the average lattice flanked by a pair of satellites situated at a distance $\pm 1/\lambda$.

Furthermore, according to the linear diffusion theory

$$\ln[I_+(t)/I_+(0)] = 2R(h)t,$$

where I(+) is the intensity as a function of time and thus a plot of $\ln[I(t)/I(0)]$ versus isothermal annealing time should be linear and the interdiffusion coefficient $\tilde{D}_B$ can be computed from the slope 2R(h) (amplification factor).

A correction is usually made for the atomic scattering at $s = 2 \sin \theta / \lambda_{x-ray}$, the Lorentz factor, the polarization factor, the absorption factor for a film, and the Debye-Waller factor [15].

It should be emphasized that this model is only valid for infinitisimal fluctuations and in cases that the differences of atomic scattering factors and the lattice parameters of the two atoms are relatively small. More rigorous models can be found in a review paper by McWhan [16] and Tsakalakos [17].

THE CONCEPT OF COHERENCY

The Concept of Coherency in Modulated Structures

A modulated structure ABAB... in a state of coherency is composed of alternate layers of A and B strained in such a way that the interatomic spacings in the planes normal to the direction of the modulation are the same in A and B. This is schematically shown in Fig. 3. The coherency strain energy has been calculated by Cahn [3] $E_c = \eta^2 Y[hkl]A^2$ for a square wave, and $E_c = \eta^2 Y[hkl]A^2/2$.

There is a possibility for the lattice to lose coherency as soon as the presence of dislocation grids at the interfaces between adjacent layers becomes more favorable energetically. In a unit volume of modulated foil the energy associated with such grids of dislocations is $E_s = 2\tau/\lambda$. Thus a total loss of coherency should be expected when

$$E_c \gtrless 2\tau/\lambda \qquad \text{or} \qquad \lambda = \frac{2\tau}{E_c}$$

Henein [18] has calculated τ for 50/50 Ag-Pd and predicted that loss of coherency for a square-like modulation will occur at wavelengths around 4.0 nm.

Philofsky and Hilliard [19] performed a very interesting experiment to determine the loss of coherency in the Cu-Pd. They measured the variation of the interdiffusivity D_λ as a function of wavelength which showed a large decrease between 2.8 nm and 3.8 nm. This decrease is due to the elimination from the driving force of the coherency strain energy contribution $2\eta^2 Y[111]$. Similar drop was also observed by Henein [18] in the Ag-Pd system.

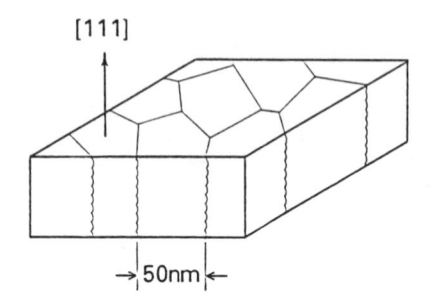

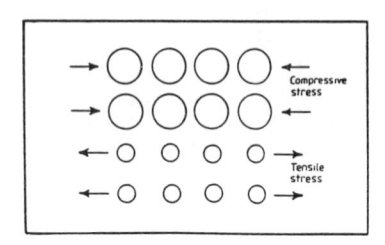

Fig. 2: Grain structure of metallic superlattice of [111] epitaxial growth.

Fig. 3: Coherency between atomic layers for an A/B composition modulated structure.

The influence that the coherency strains may have on the mechanical properties of the modulated foils could be easily understood by the presence of the enormous stresses which have to be developed to maintain the coherency. Actually half of the material is under large elastic stresses which cannot be produced by other macroscopic or microscopic phenomena.
The biaxial stresses are [20]:

$$\sigma_{11} = \sigma_{12} = \frac{1}{S_{11}' + S_{12}'} \varepsilon = Y[hkl]\varepsilon$$

where S_{11}' and S_{12}' are the elastic compliances referring to a system of coordinates where the x_3 is perpendicular to the interfaces and is the misfit strain

$$\varepsilon = \frac{1}{a}\frac{da}{dc}$$

In the case of epitaxial foils for which the x-axis (perpendicular to the foil) is parallel to the [111] or [100] cubic directions we obtain:

$$Y[111] = \frac{6\,C_{44}(C_{11}+2C_{12})}{C_{11}+2C_{11}+4C_{44}} \qquad Y[100] = C_{11}+C_{12}-2\frac{C_{12}^2}{C_{11}}$$

In general, under homogeneous strain, each initial lattice site $\underline{r}$ undergoes transformation into $\underline{r}'$ by a homogeneous distortion matrix $\underline{\underline{A}}$: $\underline{r}' = \underline{\underline{A}}\,\underline{r}$. Since the total crystal energy involves summation over the reciprocal lattice vectors, it is essential to establish the corresponding reciprocal lattice transformation. If q and q' are reciprocal lattice vectors before and after the deformation, it can be easily shown that: $\underline{q}' = \underline{B}\,\underline{q}$ where $\underline{B} = (\underline{A}^{-1})^T$. That is, the reciprocal lattice transformation corresponding to the crystal lattice rearrangement described by the matrix A is its transposed inverse. The distortion matrix corresponding to the biaxial state of stress is

$$A = \begin{bmatrix} 1+e_1 & 0 & 0 \\ & 1+e_2 & 0 \\ & & 1+e_3 \end{bmatrix}$$

where the principal strains are

$$e_1 = \varepsilon \qquad e_2 = \varepsilon \qquad e_3 = -2\frac{c_{13}}{c_{33}}\varepsilon$$

with e defined as the misfit strain.

The biaxial state of stress in layered structures transforms a cubic system into a tetragonal system. Thus, the six elastic constants are considered. The corresponding biaxial modulus for a tetragonal system is

$$Y[100] = C_{11} + C_{12} - \frac{2C_{13}^2}{C_{33}}$$

In another spectacular experiment, Yang [21] has determined the state of coherency of the nanoscale multilayered Au-Ni structure from the ratio of the low to high angle satellite intensity about the Bragg peak (Fig. 4). The transition of coherent to incoherent state occurs for bilayer thickness of around 2.5 nm. At the same time, the generalized interdiffusion coefficient versus layer thickness exhibits a discontinuity which is large enough to take the multilayered structure into the spinodal regime (Fig. 5). An additional explanation of this discontinuity has been proposed, based on screening singularities in the total crystal energy of the system [22]. These data

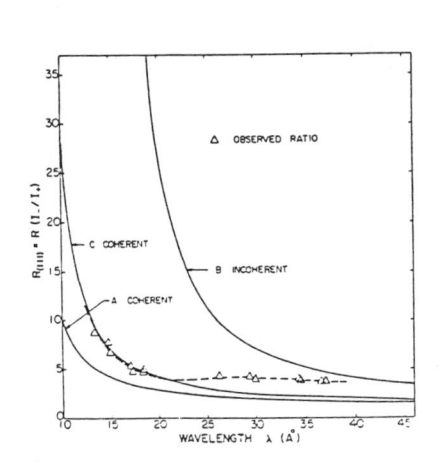

Fig. 4: The transition from coherent to incoherent modulation in Au-Ni. The observed satellite ratio of an Au-30% Ni alloy and the calculated ratios for coherent and incoherent modulations as function of λ (Yang [21]).

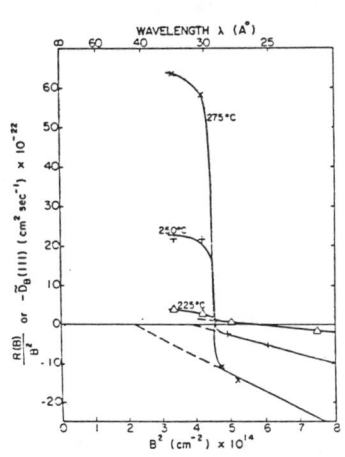

Fig. 5: Variations of the generalized interdiffusion coefficient $-\tilde{D}_B$ as a function of wavelength (upper scale) (Observed discontinuity at ~15-25 nm) Yang [21].

clearly indicate that the critical bilayer thickness for which a multilayered structure loses coherency is an extremely important materials parameter which dramatically changes the macroscopic behavior of this class of materials. Similar critical phenomena in diffusivities have been observed in Cu/Ni [20], Cu/NiFe [23] and Ag/Pd [18] systems.

CRITICAL PHENOMENA IN THE ELASTIC BEHAVIOR OF NANOSCALE MULTILAYERS

The elastic properties of one-dimensional nanoscale multilayer structures have been characterized with several experimental techniques. The elastic moduli (biaxial, Young's, flexural, shear, and torsional) of these thin films are known to be dependent upon their composition profile (amplitude, wavelength, and stoichiometry). In addition, the importance of a controlled and well-defined texture for meaningful modulus measurements is clearly demonstrated as shown in Figs. 6 and 7 [24,25].

The mechanical test to measure the biaxial modulus was performed on a bulge tester [24]. The bulge tester provides a biaxial stress load for thin foils. The deflection of a washer mounted foil due to applied gas pressure is used to calculate the biaxial stress and strain via elasticity theory. From the initial slope of the curves, the biaxial modulus is determined for each composition wave-length . The elastic non-linearity observed increases with increasing amplitude of the modulation. Similar results were noted for composition modulated Ag-Pd foils [25].

In order to qualitatively compare the measured elastic moduli for different wavelengths, the values are normalized to a selected modulation amplitude. A linear relationship between the biaxial modulus and square of the composition amplitude is shown experimentally as seen in Figure 10. Examples of the variation of biaxial elastic modulus Y[111] with wavelength are shown in Figures 1 and 2 for the cases of Cu-Ni and Ag-Pd, respectively. The enhanced modulus at wavelength ranges of 1.5 to 3.nm is clearly evidenced, with homogeneous solid solution values present for all other wavelengths.

The flexural modulus is determined from the flexural vibrations generated, as one end of a rectangular system is clamped and it is allowed to vibrate naturally. The vibrations may be generated and detected using electrostatic forces and a laser beam-position sensitive diode [13] or by acoustic drive and a capacitance method [12]. An example of the flexural modulus Y [111] observed for the Cu-Ni system [13] may be seen in Fig. 11. Enhancement is observed for the 1.5 to 3.nm wavelengths of modulation with a maximum at approximately 2.0 nm.

Other experiments using microtensile mechanical tester, vibration reed technique and acoustic waves Brillouin scattering experiments are summarized in Table 1 [27]. The table clearly demonstrates the existence of either "stiffening" or "softening" once the layer thicknesses aproach nanoscale (1-3 nm). It is also interesting to point out that most of the C_{44} softening observed by Falco and Schuller [4,5,6,27] are in dissimilar systems of fcc-bcc multilayers.

COHERENCY STRESSES AND ELASTIC MODULI OF NANOSCALE MULTILAYERS

The "supermodulus" effect is a most noteworthy characteristic in many composition modulated thin film systems. The effect has been typically observed in noble-transition binary metal systems with one-dimensional long range order (artificially created by finely controlled deposition techniques). The several hundred percent increase in the elastic moduli is attributable to a fundamental structural-bonding change. In the supermodulus systems only those modulated structures which have layer thicknesses of three to five atomic spacing generate large elastic strains, far beyond those

Fig. 6: Biaxial modulus Y[111] vs. the composition wavelength λ for the Cu/Ni composition modulated structure.

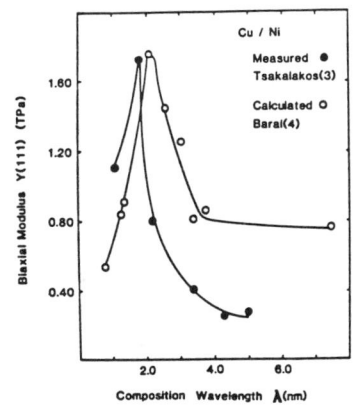

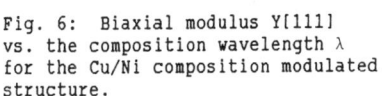

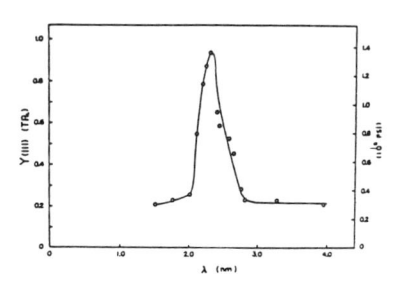

Fig. 7: The biaxial modulus Y[111] vs. the composition wavelength λ of the composition modulation in Ag-Pd foils. [25]

elastically achievable in conventional deformation processes, (by a biaxial mode of deformation) rather than dislocations in the interfaces. In smaller layer thicknesses interdiffusion during deposition results in a structure which is not atomically well defined and not capable of maintaining these large elastic strains. When the spacial periodicity becomes large enough, dislocation grids emerge at the layer interfaces. Therefore, the concept of interlayer coherency arises as a plausible explanation for the supermodulus effect. By displacing atoms from their equilibrium positions which they would occupy in a perfect, infinitely modulated structure, thereby "stiffening" the interatomic potential, one might be able to increase the elastic modulus. Theoretically the force equilibrium and competing effects between layers alternating under tension and compression in modulated structures has been considered [23]. A pseudopotential and third-order elastic constant calculation of the biaxial modulus has shown that the tetragonal lattice resulting from a biaxial distortion of a cubic structure leads to modulus enhancement for compressive strains and modulus reduction for tensial strains. The effect is not linear, however. That is, the change due to compression exceeds that due to tension. In a structure composed of an equal number of alternating layers under compression and tension, the overall effect should be influenced by compressive stresses. Correlation of the pseudopotential calculations with the Cu-Ni system [20] have predicted the correct percentage of modulus enhancement for the degree of compressive strain present in the Cu layers. An exact model of the layered structure system incorporating stress equilibrium between layers is yet to be formulated. (Stress equilibrium may not be a prerequisite for the artificially ordered structures.) Cross-sectional transmission electron microscopy on Cu-Ni composition modulated foils suggest that the enhanced modulus is a feature of coherent interfaces. Misfit dislocations were absent for only the wavelength range of foils associated with enhanced modulus [12].

Recent calculations [30] based on pseudopotential approach have demonstrated an excellent agreement between experimental and theoretical calculations of elastic constants and their pressure derivations as shown in Table II. Fig. 9 shows the elastic constants of Cu as a function of biaxial homogeneous strain on the 111 plane. The effect of compressive elastic strain of the biaxial Y[111] modulus is shown in Fig. 10. Similar calculations of Nb bcc metal shows the elastic constants change less

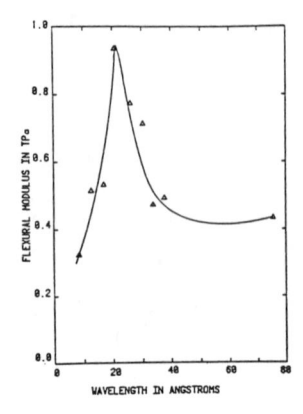

Fig. 8: Variation of the flexural modulus with the wavelength of modulation for as-deposited Cu-Ni foils of constant composition (~50 at%Cu) and [111] texture (Boral 1983)[13].

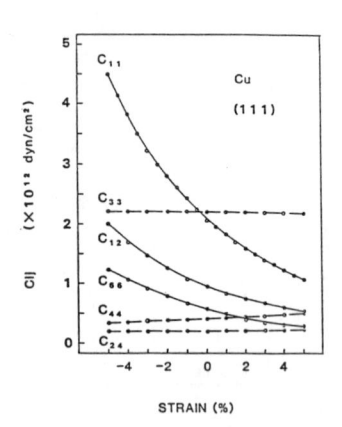

Fig. 9 : The elastic constants of Cu under a homogeneous deformation on the (111) plane.

drastically with compressive stresses, Fig. 11. From Figs. 9 and 11 one can show that in the case of Cu/Nb, a softening is expected.

It is thus concluded that the "stiffening" or "softening" of elastic constants depends on the specific system and the specific habit plane orientation. For fcc/fcc metals of a [111]//[111] orientation, a "stiffening" is expected while for the [111]//[110] fcc/bcc couples a "softening" of about 30%-40% is predicted, which is in excellent agreement with the experiments of Falco and Schuller shown in Table I.

It is also interesting to point out that the homogeneous lattice distortions have a more pronounced effect in fcc then bcc materials and that the elastic anisotropy increases substantially upon the lattice misfit strain.

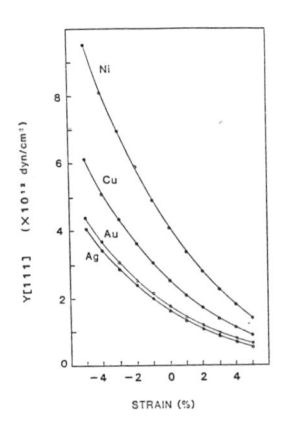

Fig. 10: Calculations of the biaxial modulus Y(111) as a function of applied strain on the (111) plane.

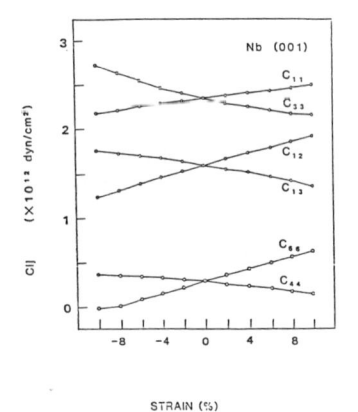

Fig. 11: The elastic constant of bcc Nb as a function of strain on the (001) plane.

TABLE I: MECHANICAL PROPERTIES OF ARTIFICIAL METALLIC SUPERLATTICES

SYSTEM	CONSTANT	RESULT	MEASUREMENT METHOD	REFERENCE
Au/Ni	Y_B	E(~225%)	BT	27
Cu/Pd	Y_B	E(~470%)	BT	27
Cu/Ni	F	E(~225%)	CR	27
	Y	E(~170%)	TT	27
	T	E(~170%)	CR	27
	Y	E(~200%)	CR	27
	T	E(~180%)	CR	27
	Y_B	E(~200%)	BT	27
	Y	N	CR	27
	BS	E(~175%)	TT	27
	YS	E(~300%)	TT	27
Ag/Pd	Y_B	E(~230%)	BT	27
Cu/Au	Y_B	N		27
Cu/Al	BS			27
Cu/NiFe	Y	E 200-300%	PT	23
CuFe/NiFe	Y	E ~200-300%	PT	23
Cu/Ni/Fe	Y	E 200-300%	PT	23
Nb/Cu	C_{44}	S(~35%)	BS	27
Mo/Ni	C_{44}	S(~40%)	BS	27
V/Ni	C_{44}	S(~40%)	BS	27
Mo/Ta	C_{44}	S(~40%)	BS	27
Au/Cr	C_{44}	E(~35%)	BS	28
WC/Co	Y	E(~200%)	CR	29

KEY: Y_B: Biaxial Elastic Modulus, F: Flexural modulus
Y: Young's Modulus, BS: Breaking stress
YS: Yield strength,
BT: Bulge test, CR: Cantilever Reed
BS: Brillouin Scattering, TT: Tensile test
PT: Piezotron tensile test

TABLE II: PRESSURE EFFECT ON ELASTIC CONSTANTS OF Cu

	Experimental	Present Calculations
C_{11} (10^{12} dynes/cm^2)	1.6570	1.6297
C_{12} "	1.1100	1.0920
C_{44} "	0.7346	0.7226
dC_{44}/dP	2.63	3.10
$d/dP[1/2(C_{44}-C_{12})]$	0.58	0.95
$d/dP[1/3(C_{44}+2C_{12})]$	5.8	5.42

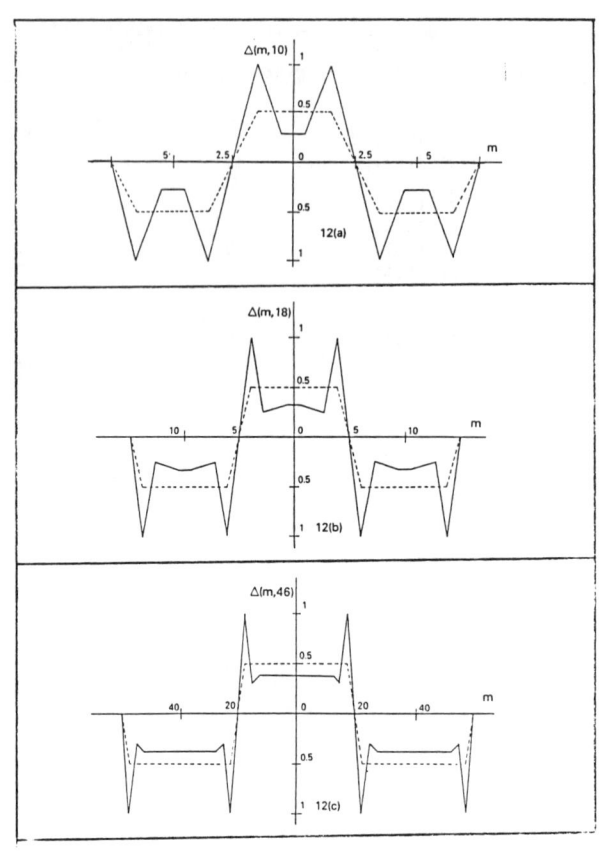

Fig. 12 (a) the percentage d-spacing change in m-th layer for a $5d_A$-$5d_B$ multilayer (solid line) $\Delta(m,10)$. The broken line corresponds to continuum elasticity model and the first-neighbor interaction microscopic (they are identical); (b) $\Delta(m,18)$; (c) $\Delta(m,46)$.

INTERFACIAL PHENOMENA

The coherency is an extremely important concept in materials science and certainly plays a central role in interfacial phenomena. The "supermodulus effect" in nanoscale multilayers, the strengthening mechanism in precipitation hardening alloys, and the optoelectronic properties of semiconducting materials represent some examples of the wide range of applicability of the coherency concept. However, when dimensions reach nanoscale, the continuous elasticity solution for the coherency stress given above fails to address the fundamental different situation.

In such dimensions the interface-interface interactions, the distortions of the binding energies of atoms at the interface, and the local disorder due to interdiffusion are factors which should be taken into account promptly. In fact, a number of interfacial phenomena cannot be understood on the basis of coherency. The loss of the supermodulus effect near the monolayer superlattice scale, the discrepancy of the critical bilayer thickness between experiment and theory and a number of experimental data of anomalous interplanar expansions and contractions near interfaces [28], are examples directly related to the nanoscale dimensionality.

Obviously, any adequate physical and mathematical treatment of this problem should be made in terms of the changes of binding energies near the interface. Similar models have been used on theoretical modeling of free surfaces using Monte Carlo, Molecular Dynamics, Embedded Atom Method, or sometimes First Principle coherent potential approximation or tight binding energy calculations [1,2,7,8,9].

However, in the present study we propose to use a modification of Khachaturyan's microscopic theory of elasticity [31] which takes into account the discrete nature of the crystal lattice. This approach provides a simple and elegant analytical solution to the problem without the use of the tedious and elaborate computer simulation analysis.

The case of elastic displacement field generated by solute atoms near the interfaces becomes substantially more important when the layer thicknesses are of the same order of magnitude as the interatomic distances. The model should be viewed as a harmonic approximation of the homogeneous force constant case. Details will be published by Tsakalakos and Khachaturyan elsewhere [32], and only the major points are below:

The elastic displacement field $\underline{u}(\underline{r})$ which is generated by a concentration wave $\Delta c(\underline{r})$ is given by

$$\underline{u}(\underline{r}) = \frac{1}{N} \sum_{\underline{k}} G_{ij}(\underline{k}) \, F_j(\underline{k}) \, \Delta c(\underline{k}) \, e^{i\underline{k}\underline{r}} \hat{e}_i \,. \tag{1}$$

where $\Delta c(\underline{k}) = \sum_{\underline{r}} \Delta c(\underline{r}) \, e^{i\underline{k}\underline{r}}$ is the Fourier transform of the concentration wave and the summation is over N points in the first Brillouin zone of the average lattice. $\underline{F}(\underline{k})$ is the Fourier Transform of the so-called Kanzanki [32] forces that describe the ability of a solute atom to distort the environmental host lattice. $G_{ij}(\underline{k})$ is the Green's tensor of the inverse of the Fourier Transform of the dynamic Born-von Karman matrix [32] which is the fundamental characteristic of the dynamic properties of a crystal because it determines the vibration frequency spectrum. In general, the Hermittian tensor $G_{ij}(\underline{k})$ can be written in terms of the eigenvectors, $\hat{e}_s(\underline{k})$, and eigenvalues, $m\omega_s^2(\underline{k})$ of the dynamic matrix, where m is the mass of the host atom, ω_s is the vibration frequency of the branch s(s=1,2,3)

$$G_{ij}(\underline{k}) = \sum_{s=1}^{3} \frac{e_s^i(\underline{k}) e_s^{*i}(\underline{k})}{m\omega_s^2(\underline{k})}$$

Thus, the calculation of the displacement field requires the knowledge of the phonon dispersion curves. On the other hand, it is possible to evaluate approximately the Green's tensor from the Born-von Karman constants. Expressions of the dynamic matrix up to eighth coordination shell are given in reference [33]. The determination of the Kazanki forces is the most difficult problem. Expressions are also given in reference [33].

Application of this model in a square superlattice (square concentration wave) is considered in the following: A[001]//[001] A/B square superlattice of fcc crystal structure can be represented by a static concentration wave

$$\Delta c(r) = \begin{array}{l} 1 - \frac{L}{2} \bar{d} \leq z \leq \frac{L}{2} \bar{d} \\ 0 - \frac{M}{2} \bar{d} \leq z < (-\frac{L}{2} - 1)\bar{d} \text{ and } (\frac{L}{2} + 1)\bar{d} < z < \frac{M}{2} \bar{d} \end{array}$$

where $\bar{d}$ is the average interplanar spacing, L+1 planes in layer A and M is the total number of planes in the superlattice period. It can be shown that:

$$\Delta c(\underline{k}) = \sin \frac{\pi(L+1)}{M} / \sin \frac{\pi}{M}$$

$$G_{33}(k) = \{8\alpha_1[1-\cos\frac{\pi}{M}]+4\alpha_2\sin^2\frac{2\pi}{M}+4\alpha_3[1-\cos\frac{2\pi}{M}]+4\alpha_3[1-\cos\frac{2\pi}{M}]\}^{-1}$$

$$F_3(k) = 8\hat{\alpha}_1(1-\cos\frac{\pi}{M})+2\hat{\alpha}_2(1-\cos\frac{2\pi}{M})+8\hat{\alpha}_3(1-\cos\frac{2\pi}{M})+16\hat{\beta}_3(1-\cos\frac{\pi}{M})$$

where $4\alpha_1 + 4\alpha_2 + 16\alpha_3 + 8\beta_3 = \alpha C_{11}$(long wavelength approximation), $\alpha_1, \alpha_2, \alpha_3, \beta_3$ are the Born-von Karman constants up to the 3rd coordination shell, and $2\hat{\alpha}_1 + 2\hat{\beta}_1 + 4\hat{\beta}_2 + 4\hat{\alpha}_3 + 20\hat{\beta}_3 = \alpha C_{44}$(long wavelength approximation) $\hat{\alpha}_1, \hat{\beta}_1, \hat{\beta}_2, \hat{\alpha}_3, \hat{\beta}_3$ are the Kazanki forces up to third neighbor approximation.

Inserting these expressions in Eq. (1), the displacement field is calculated. By defining the percentage change in the mth layer of a M-layer period supperlattice as:

$$\Delta(m,M) = 100\%[u(m)-u(m+1)]/\bar{d}$$

We can finally demonstrate the effect of interfaces on interlayer relaxations in Figs. 12(a),(b),(c). Fig. 12(a) is a plot of the $\Delta(m,10)$ (for 5 planes of A and 5 planes of B multilayer) where the square wave (curve A) shows the macroscopic coherency strain as calculated by continuous elasticity and curve B is the 3-neighbor approximation of the microscopic approach. Similarly Figs. 12(b) and (12(c) represent the interplanar relaxations for $9d_A-9d_B$ and $23d_A-23d_B$ superlattices of (001) layers.

The calculations are made for an Au-Ni superlattice for which approximate values of the constants are known. The most striking feature of these results is the large Δ' near the interface. The 200% increase of the strain near the interface is indicative of the strong long range interactions in this system (Fig. 12(b) and Fig. 12(c)). On the other hand, the 300% increase of the structural parameter Δ (Fig. 12(a)) can be understood as an interference of the relaxations from each of the neighboring interfaces. Similar behavior has been observed experimentally and predicted theoretically by Chen, Voter and Albers in very thin slabs of Al and NiAl (110) and (210) surfaces.

In the case of the Au-Ni multilayer, Imafuku et. al. [35] have used a molecular dynamics model to explain the origin of the supermodulus effect. It is interesting to point out that their model predicts structural relaxation $\Delta(m,M)$ as described above, which are used to show the elastic modulus enhancement. Such relaxations cannot be understood on the basis of coherency itself. It should be also pointed out that the first neighbor approximation of the microscopic approach coincides with the macroscopic coherency strain shown in Figs. 12(a),(b),(c).

The non-uniform contractions and expansions in each layer near interfaces is of utmost importance of a variety of mechanical, electronic, magnetic behavior of layered structures. The recent experiments by Bizanti et. al. [28] on Au/Cr superlattices have shown a dramatic change in the lattice of Au and Cr perpendicular to the superlattice plane for thickness less than 5 nm. Structural studies proved the epitaxial growth (100) plane of Cr layers and the (100) of the Au layer rotated by 45°. This habit plane orientation predicts a perfect lattice matching i.e. $a_0(Au)/\sqrt{2}=2.8838$ Å and $a_0(Cr)= 2.8839$ Å. However, an elaborate x-ray diffraction d-spacing determination showed that the Cr interplanar spacing was stretched by 8% along [001] direction, while Au contracted by 2% along the growth direction. Given the excellent agreement of lattice matching between Au and Cr, these changes cannot be attributed to bulk lattice misfit strain at the interfaces. At the same time, a 30 pct increase of the elastic constants C_{44} for the layer thickness that these d-spacing anomalies occur.

Preliminary calculations have shown that these unusual contractions and expansions can be explained by means of the Kazanki forces and the enhancement of the C_{44} is consistent with the elastic constant calculation given above. Details of these calculations will be given in a paper to be published [30].

Finally, electronic and transport properties which depend on local environment are expected to be sensitive to these interface structural relaxations. Preliminary work on NiAl epitaxial growth on Ga-As indicates a strong dependence of the growth and critical thickness characteristics on the above interfacial relaxations. Perhaps the most important example of this dependence is the magnetic behavior of Fe layers on Cu. In a spectacular experiment [36] using Mossbauer spectroscopy, it was clearly demonstrated that α-Fe transforms into γ-Fe near the interface due to the large d-spacing expansions and contractions. From the paramagnetic component (central peak) and the loss of the ferromagnetic α-Fe behavior (sextet) in the Mossbauer spectrum, it is estimated that γ-Fe is present only for the first 3-interplanar spacings of the Fe layer and the remaining interior layer remains α-Fe ferromagnetic. This is in excellent agreement with interference of relaxations shown in Fig. 12(a).

In summary, we have shown the importance of interfacial phenomena on electronic and transport properties which depend on local environment. The coherency of multilayers and its dramatic effect on macroscopic properties was demonstrated through thermodynamic, interdiffusion and elastic properties. It was shown that our fundamental understanding of elastic behavior on the basis of macroscopic elasticity is insufficient when interfaces approach each other to within interplanar distances. The stiffening or softening of elastic constants can be explained by means of contractions and expansions of the interplanar spacings. These spacing changes are not only strictly related to the lattice coherency across the interface, but also to structural relaxations because of changes of the binding energies of atoms at the interface. These structural relaxations were particularly pronounced when interface distances reached a few (3-5) interplanar spacings. A number of critical phenomena could be explained by these relaxations including interface strain induced magnetic phase transitions (such as α-Fe to γ-Fe and the development of magnetic moment in Pd layers alternating with Au layers) [37]. The d-spacing anomalies of Au-Cr layers of nanoscale dimensions could also be explained on the basis of the interference of relaxations near the interface of superlattice layers. It is also expected that a number of optoelectronic properties, in systems containing semiconductor-semiconductor or metal-semiconductor interfaces, may be sensitive to the large structural relaxations and atomic distributions proposed herewith. These effects might have an enormous impact on our understanding and subsequently on the development of our electronic and structural materials for future applications.

ACKNOWLEDGEMENT

This work was performed under the auspices of the United States Department of Energy by Lawrence Livermore National Lab. under contract#W-7405-ENG-48.
The authors wish to thank Professor A. Khachaturyan for numerous discussions and Mr. G. C. Joo, Mr. J.W. Lee and Ms. K. Finnerty for assistance in preparing the manuscript.

REFERENCES

1. "The Structure of Surfaces," ed. by M.A. Hove and S.Y. Tong (1985), Spring Series in Surface Science, Vol. 2 (Spring, Berlin).
2. "Interfaces, Superlattices, and Thin Films," ed. by J.D. Dow, I.K. Schuller and J.E. Hilliard, Mat. Res. Soc. Proc., Vol 77, (1987) Mat. Res. Soc., Pittsburgh, PA.
3. W.M. C. Yang, T. Tsakalakos and J.E. Hilliard, J. Appl. Phys. 48 (1977) 876.
4. A. Kueny, M. Grimsditch, K. Miyano, I. Banerjee, C.M. Falco, and I.K. Schuller, Phys. Rev. Lett. 48, 166 (1982).

5. M.R. Khan, C.S.L. Chun, G.P. Felcher, M. Grimsditch, A. Kueny, C.M. Falco, and I.K. Schuller, Phys. Rev., B27, 7186 (1983).
6. R. Danner, R.P. Huebener, C.S.-L. Chun, M. Grimsditch, and I.K. Schuller, Phys. Rev., B33, 3696.
7. "Modulated Structure Materials," ed. by T. Tsakalakos, Martinus Nijoff, Dordrecht, 1984.
8. "Synthetic Modulated Structures," ed. by L. Chang and B. Giessen, Academic Press, New York, 1985.
9. "Layered Structures and Epitaxy," Mat. Res. Soc., Symposium Proc. ed. J.M. Gibson, G.C. Osbourn, and R.M. Tromp, Materials Research Society, Pittsburgh, 1986.
10. Papers in this volume.
11. A. Purdes, Ph.D. Thesis, Northwestern University, 1976.
12. L.R. Testardi, R.H. Willens, J.T. Krause, D.B. McWhan and S. Nakahara, J. Appl. Phys. 52 (1981) 510.
13. D. Baral, Ph.D. Thesis, Northwestern University, Illinois (1983).
14. A. Guinier, X-ray Diffraction in Crystals, Imperfect Crystals and Amorphous Bodies, translated by P. Lorrain and D. Sainte-Mari Lorrain (Freeman, San Francisco, 1963).
15. International Tables for X-ray Crystallography (Kynoch, Birmingham 1905).
16. D.B. McWhan, in Synthetic Modulated Structures, ed. by L. Chang and B.C. Giessen (Academic, New York, 1985).
17. T. Tsakalakos, to be published.
18. G.E. Henein, Ph.D. Thesis, Northwestern University, 1979.
19. E.M. Philofsky and J.E. Hilliard, J. Appl. Phys., 43, 2040 (1972).
20. T. Tsakalakos, Ph.D. Thesis, Northwestern University, Illinois (1977).
21. W.M.C. Yang, Ph.D. Thesis, Northwestern University, Evanston, IL, 1970.
22. T. Tsakalakos, Thin Solid Films 86 (1981b) 79.
23. A. Jankowski, Ph.D. Thesis, Rutgers University, New Jersey (1984).
24. T. Tsakalakos, Thin Solid Films 75 (1981a) 293.
25. G.E. Henein and J.E. Hilliard, J. Appl. Phys. 54 (1983) 728.
26. T. Tsakalakos and J.E. Hilliard, J. Appl. Phys. 54 (1983) 734.
27. I.K. Schuller, Institute of Electrical and Electronic Engineers, 1985, Ultrasonics Symosium, ed. by B.R. McAvoy (IEEE, New York, 1970).
28. P. Bisanti, M.B. Brodsky, G.P. Felcher, M. Grimsditch and L.R. Sill, Phys. Rev. B35. 7813-7819 (1987).
29. T. Moustakas, unpublished work.
30. T. Tsakalakos and A. Jankowski, to be published.
31. A.G. Khachaturyan, Fiz. Tuerd. Tela, 9, 2595 (1967), Sov. Phys. Solid State (Engl. Transl.) 9, 2040 (1968).
32. T. Tsakalakos and A.G. Khachaturyan, to be published.
33. "Theory of Structural Transformations in Solids," A.G. Khachaturyan, John Wiley & Sons, New York, p. 464-420, 1983.
34. S.P. Chen, A.F. Voter and R.C. Albers, submitted to Physical Review Letters, 1988.
35. M. Inafuku, Y. Susajima, R. Yamamoto and M. Doyama, J. Phys. F: Met. Phys. 823-829 (1986).
36. A. Kosticas, A. Simopoulos and T. Tsakalakos, unpublished work.
37. M.B. Brodsky, private communication.

environment are expected to be sensitive to these interface structural relaxations. Preliminary work on NiAl epitaxial growth on Ga-As indicates a strong dependence of the growth and critical thickness characteristics on the above interfacial relaxations. Perhaps the most important example of this dependence is the magnetic behavior of Fe layers on Cu. In a spectacular experiment [36] using Mossbauer spectroscopy, it was clearly demonstrated that α-Fe transforms into γ-Fe near the interface due to the large d-spacing expansions and contractions. From the paramagnetic component (central peak) and the loss of the ferromagnetic α-Fe behavior (sextet) in the Mossbauer spectrum, it is estimated that γ-Fe is present only for the first 3-interplanar spacings of the Fe layer and the remaining interior layer remains α-Fe ferromagnetic. This is in excellent agreement with interference of relaxations shown in Fig. 12(a).

In summary, we have shown the importance of interfacial phenomena on electronic and transport properties which depend on local environment. The coherency of multilayers and its dramatic effect on macroscopic properties was demonstrated through thermodynamic, interdiffusion and elastic properties. It was shown that our fundamental understanding of elastic behavior on the basis of macroscopic elasticity is insufficient when interfaces approach each other to within interplanar distances. The stiffening or softening of elastic constants can be explained by means of contractions and expansions of the interplanar spacings. These spacing changes are not only strictly related to the lattice coherency across the interface, but also to structural relaxations because of changes of the binding energies of atoms at the interface. These structural relaxations were particularly pronounced when interface distances reached a few (3-5) interplanar spacings. A number of critical phenomena could be explained by these relaxations including interface strain induced magnetic phase transitions (such as α-Fe to γ-Fe and the development of magnetic moment in Pd layers alternating with Au layers) [37]. The d-spacing anomalies of Au-Cr layers of nanoscale dimensions could also be explained on the basis of the interference of relaxations near the interface of superlattice layers. It is also expected that a number of optoelectronic properties, in systems containing semiconductor-semiconductor or metal-semiconductor interfaces, may be sensitive to the large structural relaxations and atomic distributions proposed herewith. These effects might have an enormous impact on our understanding and subsequently on the development of our electronic and structural materials for future applications.

ACKNOWLEDGEMENT

This work was performed under the auspices of the United States Department of Energy by Lawrence Livermore National Lab. under contract#W-7405-ENG-48. The authors wish to thank Professor A. Khachaturyan for numerous discussions and Mr. G. C. Joo, Mr. J.W. Lee and Ms. K. Finnerty for assistance in preparing the manuscript.

REFERENCES

1. "The Structure of Surfaces," ed. by M.A. Hove and S.Y. Tong (1985), Spring Series in Surface Science, Vol. 2 (Spring, Berlin).
2. "Interfaces, Superlattices, and Thin Films," ed. by J.D. Dow, I.K. Schuller and J.E. Hilliard, Mat. Res. Soc. Proc., Vol 77, (1987) Mat. Res. Soc., Pittsburgh, PA.
3. W.M. C. Yang, T. Tsakalakos and J.E. Hilliard, J. Appl. Phys. 48 (1977) 876.
4. A. Kueny, M. Grimsditch, K. Miyano, I. Banerjee, C.M. Falco, and I.K. Schuller, Phys. Rev. Lett. 48, 166 (1982).

5. M.R. Khan, C.S.L. Chun, G.P. Felcher, M. Grimsditch, A. Kueny, C.M. Falco, and I.K. Schuller, Phys. Rev., B27, 7186 (1983).
6. R. Danner, R.P. Huebener, C.S.-L. Chun, M. Grimsditch, and I.K. Schuller, Phys. Rev., B33, 3696.
7. "Modulated Structure Materials," ed. by T. Tsakalakos, Martinus Nijoff, Dordrecht, 1984.
8. "Synthetic Modulated Structures," ed. by L. Chang and B. Giessen, Academic Press, New York, 1985.
9. "Layered Structures and Epitaxy," Mat. Res. Soc., Symposium Proc. ed. J.M. Gibson, G.C. Osbourn, and R.M. Tromp, Materials Research Society, Pittsburgh, 1986.
10. Papers in this volume.
11. A. Purdes, Ph.D. Thesis, Northwestern University, 1976.
12. L.R. Testardi, R.H. Willens, J.T. Krause, D.B. McWhan and S. Nakahara, J. Appl. Phys. 52 (1981) 510.
13. D. Baral, Ph.D. Thesis, Northwestern University, Illinois (1983).
14. A. Guinier, X-ray Diffraction in Crystals, Imperfect Crystals and Amorphous Bodies, translated by P. Lorrain and D. Sainte-Mari Lorrain (Freeman, San Francisco, 1963).
15. International Tables for X-ray Crystallography (Kynoch, Birmingham 1905).
16. D.B. McWhan, in *Synthetic Modulated Structures*, ed. by L. Chang and B.C. Giessen (Academic, New York, 1985).
17. T. Tsakalakos, to be published.
18. G.E. Henein, Ph.D. Thesis, Northwestern University, 1979.
19. E.M. Philofsky and J.E. Hilliard, J. Appl. Phys., 43, 2040 (1972).
20. T. Tsakalakos, Ph.D. Thesis, Northwestern University, Illinois (1977).
21. W.M.C. Yang, Ph.D. Thesis, Northwestern University, Evanston, IL, 1970.
22. T. Tsakalakos, Thin Solid Films 86 (1981b) 79.
23. A. Jankowski, Ph.D. Thesis, Rutgers University, New Jersey (1984).
24. T. Tsakalakos, Thin Solid Films 75 (1981a) 293.
25. G.E. Henein and J.E. Hilliard, J. Appl. Phys. 54 (1983) 728.
26. T. Tsakalakos and J.E. Hilliard, J. Appl. Phys. 54 (1983) 734.
27. I.K. Schuller, Institute of Electrical and Electronic Engineers, 1985, Ultrasonics Symosium, ed. by B.R. McAvoy (IEEE, New York, 1970).
28. P. Bisanti, M.B. Brodsky, G.P. Felcher, M. Grimsditch and L.R. Sill, Phys. Rev. B35. 7813-7819 (1987).
29. T. Moustakas, unpublished work.
30. T. Tsakalakos and A. Jankowski, to be published.
31. A.G. Khachaturyan, Fiz. Tuerd. Tela, 9, 2595 (1967), Sov. Phys. Solid State (Engl. Transl.) 9, 2040 (1968).
32. T. Tsakalakos and A.G. Khachaturyan, to be published.
33. "Theory of Structural Transformations in Solids," A.G. Khachaturyan, John Wiley & Sons, New York, p. 464-420, 1983.
34. S.P. Chen, A.F. Voter and R.C. Albers, submitted to Physical Review Letters, 1988.
35. M. Inafuku, Y. Susajima, R. Yamamoto and M. Doyama, J. Phys. F: Met. Phys. 823-829 (1986).
36. A. Kosticas, A. Simopoulos and T. Tsakalakos, unpublished work.
37. M.B. Brodsky, private communication.

MICROSTRUCTURE AND MAGNETIC BEHAVIOR
OF Fe/Ti MULTILAYERED FILMS

SATOSHI ONO, MICHIO NITTA AND MASAHIKO NAOE
Dept. of Physical Electronics, Tokyo Institute of Technology,
2-12-1 O-okayama, Meguro-ku, Tokyo 152, Japan

ABSTRACT

Fe/Ti multilayered films composed of Fe and Ti layers with various thickness (d_{Fe} of 10~1000 A and d_{Ti} of 10~200 A) have been prepared at Ar gas pressure of 2 mTorr by two pairs of Facing Targets Sputtering apparatus which can deposit very thin and continuous films on plasma-free substrates. The total thickness of Fe layers was 1000 A in all of the specimen films.
  The periodic microstructure due to compositional modulation was clearly observed not only in the low angle region of X-ray diffraction diagrams but also in the Auger electron spectroscopic depth profile.
  With decreasing d_{Fe} from 1000 to 20 A, the diffraction intensity of (110) plane in the Fe bcc phase lowered for d_{Ti} of 10~200 A and its interplanar spacing (d-spacing) increased for d_{Ti} above 50A. This diffraction peak disappeared with further decreasing d_{Ti} below 15 A. At d_{Fe} around 25 A, a different peak appeared at the angle slightly lower than that of Fe(110) peak. The diffraction intensity of (002) plane in the Ti hcp phase increased with decreasing d_{Fe}. However, its d-spacing depended little on the d_{Fe} and took the same value as that of bulk Ti except for very small d_{Ti}.
  The net saturation magnetization of Fe layers in all of the specimen films decreased gradually with decreasing d_{Fe} from 1000 to 50 A, decreased abruptly with further decreasing d_{Fe} and became nearly zero at d_{Fe} below 15A.
  The films annealed at 200 and 400 °C showed obscure periodic microstructure and had larger d-spacing of α-Fe(110) and smaller saturation magnetization than those of the as-deposited ones.

INTRODUCTION

  In recent years, much interest has been focused on magnetic multilayered films from the viewpoint not only of fundamental study on the interface magnetism[1] but also of practical application especially for the high density magnetic recording[2].
  Fe/Ti multilayered films have been fabricated by using Facing Targets Sputtering (FTS) apparatus[3] in order to pursue the possibility of developing several new types of magnetic devices with multilayered films. The very thin and continuous Fe films with large saturation magnetization can be easily prepared by using plasma-free sputtering methods such as FTS and ion beam sputtering ones. And the Ti thin films fabricated by FTS exhibited very smooth surface and homogeneous morphology[4]. Therefore, both films seem to be favorable for the formation of multilayered films.
  In this paper, the microstructure and the magnetic behavior of Fe/Ti multilayered films with various combinations in thickness of Fe and Ti layers and the annealing effect on these characteristics will be described.

EXPERIMENTAL PROCEDURE

  The principle of FTS is shown in Fig. 1. Two targets are located opposing their faces at a distance of 120 mm. A magnetic field of 220 G is applied perpendicular to the target planes by installing permanent magnets behind the targets. This magnetic field can confine γ-electrons and promote the ionization of working gas. Therefore, the substrate located at 80 mm from the midpoint between targets is free from plasma so that the excessive elevation of substrate temperature and the severe bombardment of high energy particles on growing films can be avoided.

Using two pairs of FTS equipments with 99.99 % pure Fe targets and 99.9% pure Ti ones, Fe/Ti multilayered films with various combinations in thickness of Fe and Ti layers (d_{Fe} of 10~1000 A and d_{Ti} of 10~200 A) have been deposited on thermally oxidized Si-wafers at room temperature and at Ar gas pressure of 2 mTorr. The deposition rates of Fe and Ti layers were 2.6 and 5.9 A/sec, respectively. The total thickness of Fe layers was fixed at 1000 A in all of the specimen films. The first and last layers were Ti ones so that every Fe layer was sandwiched by Ti ones.

The films with d_{Fe} of 10~1000 A and d_{Ti} of about 50 A were heated in vacuum at 200 and 400 °C for one hour in order to investigate the annealing effect.

The periodic structure due to compositional modulation was verified by low angle region X-ray diffractometry (Cu-Kα) and Auger electron spectroscopy. The crystal structure was analyzed by conventional X-ray diffractometry. And the net saturation magnetization of Fe layers at room temperature was determined on the M-H loops measured by using vibrating sample magnetometer.

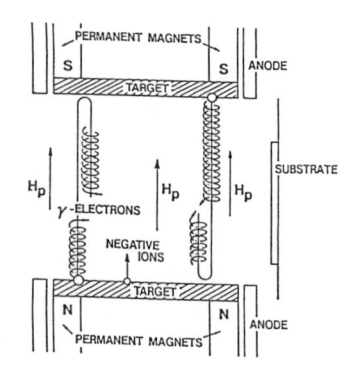

Fig.1 Principle of FTS method.

RESULTS AND DISCUSSION

Figure 2 shows the low angle X-ray diffraction diagrams of two kinds of A and B films with compositional modulation periods of Fe(13 A)/Ti(6 A) and Fe(52 A)/Ti(53 A), respectively. The first order peak for the A film and the second one for the B film were clearly observed. And the modulation lengths d_p estimated on the angles of these peaks were 19 A and 110 A for the A and B films, respectively, in good accordance with the aimed ones. It seems to be, therefore, verified that good periodic microstructure is formed in the specimen films. This was also confirmed by the Auger electron spectroscopic depth profile as shown in Fig. 3 where the correspondence in peak to valley of Fe and Ti profiles was clear.

In the high angle region of X-ray diffraction diagrams, two main peaks were observed at 2θ corresponding to (110) diffraction in the Fe bcc phase and (002) diffraction in the Ti hcp one.

Figures 4 and 5 show the dependence of their intensities and d-spacings

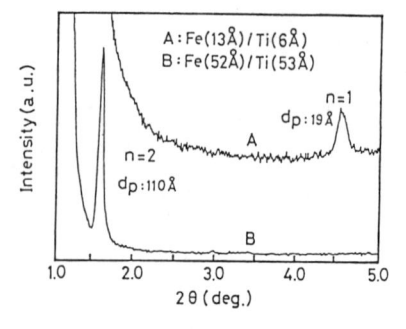

Fig.2 Low angle X-ray diffraction diagrams of the films with modulation periods of Fe(13A)/Ti(6A) and Fe(52A)/Ti(53A).

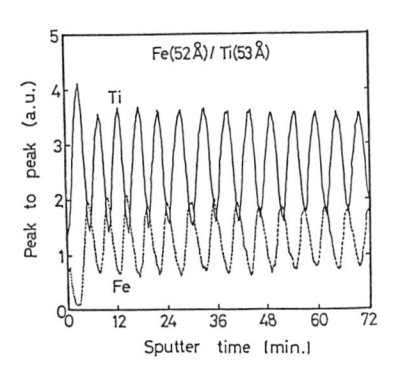

Fig.3 Auger electron spectroscopic depth profile of the film with modulation period of Fe(52A)/Ti(53A).

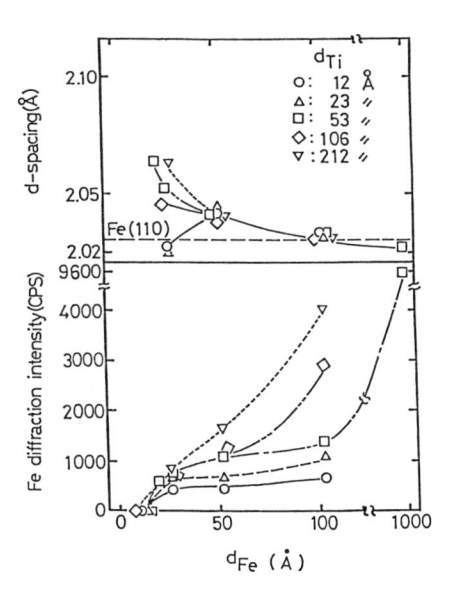

Fig.4 Dependence of Fe diffraction intensity and its d-spacing of the films with various d_{Ti} on d_{Fe}.

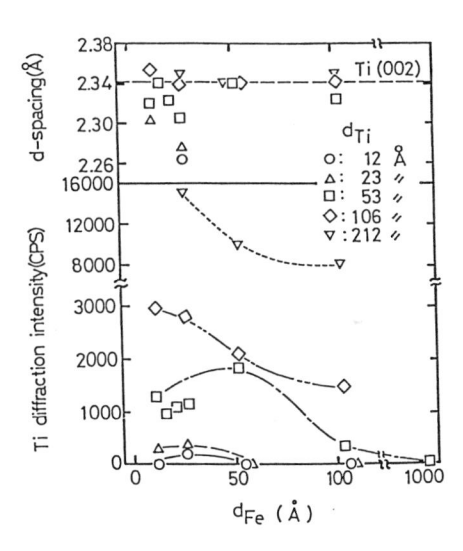

Fig.5 Dependence of Ti diffraction intensity and its d-spacing of the films with various d_{Ti} on d_{Fe}.

on the d_{Fe}. The intensity of Fe(110) decreased with decreasing d_{Fe} and became zero at d_{Fe} of about 15 A for d_{Ti} of 10~200 A, while it tended to become higher with increasing d_{Ti}. Its d-spacing was almost same as that of bulk Fe at d_{Fe} above 100 A and increased with further decreasing d_{Fe} for d_{Ti} above about 50 A. These results indicate that the transition layers of amorphous-like interdiffusion phase are formed inevitably at the interface between Fe and Ti layers and become thicker with decreasing d_{Fe}. The intensity of Ti(002) increased with decreasing d_{Fe} and with increasing d_{Ti} though the low peak of Ti(002) appeared even at d_{Fe} below 25 A for d_{Ti} below 25 A. Its d-spacing was almost same as that of bulk Ti at d_{Ti} above 50 A, while it became smaller than that of bulk Ti at d_{Ti} below 25 A.

Figure 6 shows several X-ray diffraction diagrams of the specimen films with various d_{Fe} at d_{Ti} of about 50 A. The film with d_{Fe} of

Fig.6 High angle X-ray diffraction diagrams for various d_{Fe} at d_{Ti} of 53A.

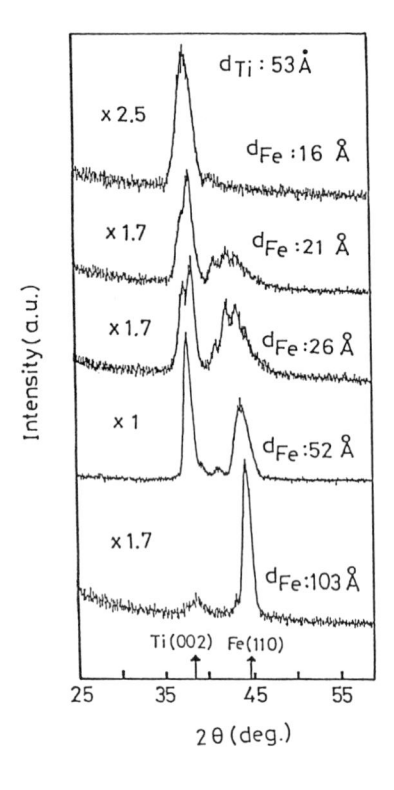

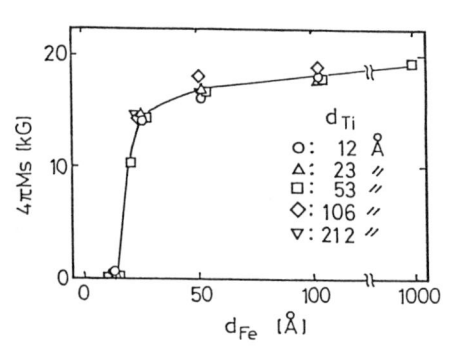

Fig.7 Dependence of $4\pi Ms$ of the films with various d_{Ti} on d_{Fe}.

about 50 A exhibited a different diffraction peak at 2θ in the midway between those of Fe(110) and Ti(002). This peak seems to be one of the satellites. The films with d_{Fe} of 20~25 A exhibited two other peaks at the angles slightly lower than those of Fe(110) and Ti(002). These peaks were also observed in the other films with different d_{Ti}. It is not yet clear whether they are satellites or appear due to the lattice mismatch through the interface between Fe and Ti layers.

Figure 7 shows the dependence of the net saturation magnetization $4\pi Ms$ of Fe layers on the d_{Fe}. Little effect of d_{Ti} on the $4\pi Ms$ was recognized. The $4\pi Ms$ at d_{Fe} of about 1000 A was about 20 kG, the value of which is nearly equal to that of bulk Fe. It decreased gradually with decreasing d_{Fe} from 1000 to 50 A, decreased abruptly at d_{Fe} below 25 A and became almost zero at d_{Fe} of about 15 A. This magnetic behavior could be attributed to the non-magnetic interdiffusion phase at the interface and to the fine grain size resulted from very small d_{Fe}, that is, the superparamagnetic behavior.

The films with various d_{Fe} at d_{Ti} of about 50 A annealed at 200 and 400 °C for one hour in vacuum exhibited no diffraction peak in the low angle region of X-ray diffraction diagrams. And the separation of Fe and Ti peaks in the Auger electron spectroscopic depth profile became more obscure with elevation of annealing temperature. The periodic microstructure could be, therefore, destroyed by the interdiffusion of Fe and Ti atoms through the interface which should be promoted by annealing.

Figure 8 shows the annealing effect on the diffraction intensity and the d-spacing of Fe(110). The annealing increased the intensity of Fe(110) in the films with d_{Fe} of 1000 A, which may be due to the improvement of crystallinity of Fe layers. The film annealed at 200°C showed the same behavior of intensity as the as-deposited one at d_{Fe} below 100 A. On the other hand, the intensity of the film annealed at 400 °C decreased with decreasing d_{Fe} from 1000 to 100 A in the same way as those of the films as-deposited and annealed at 200°C, but it increased rapidly with further decreasing d_{Fe} below 50 A. There was little effect of annealing on the d-spacing of Fe(110) at d_{Fe} above 100 A. However, this d-spacing increased with elevation of annealing temperature at d_{Fe} below 50 A. The diffraction peak observed at the angle slightly lower than that of Fe(110) in the films as-deposited and annealed at 200 °C did not appear in the film annealed at 400°C. The behavior of intensity and d-spacing of Fe(110) in the film annealed at 400 °C indicates that the intermetallic compound of cubic FeTi having (110) orientation with d-spacing of 2.097 may be formed.

Figure 9 shows the annealing effect on the Ti(002) diffraction. The intensity of Ti(002) of the films annealed at 200°C was almost same as that

Fig.8 Annealing effect on Fe diffraction intensity and its d-spacing of the films with various d_{Fe} at d_{Ti} of 53A.

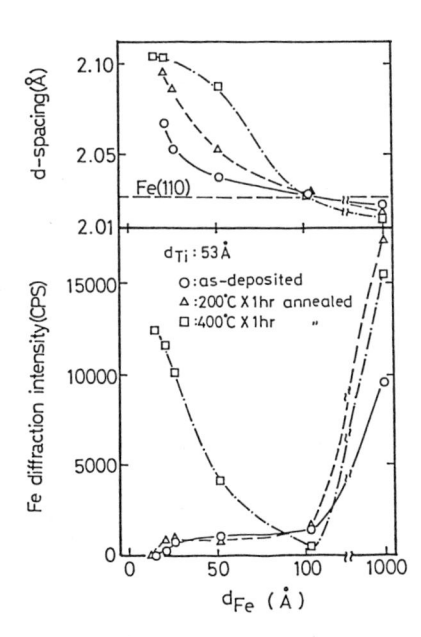

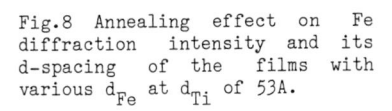

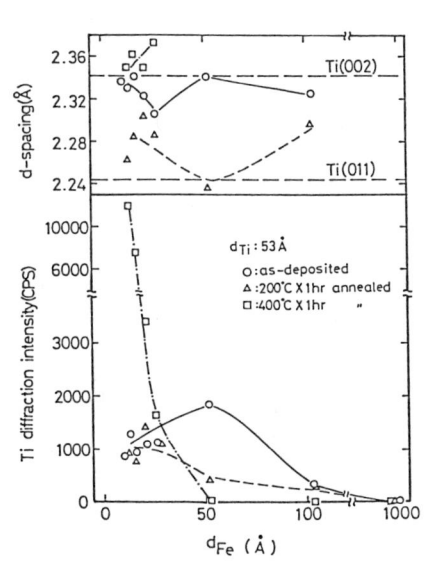

Fig.9 Annealing effect on Ti diffraction intensity and its d-spacing of the films with various d_{Fe} at d_{Ti} of 53A.

of the as-deposited one. But its d-spacing was smaller than that of the as-deposited one. In the film annealed at 400°C, no peak of Ti layer was observed at d_{Fe} above 50 A. The intensity of Ti(002) observed at d_{Fe} below 25 A increased with decreasing d_{Fe}, while its d-spacing was same as that of bulk Ti.

Figure 10 shows the annealing effect on the net $4\pi Ms$ of Fe layers. No annealing effect on it was observed at d_{Fe} of 1000 A. However, the $4\pi Ms$ at d_{Fe} below 100 A decreased with elevation of annealing temperature probably due to the increase in thickness of non-magnetic interdiffusion layers caused by annealing.

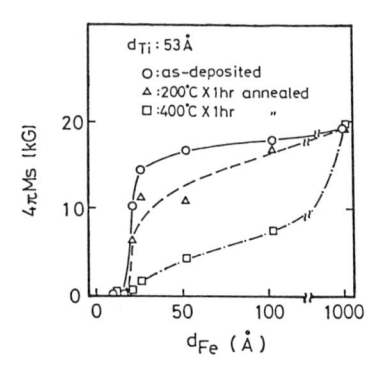

Fig.10 Annealing effect on $4\pi Ms$ of the films with various d_{Fe} at d_{Ti} of 53A.

CONCLUSIONS

Fe/Ti multilayered thin films have been prepared by Facing Targets Sputtering, and their microstructure and magnetic behavior were investigated.
(1) The periodic microstructure due to compositional modulation was clearly observed not only in the low angle region of X-ray diffraction diagrams but also in the Auger electron spectroscopic depth profile.
(2) The decrease of d_{Fe} lowered the intensity of (110) diffraction in the Fe bcc phase and increased its d-spacing for d_{Ti} above 50 A. This diffraction peak disappeared with further decreasing d_{Fe} below 15 A.
(3) The intensity of (002) diffraction in the Ti hcp phase increased with decreasing d_{Fe}. Its d-spacing depended little on the d_{Fe} and took the same value as that of bulk Ti except for very small d_{Ti}.
(4) The net saturation magnetization of Fe layers decreased gradually with decreasing d_{Fe} from 1000 to 50 A, decreased abruptly with further decreasing d_{Fe} and became nearly zero at d_{Fe} below 15 A for all of the specimen films.
(5) The films annealed at 200 and 400 °C exhibited obscure periodic microstructure and had larger d-spacing of Fe(110) and smaller saturation magnetization than those of the as-deposited ones.

REFERENCES

[1] K. Terakura, in METALLIC SUPERLATTICES, edited by T. Shinjo and T. Takada (Elsevier Science Publishers, New York, 1987), P.213
[2] T. Katayama, H. Awano and Y. Nishihara, J. Phys. Soc. Jpn. 55, 2539(1986)
[3] M. Naoe, S. Yamanaka and Y. Hoshi, IEEE Trans. Magn. MAG-16, 646(1980)
[4] S. Ono and M. Naoe, Supplement to Trans. JIM. 29, 57(1988)

ELECTRODEPOSITED METALLIC SUPERLATTICES

D.S. LASHMORE, ROBERT OBERLE, MOSHE P. DARIEL*, L. H. BENNETT AND LYDON SWARTZENDRUBER

Institute for Materials Science and Engineering
National Institute of Standards and Technology (formerly NBS)
Gaithersburg, Maryland 20899
* Ben-Gurion University, Beer-Sheva, Israel

ABSTRACT

Electrochemical deposition of artificial compositionally modulated superlattices is described. It is shown that the quality of these alloys is comparable or superior to materials produced by vapor deposition or sputtering. The ambient temperature process described by Yahalom** has been modified to include a feedback and control system in order to compensate for natural convective disturbances in the electrolyte. Data is presented for copper-nickel samples of varying wavelengths down to 2 nm which suggests that magnetic properties of thin nickel layers are comparable with bulk nickel. Alloys of other types whose properties can be tailored on a near atomic scale will also be discussed along with potential applications.

** U.S. Pat. 4,653,348 (1987)

INTRODUCTION

The production of artificial superlattices using vapor deposition techniques was initiated in order to verify Cahn's diffusion equation [1]. Studies of these early artificially produced alloys led to the observation of their unique and valuable properties. Among these properties are their unusual flow stress, first predicted by Koehler [2] and verified by Hilliard, Tsakalakos and Jankowski [3,4], along with the observation that, at a wavelength of 2 nm to 3 nm, several of the artificial superlattices exhibited a very high biaxial modulus. The magnetic properties of artificial superlattices studied by a number of investigators [5-12] show a marked deviation from bulk alloy behavior. This paper describes new electrochemical techniques used to produce artificial superlattices. The magnetic properties of the electrodeposited alloys suggest reduced alloying between layers compared with the amount of alloying of sputtered alloys of the same wavelength. Saturation magnetization values are higher than previously reported.

PROCESS DECRIPTION

In 1921 Blum produced the first microlayered alloys of copper and nickel deposited from two separate electrolytes [13]. He demonstrated that these layered alloys exhibited an enhanced flow stress and attributed this enhancement to grain refinement strengthening. The layer thicknesses were too large to show epitaxy. Later work by Brenner on copper bismuth [14] demonstrated the possibility of reducing layer thickness to between 300 and 600 nm. The intended application for the Cu-Bi alloy was for optical diffraction gratings and the mechanical properties were not studied. Cohen, Koch, and Sard investigated the behavior of electrochemically produced Ag Pd layered alloys [15] to improve the wear performance of electrical contacts. Their work led to layers as thin as 30 nm. The first patent on electrochemically produced layer structures was by Yahalom and Zadok [16] who made use of a single electrolyte in which the more noble element was present in dilute concentration. Using a similar process, Tench and White [17], in a study of copper nickel demonstrated a strength enhancement at layer thicknesses of 100 to 300nm. Ogden showed [18] that the microhardness was affected by wavelength. Our previous studies, using other variants of the process patented by Yahalom, demonstrated the first electrochemically produced superlattices

[19], whose magnetic properties yield a clear magnetic aftereffect and, for some wavelengths, a magnetic vector perpendicular to the plane of the foil [20]. This paper describes an improvement in electrochemical deposition process, using a feedback and control system. This system consists of a microcomputer controlling and communicating with a digital coulometer and a potentiostat. The potentiostat is connected to the electrochemical deposition cell. A schematic diagram of this system is presented in Fig. 1. A computer controls the process and changes the potential according to the charge passed. The wavelength, presently limited only by the composition of the electrolyte, is about 0.5 nm for the copper and about 2 nm for the nickel. Most of the communication between instruments takes place during the copper pulse. The cell consists of an anodic chamber separated from the cathodic chamber by a NAFION membrane to keep anodic reaction products from being incorporated into the alloy. The temperature is maintained at 30 ± 1 °C with no agitation.

STRUCTURAL CHARACTERIZATION

Metallic multilayer structures exhibit an epitaxial relationship between the layers which depends on the misfit between the crystal structure and the thickness of the layers. The existence of epitaxy can be inferred from the presence of a single Bragg peak. The spacing of the layers can be accurately determined by the position of the satellite peaks. A comparison of the x-ray data of samples made with the present system and samples made with the previously used system [19], shown in Fig. 2, indicates that the new system provides improved alloys. The improvement shown by the enhanced satellite intensity is due to reduced alloying at the interface and/or to improvement in the reproducibility of the layer thickness as discussed by McWhan[21]. The magnetic evidence presented below suggests that alloy formation at the interfaces is reduced. The maximum thickness before epitaxy is lost was discussed by Jesser and Kuhlmann-Wilsdorf [22]. For copper on nickel, Jesser and Kuhlmann-Wilsdorf showed that this critical thickness is about 1.1 nm. The conditions for epitaxy in a multilayer system are much different, as has recently been discussed by Jesser and Van Der Merwe [23-24]. The occurrence of epitaxy depends not only on the thickness of each individual layer but also on the ratio of the thicknesses of the layers. In electrodeposited multilayers produced from a single electrolyte, the more noble reduction potential results in a pure element and the less noble potential produces an alloy with the more noble element. The composition of the alloy is determined by the composition of the solution, the transport conditions, and the electrochemical kinetics. In the deposition process described here the rate of deposition of the more noble element is always transport limited. An exact treatment of the compositional dependence on electrochemical parameters was recently provided by Despic and Jovic [25]. The composition near the interface may be quite different from the average and is linked to the deposition parameters. A comparison of the samples, which are identical except for the potential of the nickel pulse, is shown in Fig.3. Clearly, as the cathodic overpotential increases, the degree of alloying decreases and/or the repeatability of the layer thickness improves. It remains unclear which effect is taking place; however, the magnetic data suggests that alloying is reduced as the potential increases.

A typical electrochemical current transient, shown in Fig. 4. illustrates the effect of double layer charging, and the $T^{-1/2}$ diffusion of reactants to the cathode. As the potential returns to the level to deposit the more noble element the current is forced anodic for a short time. During this time, some of the less noble element is dissolved and some hydrogen may be oxidized.

The perfection of the alloy is clearly related to the level of the less noble reduction potential as was shown in Fig. 3. The epitaxy can be lost by increasing the potential beyond a certain limit. This effect is most pronounced on the high index planes, as shown in Fig. 5a, but epitaxy can be regained by reducing the wavelength as shown in Fig. 5b. On a single crystal copper substrate, the epitaxy can be determined only by the orientation of the substrate and, with the system described here, no other crystal orientations are grown. This is contrary to previous work [19] where some other orientations usually occur in addition to the orientation of the substrate.

MAGNETIC CHARACTERIZATION

A typical hysteresis curve is shown in Fig. 6 with $\mathbf{M_s}$ defined on the Figure. If the applied field is taken to a very high forward value, say 10 KOe, then reversed to various

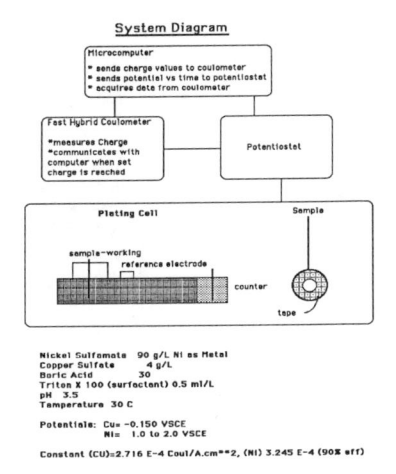

Figure 1. A Block Diagram of the system used to electrochemically produce artifical superlattices

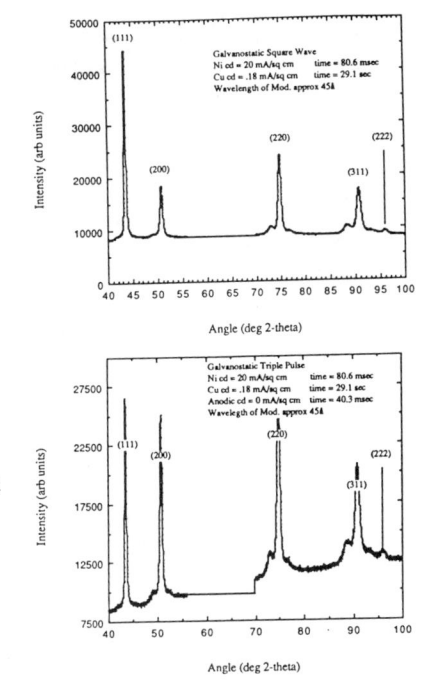

Figure 2. Comparison of x-ray diffraction data with two different electrochemical processing techniques.

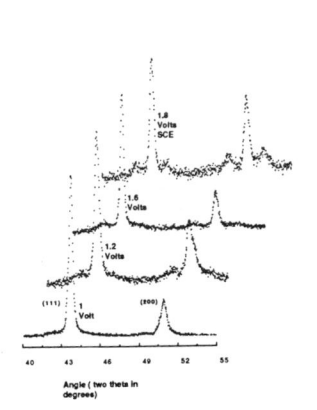

Figure 3. The effect of nickel reduction potential on development of x ray satellites.

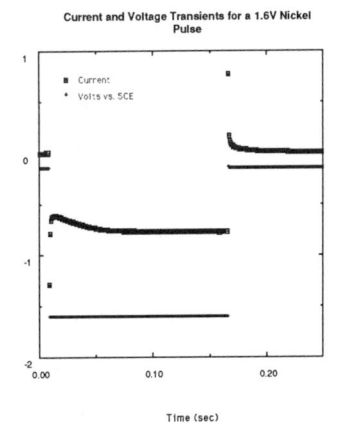

Figure 4. A typical electrochemical current transient for a square applied potential pulse.

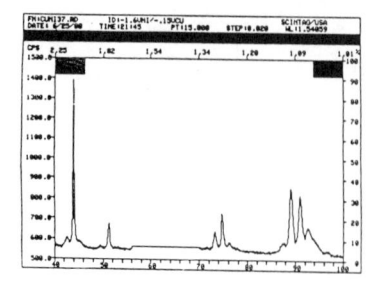

Figure 5a. x-ray data showing that epitaxy is lost if nickel reduction potential is too cathodic.

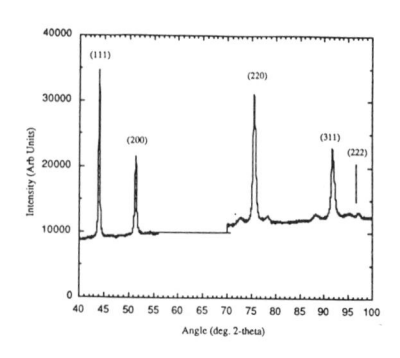

Figure 5b X-ray data showing that epitaxy is regained if wavelength is reduced.

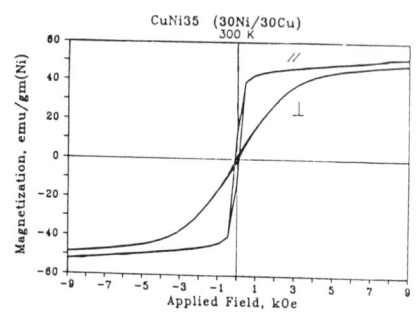

Figure 6. A typical hysteresis curve for a 6nm period Cu-Ni superlattice.

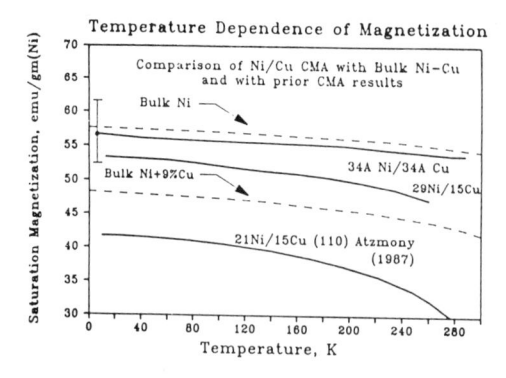

Figure 7. Saturation magnetization vs. temperature for various samples.

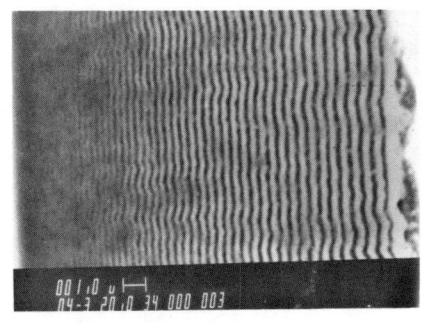

Figure 8a. An SEM micrograph of a cross section of a graded Cu-Ni alloy with the wavelength varying from 30nm to 300nm.

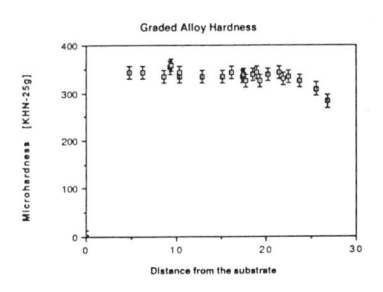

Figure 8b. The microhardness of the above alloy as a function of position.

levels, one would expect the induced magnetization to decay to its equilibrium value very rapidly (<1s). In practice the decay is on the order of 10's of seconds. This effect, described in detail elsewhere [26] is unexpected for layered alloys. The variation of M_s with temperature provides a great deal of information on not only the magnetic behavior but also the degree of alloying in the sample. A comparison of M_s of samples produced using various process techniques is shown in Fig. 8. The unusually high values of M_s, has been discussed in detail by Bennett et. al. [27].

MICROTAILORING OF PROPERTIES

The capability of controlling the wavelength on a near atomic scale, makes it possible to tailor, on a near atomic scale, such properties as M_s, flow stress, modulus, hardness, and corrosion performance which depend on the wavelength. For example, electrical contacts may be designed to maximize their surface conductivity -by using multilayers of gold and cobalt with an outer layer of gold and, and to maximize their resistance to deformation from the contact stress - by making the layers very thin at the point of maximum Hertzian stress. This is possible because of the strong dependence of the flow stress on wavelength [2,17]. Another example of this microtailoring capability is the in the design of interface regions in metal matrix composites which must distribute plastic deformation over a large volume of material in order to avoid cracking at the interface. Another example of an application is in corrosion protection. By continuously changing the ratio of one layer thickness to another, the chemical potential of the alloy can be tailored with the overall thickness in a nearly continuous manner, so that pitting corrosion can be effectively suppressed. An example of a graded copper nickel alloy with the wavelength graded from 30 nm to 300 nm is shown in Fig. 8 a, and the microhardness of the same sample is Fig 8b.

SUMMARY

1. A new process has been developed to produce artificially modulated structures which at small wavelengths, are superlattices. This process utilizes a feedback and control system to compensate for unexpected variations in ionic transport.
2. A strong and unexpected dependence on degree of alloying at the interface on the deposition potential of the less noble element has been observed.
3. The magnetic properties of the artificial superlattices are superior to those of alloys produced using other materials processing techniques. In particular the M_s, is higher than that of any copper nickel superlattice yet reported.

ACKNOWLEDGEMENTS

The authors gratefully acknowledge the financial support of the IBM corporation. We also acknowledge the help of Sandra Claggett for sample preparation, Rosetta Drew for making many of the magnetic measurements, and Alexander Shapiro for assistance in composition analysis.

REFERENCES

1. J.E. Hilliard (1979) in "Modulated Structures", (J. M. Crowley, J. B. Cohen, M. B. Salamon and B. J. Wensch, eds.) pg 407 AIP Conf. Proceedings 53, Am. Inst. of Physics, New York
2. J.Koehler, *Phys. Rev. B.*, **2**, 6,(1970) 547
3. T. Tsakalakos, Hilliard, J. E. *J. Appl. Phys.* **54,** 734 (1983)
4. A.Jankowski.,T.Tsakalakos, *J. Appl. Phys.* **57**, 1835 (1985)
5. J. B. Thaler, C. M. Ketterson and J. E. Hilliard, *Phys. Rev. Lett.* **41**, 336 (1978)
6. E.M. Gyorgy, J.F. Dillon Jr..,D. B. McWhan, L.W. Rupp Jr., L. R.Testardi.and P. Flanders P. *Phys. Rev. Letter* **45**, 57(1980)
7. J. F. Dillon Jr., E. M. Gyorgy, L. W. Rupp, Jr., Y. Yafet and L. R. Testardi, *J. Appl. Phys.* **52**, 2256 (1981)
8. E. M. Gyorgy, D. B. McWhan, J. F. Dillon, Jr., L. R. Walker, and J. V. Waszczak, *Phys. Rev. B* **25**, 6739 (1982)
9. E. M. Gyorgy, D. B. McWhan, J. F. Dillon Jr., L. R. Walker , J. V. Waszczak , Musser, D. P., and R. H. Willens, *J. Magn. Magn. Mat* **31-34**(1983) 915
10. J. Q. Zheng, J. B. Falco, C. M. Ketterson, and I. K. Schuller, *Appl. Phys. Lett.* **38**, 424 (1981)
11. J. Q. Zheng, C. M. Ketterson,J. B. Falco, and Schuller, I. K., *J. Appl Phys.* **53**, 3150 (1982)
12. Flevaris, N. K., C. M. . Ketterson, and J. E. Hilliard, *J. Appl. Phys.* **53**, 2439 (1982)
13. W. Blum, "The Structure and Properties of Alternately Electrodeposited Metals", *Trans. Am. ELectochem. Soc.* **40**, (1921) 307
14. A. Brenner in *Electrodeposition of Alloys, Principoles and Practice*, Vol. 2, Academic Press, New York (1963).
15. U. Cohen, F. B. Koch, R. Sard, *J. Electrochemical Soc.* **130**, 10 (1987)
16. J. Yahalom, O.Zadok, *J. Mat. Sci* **22** ,6,(1987) 499, U. S. Pat. 4,652,348 (1987)
17. D. Tench, J. White, *Met. Trans.* **15A** (1984) 2039
18. C. Ogden, *Plating and Surface Finishing 73*, (1986) 130
19. D.S. Lashmore, M. P. Dariel, *J. of the Electrochem. Soc.* **135**,5 (1988) 1218
20. L.H. Bennett , D.S. Lashmore, M. P. Dariel, Kaufman, M. J., Rubinstein, M., P. Lubitz, O. Zadok,. and J.Yahalom *J. of Magnetism and Magnetic Materials* **67** (1987) 239
21. D. B. McWhan To be Published in Physics, Fabrication and Applications of Multilayer Structures, P. Dhez, Ed. by Plenum, New York
22. W.A. Jesser, D. Kuhlmann-Wilsdorf , *Phys. Stat. Sol.* **19**, 95 (1967)
23. W. A. Jesser , J. H. van der Merwe, "An Exactly Solvable Model for Calculating Critical Misfit and Thickness in Epitaxial Superlattices", I. Layers of Equal Elastic constants and Thickness and II. Layers of Unequal Elastic Constants and Thicknesses (to be published)
24. W. A. Jesser, and J.H.van der Merwe, Critical Misfit and Thickness: An Overview (to be published)
25. A. R. Descpic, and V. D. Jovic, *J. Electrochem. Soc.* **134**,(1987)3004
26. U. Atzmony, L. J. Swartzendruber, L. H. Bennett, M. P. Dariel, D. S. Lashmore, M. Rubinstein, and P. Lubitz, J. Magn.and Magn Matls. **69** (1987) 237
27. L.H. Bennett, L. J. Swartzendruber, D.S. Lashmore, R. R. Oberle, U. Atzmony, M. P. Dariel and R. E. Watson (to be published)

PREPARATION AND STRUCTURE OF Cu-W MULTILAYERS

K.M. UNRUH,[*] B.M. PATTERSON,[*] S.I. SHAH,[**] G.A. JONES,[**] Y.-W. KIM,[***] and J.E. GREENE[***]
[*] Department of Physics and Astronomy, University of Delaware, Newark, DE 19716.
[**] Central Research and Development Department, Experimental Station, E.I. du Pont de Nemours and Co., Wilmington, DE 19880.
[***] Department of Materials Science, the Coordinated Science Laboratory, and the Materials Research Laboratory, University of Illinois at Urbana-Champaign, Urbana, Il 61801.

ABSTRACT

Sputtered multilayer samples of W and Cu have been prepared on a variety of substrates with nominal individual layer thicknesses ranging from about 5 to about 100 A. High angle X-Ray Diffraction (XRD) data have been obtained at room temperature on all of these samples. In addition, selected samples have been studied by Transmission Electron Microscopy (TEM) and by low and high angle XRD from room temperature to above the melting point of bulk Cu. XRD and TEM data indicate that the as-deposited multilayer samples are comprised of well-defined individual layers due, in part, to the very small mutual solubility of W and Cu. At high temperatures the existence of low angle scattering peaks as well as satellites about Bragg peaks indicates that the layered structure is not lost. As a result, the W-Cu multilayer system seems to be an interesting candidate for the study of the melting behavior of thin Cu layers.

INTRODUCTION

The fabrication and study of artificially structured materials (ASM) has attracted considerable recent attention, driven by an interest in their fundamental physical and chemical properties as well as their technological importance. Layered materials consisting of alternating, quasi two-dimensional sheets of various elements and/or alloys are an important class of ASM. In this work we report on the preparation and structure of multilayer materials consisting of alternating layers of crystalline W and Cu with nominal individual layer thicknesses ranging from about 5 to about 100 A. Our motivation for this particular choice of materials was threefold. In the first instance, although extended solubility limits have been obtained by vapor deposition and ion irradiation methods [1-3], bulk W and Cu are essentially immiscible as solids and, even as liquids, the mutual solubility is extremely small [4]. As a result, a negligible amount of interdiffusional mixing is expected between the layers during deposition and post deposition annealing. Secondly, W and Cu are similar to the extensively studied prototypical Nb-Cu multilayer system in that the two crystal structures are dissimilar (BCC for W and FCC for Cu) and there is a large lattice mismatch of about 7% between the [110] planes of W and the [111] planes of Cu [5]. Finally, the melting temperatures of the two bulk elements are drastically different with $T_m(W)=3680$ K and $T_m(Cu)=1357$ K. Therefore, if the lower melting point Cu layers retain a layered structure at temperatures above the melting point of Cu it should be possible to study the melting behavior of the Cu layers as a function of the layer thickness. Although the rather high melting temperature of Cu complicates these measurements, the potential benefit of a well defined structure should be contrasted to previously reported melting studies involving Pb-Ge multilayers in which the layered Pb structure was lost at elevated temperatures as the amorphous, as-deposited, Ge layers crystallized [6-9].

EXPERIMENTAL

All of the Cu-W multilayer samples studied in this work were prepared in a dual gun magnetron sputtering system from elemental targets. A layered structure was obtained by alternatively rotating a computer controlled substrate platform for the desired period of time over each sputtering source. Base system pressures were typically about 10^{-7} Torr prior to throttling the high vacuum pump and about 5×10^{-7} Torr following throttling. With the high vacuum pumped throttled, high purity Ar gas was dynamically pumped through the deposition chamber so as to maintain a constant sputtering pressure of 5 mTorr. The sputtering gun deposition rates were monitored and feedback stabilized based on the output of a quartz crystal oscillator positioned over each sputtering gun. In each case the input power to the sputter guns and the time of deposition were adjusted so as to produce equal W and Cu layer thicknesses.

Room temperature XRD patterns were obtained at high angles using a Cu K_α Philips theta-two theta diffractometer and at low angles and high temperatures using a Rigaku theta-theta vertical mode diffractometer. High temperature XRD patterns were obtained in a vacuum of about 10^{-5} Torr and to temperatures up to 1473 K, well above the melting point of pure Cu.

Cross-sectional TEM micrographs were obtained from selected samples prepared by standard techniques and ion milled using 5 keV Ar ions at a sputtering current of one half ampere.

RESULTS AND DISCUSSION

Figures 1 and 2 illustrate a series of typical high angle XRD patterns on a selection of samples with individual layer thicknesses ranging from about 5 to about 100 A. The actual modulation wavelength determined from the positions of the scattering peaks gave values of the layer thicknesses to within about 10% of the nominal values; hereafter the various multilayer samples will be referred to by their nominal layer thickness. At relatively large layer thickness, as in the case of Figure 1, an intense Bragg peak corresponding to W[110] reflections is observed at a value of two-theta equal to 40.24 degrees and a less intense Bragg peak associated with Cu[111] reflections at an angle of 43.13 degrees. The associated W lattice parameter of 3.167 A is in good agreement with the published value of 3.165 A while the measured Cu lattice parameter of 3.630 A is somewhat larger than the published value of 3.615 A. Sideband reflections, although substantially less intense, are also observed

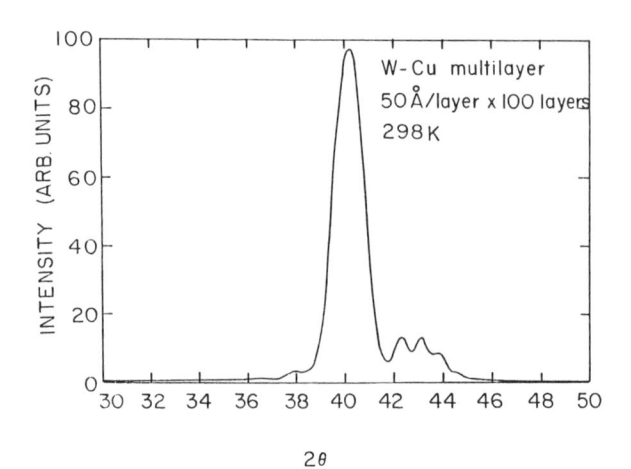

Figure 1: High angle XRD pattern from a W-Cu multilayer (50 A/layer).

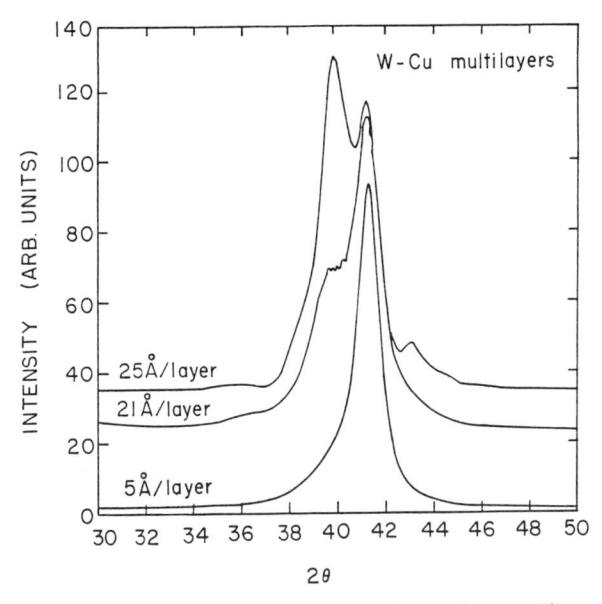

Figure 2: High angle XRD patterns from selected W-Cu multilayers.

in Figure 1. As the individual layer thickness is reduced, however, a scattering peak appears at an angular position between the original two peaks until, as can be seen in Figure 2, only this peak remains. Scattering patterns of this kind have been obtained in the past and can be understood, at least qualitatively, in terms of the scattering from a z-axis modulated multilayered structure [10].

A room temperature low angle XRD scattering pattern from a sample with individual layer thicknesses of about 50 A is shown in Figure 3. The presence of low angle scattering peaks (as well as high angle satellites) indicates that these multilayer samples are of reasonable quality, that is, the individual layers are reasonably flat and of similar thicknesses from layer to layer. An additional indication of the layer quality can also be directly seen from the cross-sectional TEM micrograph of Figure 4.

High angle XRD patterns from the sample of Figure 3 at temperatures of 1023, 1273, 1373, and 1473 K are shown in Figure 5. The presence of satellites around the W[110] and Cu[111] Bragg peaks indicates that the layered structure of the sample is not destroyed, even at temperatures above the melting point of bulk Cu. It is interesting to note that the Cu Bragg peak seems to still exist, even at these temperatures. As noted earlier, this behavior should be contrasted to the behavior of previously studied Pb-Ge multilayers in which, despite a well-layered as-deposited structure, annealing and crystallization of the Ge layers destroys the layering [7].

SUMMARY AND CONCLUSIONS

Multilayered materials consisting of alternating layers of W and Cu have been prepared by a magnetron sputtering technique. Well defined layers have been observed by low and high angle XRD measurements and by TEM. The layered structure has been found to be preserved following heat treatments at temperatures as high as 1373 K. As a result, this multilayer system seems to be a reasonable candidate for a study of the melting behavior of thin Cu films.

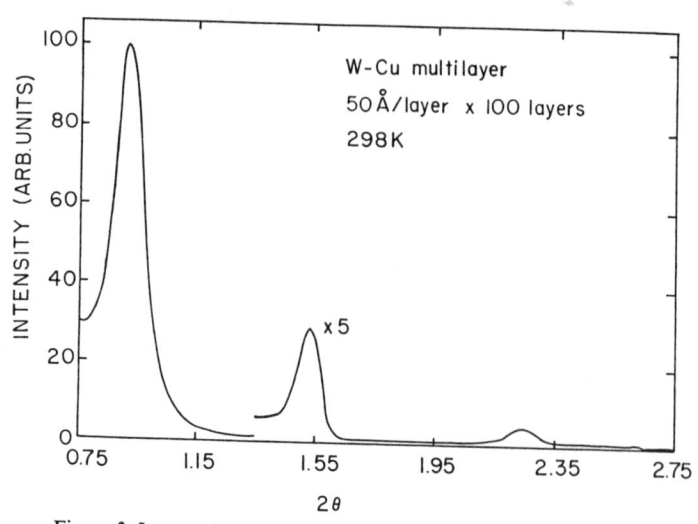

Figure 3: Low angle XRD pattern from a W-Cu multilayer (50 A/layer).

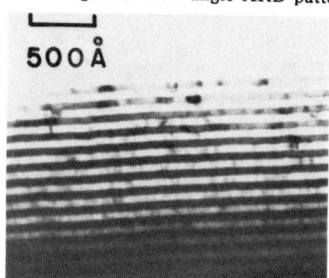

Figure 4(left): TEM cross-sectional micrograph of a W-Cu multilayer (50 A/layer).

Figure 5(bottom): High angle XRD patterns from the W-Cu mulyilayer of Figure 3 at various temperatures.

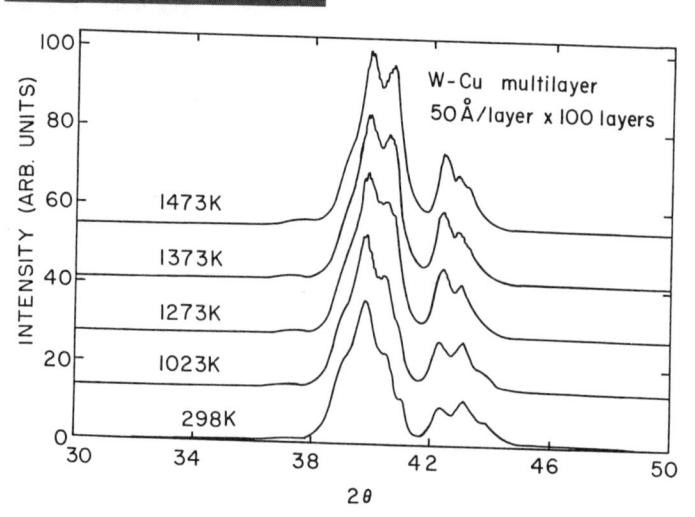

ACKNOWLEDGMENTS

This work was supported in part under ONR contract number N00014-88-K003.

REFERENCES

1) A.G. Dirks and J.J. van den Broek, J. Vac. Sci. Technol. A3, 2618 (1985).

2) M. Natasi, F.W. Saris, L.S. Hung, and J.W. Mayer, J. Appl. Phys. 58, 3052 (1985).

3) J.F.M. Westendorp, U. Littmark, and S.W. Saris, Nucl. Instrum. Meth. B18, 54 (1986).

4) F.A. Shunk, Composition of Binary Alloys, Second Supplement (McGraw-Hill, New York, 1969).

5) I.K. Schuller, Phys. Rev. Lett. 44, 1597 (1980).

6) W. Sevenhans, H. Vanderstraeten, J.-P. Locquet, Y. Bruynseraede, H. Homma, and I.K. Schuller, (preprint) to appear in Mat. Res. Soc. Sym. Proc. 1988.

7) W. Sevenhans, J.-P. Locquet, Y. Bruynseraede, H. Homma, and I.K. Schuller, Phys. Rev. B38, 4974 (1988).

8) R.H. Willens, A. Kornblit, L.R. Testardi, and S. Nakahara, Phys. Rev. B25, 290 (1982).

9) G. Devoud and R.H. Willens, Phys. Rev. Lett. 57, 2683 (1986).

10) See e.g. D.B. McWhan, in Synthetic Modulated Structures, edited by L.Y. Chang and B.C. Giessen (Academic Press, Inc., Orlando, 1985), Chapter 2.

X-RAY DIFFRACTION ANALYSIS OF Au/Ni MULTILAYERS

J. CHAUDHURI[*], S. SHAH[*] AND A. F. JANKOWSKI[**]
* The Wichita State University, Wichita, KS 67208
** Lawrence Livermore National Laboratory, Livermore, CA 94550

ABSTRACT

X-ray diffraction is a useful method to measure the microscopic strain profile in multilayered materials. Depth profiles of strain in the modulation direction are obtained by an iterative fitting of the experimental diffraction pattern with a kinematic model. This approach was used to characterize the coherency strain profile in Au/Ni superlattices.

The accommodation of coherency strain through the superlattice is dependent upon the atomic misfit between the component materials and the thickness of each layer. The depth profile of strain was determined for multilayers with repeat periodicities of 2.92 nm and 4.26 nm. A significant volume fraction of interfaces is present in these nanometric dimensioned laminates.

INTRODUCTION

Considerable effort has been directed over the past decade to the growth of artificial metallic superlattices. These thin film structures are fabricated by alternately depositing material layers whereby a one dimensional composition modulation is produced.

In order to investigate the multilayer microstructures, X-ray diffraction results are often used. Several models have been proposed to interpret the X-ray diffraction data [1-5].

In this article, a kinematic model, as proposed by Speriosu [6,7], was used to analyze the microstructures of Au/Ni metallic superlattices. Speriosu used this model to determine strain in the individual layers of GaAlAs/GaAs superlattices. In addition, quantitative measurements of transition regions at the interface were possible due to the sensitivity of this technique. In the present investigation, the strain at the Au/Ni interface was considered to be coherent. The one dimensional composition modulation was treated to be a square wave.

KINEMATIC MODEL [6,7]

A brief description of the diffraction analysis used to determine the strain profile of superlattices is presented. Details of the kinematic model are given elsewhere by Speriosu [6,7]. For diffraction calculations, a uniform epitaxial layer is described by its thickness t, structure factor F, normal absorption coefficient μ and perpendicular strain $\epsilon_\perp$, if the lattice is strained in a direction perpendicular to the surface only. The direction cosines of the incident and diffracted waves with respect to the surface normal are γ_0 and γ_H, respectively. The variables A and Y, associated with the diffracted amplitude, are given by :

$$A = \frac{r_e \lambda |F| t}{V \sqrt{|\gamma_0 \gamma_H|}}, \quad (1)$$

$$Y = - \frac{\sqrt{\gamma_0} \pi V \sin 2\theta_B}{\sqrt{|\gamma_H|} r_e \lambda^2 |F|}(\Delta\omega), \quad (2)$$

where r_e is the classical electron radius, λ is the X-ray wavelength, V is the volume of the unit cell and θ_B is the Bragg angle. If the angle between the diffracting plane and the surface is zero, then the differential angle $\Delta\omega$ is

$$\Delta\omega = \theta - \theta_B + \epsilon_\perp \tan \theta_B, \quad (3)$$

where θ is the angle of incidence with respect to the diffracting planes.

An arbitrary depth profile of strains can be represented by a discrete structure of N laminae. The normalized amplitude diffracted by such a structure is

$$E_N = i \frac{\sqrt{\gamma_0}}{\sqrt{|\gamma_H|}} \sum_{j=1}^{N} a_j e^{-i(A_j Y_j + \phi_j)} \frac{\sin A_j Y_j}{Y_j}, \quad (4)$$

where,

$$a_j = \exp\left(-\mu \frac{\gamma_0 + |\gamma_H|}{2|\gamma_0 \gamma_H|} \sum_{i=j+1}^{N} t_i\right),$$

$$a_N = 1,$$

$$\phi_j = 2 \sum_{i=1}^{j-1} A_i Y_i,$$

and $\phi_1 = 0$,

in which lamina j has its own A_j and Y_j.

A superlattice is a special case of arbitrary laminar structure. In its simplest form, the superlattice period consists of two layers, labeled a and b, each with its own thickness, strain, structure factor, and the corresponding A_a, Y_a and A_b, Y_b. The amplitude of the nth-order-peak is proportional to the superlattice structure factor F_{sn}, where

$$F_{sn} = (\sin A_a Y_{an}) \left(\frac{1}{Y_{an}} - \frac{1}{Y_{bn}} \right) \qquad (5)$$

Au/Ni SUPERLATTICE

The above kinematic model was applied to the Au/Ni superlattice system to determine the strain within the individual layers. The (111) composition modulated foils were prepared on (001) Si substrate by magnetron sputter deposition of pure gold and nickel metals. The experimental X-ray diffraction curves ((111) reflection) were obtained by using the CuK$_\alpha$ radiation and a Philips Electronic diffractometer with a graphite monochromator placed before the detector. The average repeat period, the total number of layer pairs and the total thickness of the two samples used in this study are reported in Table I. Since these Au/Ni superlattices are highly textured along the [111] direction and a buffer layer of 20 nm of Au was deposited, the pure gold (111) peak was treated as a substrate peak. For modeling purposes, the following assumptions were made:
1. The strain distribution from period to period does not fluctuate.
2. The interplanar spacing, d, within the Au layer is d_{Au}, within the Ni layer is d_{Ni}, while at the interface it is $0.5(d_{Au} + d_{Ni})$.
3. Similarly, the structure factor, F, within the Au and Ni layers are the structure factors of Au and Ni, respectively, while the structure factor at the interface is given by Eq. (5).
4. Since the elastic modulus of Ni is much higher as compared to that of Au, the strain will be carried down further from the interface in Ni layers than in Au layers [8].

Table I. Dimensions and Strain Values in Au/Ni Superlattices

Sample	Period (nm)	Total Periods	Total Superlattice Thickness (nm)	Ave. Normal Strain $<\epsilon_\perp>$	Strain $\epsilon_\perp$
1	4.26	61	260	2.57 %	5.31 %
2	2.92	63	184	6.35 %	7.52 %

The total number of layer pairs in sample no 1. is 61. Each layer pair was divided into 8 sublayers; thus altogether 488 laminae were considered. Similarly the total number of laminae considered for the sample no. 2 were 504. The width of each sublayer and the strain in each sublayer were varied until a good match was obtained between the experimental and calculated diffraction patterns. To calculate the structure factors, the corresponding values of atomic scattering factors and dispersion corrections were obtained from reference [9].

RESULTS AND DISCUSSIONS

Figures 1 and 2 show the measured and calculated X-ray diffraction patterns for the two samples. The average strain values, $<\epsilon_\perp>$, were obtained from the angular separation between the 0th-order-peak and the substrate peak and are listed in Table I. The Au substrate peak was obtained from an additional diffraction pattern. Using the ratio of the amplitude of the +1 peak to that of the 0th peak and Eq. (5), strains of 5.31% and

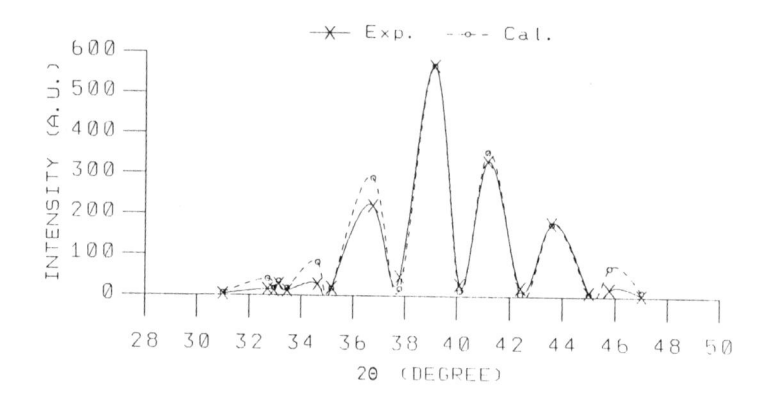

Figure 1. Measured and Calculated CuK_α (111) X-ray Diffraction Pattern Of Au/Ni Superlattice Sample 1.

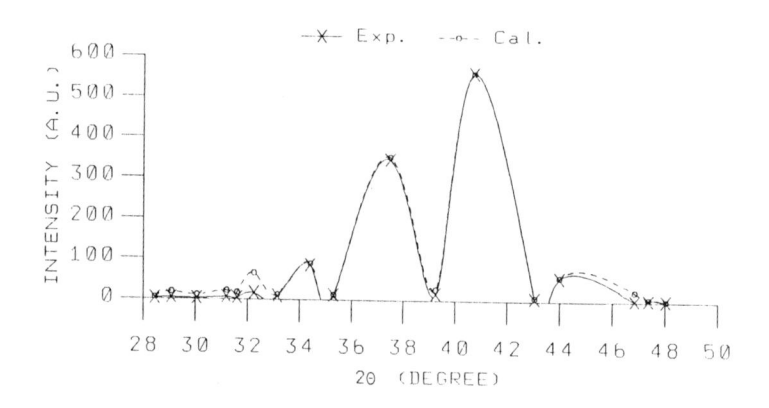

Figure 2. Measured and Calculated CuK_α (111) X-ray Diffraction Pattern Of Au/Ni Superlattice Sample 2.

7.52% were obtained for the samples with periods of 4.26 nm and 2.92 nm, respectively (Table I). The same values of strain were obtained by using the ratio of amplitude of the -1 peak to that of the 0th-order peak. These initial strain values were used to calculate the diffraction pattern.

The calculated diffraction pattern reproduces, very well, the locations as well as the intensities of the experimental peaks. The calculated curves were accepted after a trial-and-error procedure involving about 100 iterations for the first sample (with fewer iterations for the second sample). The strain profile within the first two layers is shown in Figures 3 and 4 for samples no. 1 and 2, respectively. As it was mentioned earlier, the strain profile from one layer pair to the next was considered the same. The diffraction patterns were found to be

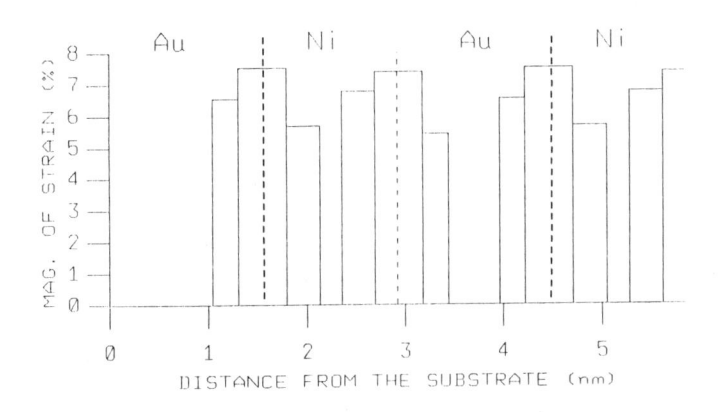

Figure 3. Depth Profile Of Perpendicular Strain In Two Layer Pairs Of Au/Ni Superlattice Sample 1.

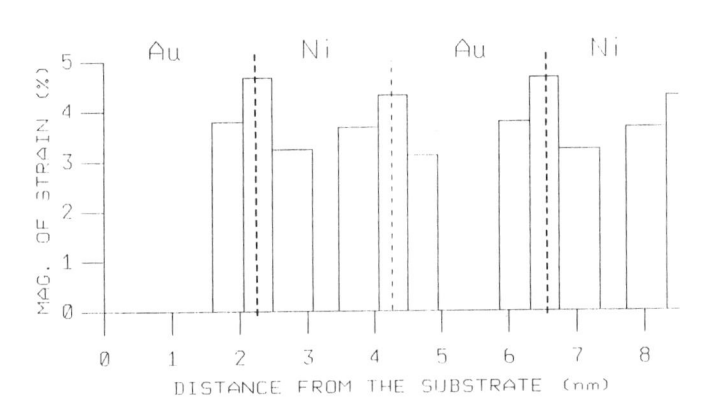

Figure 4. Depth Profile Of Perpendicular Strain In Two Layer Pairs of Au/Ni Superlattice Sample 2.

extremely sensitive to the total number of atomic layers considered at the interface. The influence of interface thickness on the X-ray spectra is also reported by Mitura et al [5]. In addition, the amount of strain considered in each sublayer has a large impact on the diffraction pattern. The change in the magnitude of strain to the fifth decimal place changes the diffraction pattern considerably.

The present analysis indicates that the strain in the thicker layer pair sample is less, which could result from a partial loss of coherency possibly due to the presence of non-perpendicular strain at the interface [7]. Yang has shown that the composition modulations in Au/Ni are coherent for repeat periodicities less than 3.0 nm. At longer repeat periodicities, there is a progressive loss in coherency which was confirmed in the diffusion measurements by a rapid increase in diffusivity due to the decrease in strain energy contribution to the driving force [10].

CONCLUSIONS

It was shown that a kinematic X-ray diffraction model can be used to reveal the depth profile of strain in the Au/Ni superlattice system. The strain values were found to be maximum at the interfaces and minimum within the Au and Ni layers. It is clear that the coherent strain at the interface plays an important role in these superlattices. The Au/Ni multilayer system exhibits a supermodulus effect [11]. In addition to several other models, the concept of coherency strain at the interface is employed in an explanation of the origin of the supermodulus effect in metallic multilayers [8]. Results for the strain profiles obtained in this study, shapes of which are more sawtooth type than square type, are in agreement with apriori modeling assumptions as used in reference [8]. Future study will be done to correlate the supermodulus effect with the strain distribution obtained from this kinematic model. Also, this model will be applied to investigate any possible existence of non-perpendicular strain leading to a partial loss of coherency at longer repeat periodicities.

ACKNOWLEDGMENT

This work was funded in part by the National Science Foundation Grant #DMR-8605564.

REFERENCES

1. I. K. Schuller, Phys. Rev. Lett. 44, 1597 (1980).
2. E. M. Gyorgy, D. B. Mcwhan, J. F. Dillon Jr., L. R. Walker and J. V. Wasczak, Phys. Rev. B 25, 6739 (1982).
3. M. R. Khan, C. S. L. Chun, G. P. Felcher, M. Grimsditch, A. Kueny, C. M. Falco and I. K. Schuller, Phys. Rev. B 27, 7186 (1983).

4. Y. Fujii, T. Ohnishi, T. Ishihara, Y. Yamada, K. Kawaguchi, N. Nakayama and T. Shinjo, J. Phys. Soc. Japan 55, 251 (1986).
5. Z. Mitura and P. Mikolajczak, J. Phys. F: Met. Phys.18, 183 (1988).
6. V. S. Speriosu, J. Appl. Phys. 52(10), 6094 (1981)
7. V. S. Speriosu and T. Vreeland, Jr., J. Appl. Phys. 56(6), 1591 (1984).
8. A. F. Jankowski, J. Phys. F: Met. Phys. 18, 413 (1988).
9. J. A. Ibers and W. C. Hamilton, eds., International Tables for X-Ray Crystallography, Vol. IV (Kynoch, Birmingham, 1974)
10. W. M. C. Yang, Ph. D. Thesis, Northwestern University, Evanston, Il (1971).
11. W. M. C. Yang, T. Tsakalakos and J. E. Hillard, J. Appl. Phys. 48, 876(1977).

Author Index

Anderson, Marc A., 41
Asami, Minuo, 105
Averback, R.S., 35

Bennett, Lawrence H., 179, 219
Bier, Th., 35
Birringer, R., 149
Breval, E., 93

Cerezo, Alfred, 155
Chaudhuri, J., 231
Chen, C.H., 191
Chen, H.S., 191
Chen, Zhe, 119
Chien, C.L., 143, 161, 185, 191
Childress, J., 161
Chow, Gan-Moog, 61
Chui, S.T., 119
Chun, Jung-Hoon, 87

Dariel, Moshe P., 219
Devaty, R.P., 99
Disko, M.M., 79

Eastman, J.A., 3, 15, 21, 27
Epperson, J.E., 15, 21

Fecht, H.J., 137
Franz, H., 149

Gallois, B.M., 49
Garland, C., 137
Gavrin, A., 143
Gleiter, H., 149
Greene, J.E., 225
Grovenor, Chris R.M., 155

Hahn, H., 21, 35
Haubold, T., 149
Hellstern, E., 137
Hetherington, Mark G., 155
Hirata, T., 167
Höfler, H.J., 35

Ito, H., 167

Jankowski, A.F., 199, 231
Jayanth, C.S., 79
Jeong, I.S., 173
Johnson, W.L., 137
Jones, G.A., 225
Jorra, E., 149

Kear, B.H., 67
Kim, B.K., 67
Kim, D.Y., 173
Kim, Y.-W., 225
Klippert, T.E., 15, 21
Kortan, A.R., 191
Kuan, Teh S., 73

Landis, Abraham L., 73
Lashmore, D.S., 219
Lee, Arthur K., 87
Lee, S.R., 49
Levy, A., 161
Liou, S.H., 191
Logas, J., 35
Luton, M.J., 79
Lyons, Alan M., 111

Mackenzie, John D., 105
MacMillan, M.F., 99
Mantese, J.V., 99
Mathur, R., 49
Matras, S., 79
McCandlish, L.E., 67

Nagelberg, A.S., 93
Nakahara, S., 111
Naoe, Masahiko, 167, 213
Narayanasamy, A., 15, 21
Nitta, Michio, 213

Oberle, Robert, 219
Ono, Satoshi, 213

Patterson, B.M., 225
Pearce, E.M., 111
Peisl, J., 149
Petry, W., 149
Pope, Edward J.A., 105

Ramasamy, S., 21
Ritter, Joseph J., 179

Sanchez-Caldera, Luis E., 87
Shah, S., 231
Shah, S.I., 225
Shapiro, Alexander J., 179
Sheng, Ping, 119
Shollock, Barbara A., 155
Shull, Robert D., 179
Siegel, R.W., 3, 15, 21
Smith, George D.W., 155
Spaepen, Frans, 127
Streitz, F.H., 185

Strutt, Peter R., 61
Suh, Nam P., 87
Swartzendruber, Lydon J., 179, 219

Trouw, F., 15, 21
Tsakalakos, T., 199

Unruh, K.M., 225

Valanju, A.P., 173
Vallone, J., 79

Wallner, G., 149
Walser, R.M., 173
White, J.W., 15, 21
Wu, L.W., 67

Xu, Qunyin, 41

Yamato, Deanne P., 73
Yoo, J.Y., 49

Zhou, Min-Yao, 119

Subject Index

age-hardening, 155
Al, 61, 79, 93
Al_2O_3, 3
  filler, 93
alloys
  Al-Cu-Mg-Ag, 155
  Al-Ru, 137
  Alnico, 155
  aluminum, 155
  aluminum-transition metal, 127
  Au-Ni, 231
  $Co_{61}B_{39}$, 175
  composition modulated, 199
  Cu-Ni, 219
  dispersion strengthened, 79, 87
  Fe-Ni, 143
  Fe-Ti, 213
  graded, 219
  granular, 143
  Tb-Fe-Co, 167
  W-Co-C ternary, 67
  W-Cu, 225
alumina, 3
amorphous, 127
  silica fibers, 61
  thin films, 167, 175
  TiO_2, 27
antiferromagnetism, 179
Arrhenius relaxation law, 185, 191
atom probe microanalysis, 155

ball milling, 137
blocking temperature, 185, 191
boride dispersion strengthened copper, 87
butyl acrylate, 105

calorimetry, 127
carbides, 49, 67
carbon, 49
  activity, 67
  film, 49
carburization
  kinetics, 67
  using CO/CO_2, 67
ceramic, 3, 15, 21, 27, 35, 41, 49, 93
  membrane, 41
cermet, 67, 99
$Co(en)_3WO_4$, 67
coercivity, 143, 167
  increase in, 185, 191

colloidal stability, 41
columnar structure, 49
composite, 49, 61, 67, 73, 79, 87, 93, 99, 105, 111, 119, 137, 161, 213, 219, 225, 231
  $(Fe_{50}Ni_{50})_x(Al_2O_3)_{1-x}$, 143
  Al-Al(O,N), 79
  Al-Ru, 137
  Al_2O_3-metal, 93
  Au-Ni, 231
  coating, 49
  Cu-Ni, 219
  $Fe-Al_2O_3$, 161, 185, 191
  Fe-mullite, 73
  $Fe-SiO_2$, 161, 185, 191
  Fe-Ti, 213
  $Fe_x(SiO_2)_{1-x}$, 143
  filament reinforced, 49
  granular film, 99
  metal-metal, 127
  $Pt-Al_2O_3$, 99
  silica gel-polymer, 105
  TiN-TiC, 49
  transparent, 105
  W-Cu, 225
  WC-Co, 67
compositional modulation, 199, 213, 219, 225, 231
contrast variation method, 21
critical phenomena, 199
cryomilling, 79
Cu, 111
CVD, 49
  computer controlled, 49
  hot wall, 49
  RF-plasma assisted, 49

Debye temperature, 161
dimethylbutadiene, 105
dispersion strengthening, 79, 87
  yield strength, 79
DMB, 105
dried gels, 73

effective medium theory, 99, 119
effective volume anisotropy, 185, 191
electrochemical deposition, 219
electron diffraction, 167

Fe, 161

field evaporation, 155
Fourier transform
  digital, 175
fractal, 21

gas-condensation method, 3, 27
gel
  Fe + silica, 179
  porous silica, 105
gel structures, 41
grain boundaries, 137
grain growth, 35
grain size, 3, 15, 21, 35, 41, 185, 191
granular
  Fe-Al$_2$O$_3$, 185, 191
  Fe-SiO$_2$, 185, 191
  film, 99
  materials, 161
growth morphologies, 175

H$_2$WO$_4$, 67
hardness, Vicker's, 3, 87
heavy deformation, 137

image processing, digital, 175
in situ chemical reaction, 15, 21, 27, 67, 87, 111
infrared, 99, 119
interfacial phenomena, 199
ionic strength, 41

Kerr rotation angle, 167

laser-induced evaporation, 61
lattice dynamics, 161

magnesium oxide, 3
magnetic
  hyperfine field, 179
  material, 155
  properties, 73, 175, 185, 191, 213, 219
  susceptibility, 179
magnetization, 73, 161, 167, 213, 219
magneto-optical properties, 167
maximum entropy analysis, 15
mechanical alloying, 137
mechanism of gelation, 41
membrane, 41
metastable phases, 67
MgO, 3
microcrystalline, 127
misfit strain, 199
Mixalloy process, 87
Mössbauer spectroscopy, 161, 179
multilayers

Ag-Pd, 199
Au-Ni, 199, 231
coherency in, 199
Cu-Ni, 199, 219
elastic behavior, 199
Fe-Ti, 213
melting behavior, 225
strain profile, 231
W-Cu, 225

nanocomposites, 67, 73, 111, 179, 199
nanocrystals, 27, 35, 67, 137, 149, 185, 191
nanophase, 3, 15, 21, 27, 67, 79
nitrides, 49
non-equilibrium reaction pathway, 67
non-periodic models, 127

optical hosts, 105
optical properties, 105, 119, 167
organic dye, 105
oxidation, 27
  directed, 93

particle size, 73, 149
particles
  copper, 111
  dispersed, 119
  iron, 161
  primary, 41
  secondary, 41
  single domain, 185, 191
Pd, 27, 149
peptization, 41
permeability
  of He and Ar, 35
phase control, 27
poly(2-vinylpyridine), 111
polymethylmethacrylate (PMMA), 105
polystyrene, 105
pore size, 41
pore volume, 35
position sensitive detector, 155
precursor, alkoxide, 41
precursor molecule, 67, 111
properties, microtailoring of, 219

quasicrystalline, 127

Raman spectroscopy, 49
rapid condensation, 61
reductive pyrolysis, 67
reflectivity, 167

Rutherford backscattering, 3
rutile, 3

shear bands, 137
Si, 93
silicone, 105
sintering, 15, 35
small angle
  neutron scattering, 3, 15, 21
  x-ray scattering, 149
sol-gel, 41, 73, 105, 179
sonication, 41
spin glass, 179
sputtering
  facing targets, 167, 213
  magnetron, 225
  plasma free substrate, 167
  RF, 185, 191
  RF sequential, 173
superlattices
  Au-Ni, 199, 231
  Cu-Ni, 199, 219
  metallic, 199
  W-Cu, 225
supermodulus effect, 199, 231
superparamagnetism, 143, 179, 185, 191, 213
surface area, 35, 41

TEOS, 73
thermal annealing, 149
thermogravimetric analysis, 41, 67
Ti, 27
TiB_2, 87
TiC, 49
TiN, 49
TiO, 21
TiO_2, 3, 15, 21, 27, 35, 41

vapor-liquid-solid mechanism, 49, 61
vibrating sample magnetometer, 73

W, 61
W-Co-C, 67
WC-Co, 67

x-ray diffraction, 73, 179, 185, 191, 199, 213, 219, 225, 231